Lecture Notes in Computer Science 14494

Founding Editors

Gerhard Goos
Juris Hartmanis

Editorial Board Members

Elisa Bertino, *Purdue University, West Lafayette, IN, USA*
Wen Gao, *Peking University, Beijing, China*
Bernhard Steffen, *TU Dortmund University, Dortmund, Germany*
Moti Yung, *Columbia University, New York, NY, USA*

The series Lecture Notes in Computer Science (LNCS), including its subseries Lecture Notes in Artificial Intelligence (LNAI) and Lecture Notes in Bioinformatics (LNBI), has established itself as a medium for the publication of new developments in computer science and information technology research, teaching, and education.

LNCS enjoys close cooperation with the computer science R & D community, the series counts many renowned academics among its volume editors and paper authors, and collaborates with prestigious societies. Its mission is to serve this international community by providing an invaluable service, mainly focused on the publication of conference and workshop proceedings and postproceedings. LNCS commenced publication in 1973.

Reneta P. Barneva · Valentin E. Brimkov ·
Claudio Gentile · Aldo Pacchiano
Editors

Artificial Intelligence and Image Analysis

18th International Symposium
on Artificial Intelligence and Mathematics, ISAIM 2024
and 22nd International Workshop
on Combinatorial Image Analysis, IWCIA 2024
Fort Lauderdale, FL, USA, January 8–10, 2024
Revised Selected Papers

Editors
Reneta P. Barneva
State University of New York at Fredonia
Fredonia, NY, USA

Claudio Gentile
Google LLC
New York, NY, USA

Valentin E. Brimkov
SUNY Buffalo State University
Buffalo, NY, USA

Institute of Mathematics and Informatics,
Bulgarian Academy of Sciences
Sofia, Bulgaria

Aldo Pacchiano
Boston University
Boston, MA, USA

ISSN 0302-9743   ISSN 1611-3349 (electronic)
Lecture Notes in Computer Science
ISBN 978-3-031-63734-6   ISBN 978-3-031-63735-3 (eBook)
https://doi.org/10.1007/978-3-031-63735-3

© The Editor(s) (if applicable) and The Author(s), under exclusive license
to Springer Nature Switzerland AG 2024

This work is subject to copyright. All rights are solely and exclusively licensed by the Publisher, whether the whole or part of the material is concerned, specifically the rights of translation, reprinting, reuse of illustrations, recitation, broadcasting, reproduction on microfilms or in any other physical way, and transmission or information storage and retrieval, electronic adaptation, computer software, or by similar or dissimilar methodology now known or hereafter developed.
The use of general descriptive names, registered names, trademarks, service marks, etc. in this publication does not imply, even in the absence of a specific statement, that such names are exempt from the relevant protective laws and regulations and therefore free for general use.
The publisher, the authors and the editors are safe to assume that the advice and information in this book are believed to be true and accurate at the date of publication. Neither the publisher nor the authors or the editors give a warranty, expressed or implied, with respect to the material contained herein or for any errors or omissions that may have been made. The publisher remains neutral with regard to jurisdictional claims in published maps and institutional affiliations.

This Springer imprint is published by the registered company Springer Nature Switzerland AG
The registered company address is: Gewerbestrasse 11, 6330 Cham, Switzerland

If disposing of this product, please recycle the paper.

Foreword

Artificial Intelligence and Image Analysis – ISAIM 2024 and IWCIA 2024

On January 8–10, 2024, ISAIM and IWCIA collocated for a joint meeting in Fort Lauderdale, Florida. As a boutique conference, participants actively worked together, establishing personal connections within our research community and new collaborations. All together, there were 46 talks given in the regular track, special sessions and keynote lectures.

The ISAIM series "International Symposium on Artificial Intelligence and Mathematics" has been held every two years in Fort Lauderdale since 1990. The general scope of the series is quite broad, fostering interactions between all aspects of AI, theoretical computer science, mathematics and their application. The series was founded by Martin Charles Golumbic, Peter L. Hammer and Frederick Hoffman, with the editorial board of the *Annals of Mathematics and Artificial Intelligence* (AMAI) serving as the permanent Advisory Committee. Traditionally, the Symposium attracts participants from a variety of disciplines, thereby providing a unique forum for scientific exchange. The three-day symposium includes invited speakers, presentations of technical papers, and special topic sessions.

The IWCIA series "International Workshop on Combinatorial Image Analysis" has taken place annually since 1991. Image analysis is a scientific discipline providing theoretical foundations and methods for solving problems appearing in a wide range of areas, as diverse as medicine, robotics, defense, and security. As a rule, the processed data are discrete; therefore, the "discrete approach" to image analysis appears to be a natural one and has an increasing importance. It is based on studying combinatorial properties of the considered digital data sets. Combinatorial image analysis often features various advantages (in terms of efficiency and accuracy) over the more traditional approaches based on continuous models requiring numeric computation.

The joint ISAIM-IWCIA 2024 conference was held in person, organized and sponsored by Florida Atlantic University, the University of Oklahoma, and the Caesarea Rothschild Institute at the University of Haifa. Program Co-chairs for ISAIM were Claudio Gentille and Aldo Pacchiano, and Program Co-chairs for IWCIA were Reneta Barneva and Valentin Brimkov. The full program can be found at the website: https://isaim2024.cs.ou.edu/.

Keynote Speakers and Special Sessions

Edith Elkind (Oxford University) gave a keynote lecture on *Group Fairness in Collective Decisions: From Multiwinner Voting to Participatory Budgeting*. Many cities around the world allocate a part of their budget based on residents' votes, following a process known as participatory budgeting. It is important to understand which outcomes of this process should be viewed as fair, and whether fair outcomes could be computed efficiently. In her talk, Prof. Elkind presented an overview of recent progress on this topic. Focusing on a special case of participatory budgeting where all candidate projects have the same cost (known as multi-winner voting), she formulated progressively more demanding notions of fairness for this setting, and identified efficiently computable voting rules that satisfy them. She concluded with a discussion of the challenges of extending these ideas to the general model.

Susan Schneider (Florida Atlantic University) shared some of her thinking on the topic of the *Global Brain Argument*. In her keynote lecture, she discussed the likely emergence of one or more hyperintelligent "global brains" in which humans are nodes, connected via brain-computer interfaces and/or wearable devices. She also described her vision for the newly created "Center for the Future Mind", of which she is the founding director.

Geoff Sutcliffe (University of Miami) gave a keynote lecture on *The TPTP World – Infrastructure for Automated Reasoning*. The TPTP World is a well known and established infrastructure that supports research, development, and deployment of Automated Theorem Proving (ATP) systems for classical logics. The data, standards, and services provided by the TPTP World have made it increasingly easy to build, test, and apply ATP technology. In his talk, Prof. Sutcliffe reviewed the core features of the TPTP World, describing key service components of the TPTP World, and presenting some successful applications in mathematics. Recent work in non-classical logics, and future work verifying generative AI system results, were also discussed.

John Hooker, Tae Wan Kim and Derek Leben organized a special track on *Alternative Models for Fairness in AI*. Fairness in AI has become a topic of intense discussion and is likely to continue to grow in importance as AI permeates society. The aim of this track was to assemble talks on a wide range of fairness models that are potentially relevant to AI, including models that go beyond the standard statistical parity metrics and may be unfamiliar to the AI community. Topics included optimization models for fairness, axiomatic and bargaining derivations of social welfare functions, logic-based models for ethical obligation, and game-theoretic approaches.

Abhishek Gupta and Zhaoran Wang organized a special track on *Deep Reinforcement Learning*. Deep Reinforcement Learning (DRL) is at the forefront of AI research, driving advancements in areas from robotics to finance. While the potential of DRL is vast, there are inherent challenges that need to be addressed, both in theory and in practice. This workshop aimed to provide participants with a comprehensive understanding of these challenges, backed by hands-on experiences and in-depth discussions. By promoting an understanding and appreciation of the latest advancements in DRL, the DRL special track addressed the existing challenges facing the field. Topics included sample efficiency, exploration versus exploitation, transfer learning, stability and convergence, as well as real-world application barriers.

Acknowledgments

We would like to thank the program chairs and special track chairs who put together the excellent program of lectures. We are especially pleased to acknowledge the hard work by all authors who submitted papers to this LNCS proceedings consisting of papers from the regular track. Our appreciation goes to the referees for their evaluations and recommendations that improved the selection of research works to be enjoyed by readers. Finally, a huge round of applause goes to Maria Provost, the organization chair from Florida Atlantic University, and to the computing team of the University of Oklahoma for much of the technical support for the website.

Dimitrios I. Diochnos
Martin Charles Golumbic
Frederick Hoffman

Preface

We are pleased to welcome you to the Artificial Intelligence and Image Analysis proceedings. The volume contains 14 papers from among those presented at the co-located scientific forums – the International Symposium on Artificial Intelligence and Mathematics (ISAIM 2024) and the International Workshop on Combinatorial Image analysis (IWCIA 2024), held on January 8–10, 2024 in Fort Lauderdale, FL, USA.

There were 25 submissions from authors based in 12 different countries on four continents. The submissions were reviewed by the members of the respective program committees and 14 contributed papers were selected for inclusion in these proceedings. EasyChair and OpenReview provided a convenient platform for smoothly carrying out the rigorous review process. The most important selection criterion for acceptance or rejection of a paper was the overall score received. Other criteria included relevance to the conference topics, correctness, originality, mathematical depth, clarity, and presentation quality. We believe that as a result, only papers of high quality were accepted for publication in this volume.

Many individuals contributed to the success of ISAIM and IWCIA. We are grateful to all participants and especially to the contributors of this volume. We are indebted to the General Chair, Martin Charles Golumbic, for his leadership and dedication, which were instrumental in coordinating and managing this memorable scientific forum. Our most sincere thanks go to Dimitrios I. Diochnos, who served as Publications and Publicity Chair of both conferences and maintained the conference website. We express our heartfelt gratitude to the members of the Program Committees whose cooperation in carrying out high-quality reviews was essential in establishing a strong scientific program. We extend our deepest appreciation to the keynote speakers, Edith Elkind from Oxford University, Susan Schneider from Florida Atlantic University, and Geoff Sutcliffe from the University of Miami for the excellent talks and overall contribution to the conference program. The success of the workshop would not have been possible without the hard work of the local organizers and particularly of Frederick Hoffman and Maria Provost. We are very grateful to them and to the host organization – Florida Atlantic University – for the support. Finally, we wish to thank the team at Springer for the efficient and kind cooperation in the timely production of this book.

We believe that ISAIM-IWCIA 2024 provided excellent conditions for professional development and the proceedings will be of interest to a broad audience of scholars and educators.

March 2024

Reneta P. Barneva
Valentin E. Brimkov
Claudio Gentile
Aldo Pacchiano

Organization

The 18th International Symposium on Artificial Intelligence and Mathematics, ISAIM 2024, and the 22nd International Workshop on Combinatorial Image Analysis, IWCIA 2024, were organized by the Florida Atlantic University and were held in Fort Lauderdale on January 8–10, 2024.

Keynote Speakers

Edith Elkind	University of Oxford, UK
Susan Schneider	Florida Atlantic University, USA
Geoff Sutcliffe	University of Miami, USA

ISAIM Chairs

Martin Charles Golumbic (General Chair)	University of Haifa, Israel
Frederick Hoffman (Conference Chair)	Florida Atlantic University, USA
Claudio Gentile (Program Co-chair)	Google Research, USA
Aldo Pacchiano (Program Co-chair)	Boston University, USA
Maria Provost (Local Organizer)	Florida Atlantic University, USA
Dimitrios I. Diochnos (Publications & Publicity Chair)	University of Oklahoma, USA

ISAIM Special Session Chairs

Boolean and Pseudo-Boolean Functions

Endre Boros	Rutgers University, USA
Yves Crama	University of Liège, Belgium

Alternative Models for Fairness in AI

John Hooker	Carnegie Mellon University, USA
Tae Wan Kim	Carnegie Mellon University, USA
Derek Leben	Carnegie Mellon University, USA

Deep Reinforcement Learning: Bridging Theory and Practice

Abhishek Gupta	University of Washington, USA
Zhaoran Wang	Northwestern University, USA

ISAIM Program Committee

Salem Benferhat	Université d'Artois, France
Endre Boros	Rutgers University, USA
Berthe Choueiry	University of Nebraska-Lincoln, USA
Yves Crama	University of Liège, Belgium
Chris Dann	Google Research, USA
Dimitrios Diochnos	University of Oklahoma, USA
Benjamin Fish	University of Michigan, USA
Claudio Gentile	Google Research, USA
Martin Golumbic	University of Haifa, Israel
Steve Hanneke	Purdue University, USA
Thomas Lidbetter	Rutgers University, USA
Debasis Mitra	Florida Tech School of Computing, USA
Aditya Modi	Microsoft, USA
Leora Morgenstern	Systems & Technology Research, USA
Mirco Mutti	Politecnico di Milano, Italy
Gergely Neu	Universitat Pompeu Fabra, Spain
Aldo Pacchiano	Boston University, USA
Matteo Papini	Politecnico di Milano, Italy
Kamal Premaratne	University of Miami, USA
Lev Reyzin	University of Illinois at Chicago, USA
Jörg Rothe	Heinrich-Heine-Universität Düsseldorf, Germany
György Turán	University of Illinois at Chicago, USA and University of Szeged, Hungary
Kristen Brent Venable	University of West Florida and IHMC, USA
Andrea Zanette	University of California at Berkeley, USA
Xuezhou Zhang	Boston University, USA

IWCIA Chairs

Reneta P. Barneva　　　　　　SUNY Fredonia, USA
Valentin E. Brimkov　　　　　 SUNY Buffalo State, USA

IWCIA Steering Committee

Gabor Herman　　　　　　　　CUNY Graduate Center, USA
Valentin E. Brimkov　　　　　 SUNY Buffalo State, USA
Giorgio Nordo　　　　　　　　 University of Messina, Italy
Tibor Lukić　　　　　　　　　 University of Novi Sad, Serbia
Renato M. Natal Jorge　　　　 University of Porto, Portugal
Joao Manuel R. S. Tavares　　 University of Porto, Portugal

IWCIA Program Committee

Eric Andres　　　　　　　　　 Université de Poitiers, France
Buda Bajić　　　　　　　　　　University of Novi Sad, Serbia
Péter Balaźs　　　　　　　　　University of Szeged, Hungary
George Bebis　　　　　　　　　University of Nevada at Reno, USA
Partha Bhowmick　　　　　　 Indian Institute of Technology Kharagpur, India
Arindam Biswas　　　　　　　 Indian Institute of Engineering Science and
　　　　　　　　　　　　　　　Technology, Shibpur, India
Boris Brimkov　　　　　　　　 Slippery Rock University, USA
Alfred M. Bruckstein　　　　　Technion, I.I.T., Israel
Li Chen　　　　　　　　　　　 University of the District of Columbia, USA
Lidija Čomić　　　　　　　　　University of Novi Sad, Serbia
Mousumi Dutt　　　　　　　　 St. Thomas College of Engineering and
　　　　　　　　　　　　　　　Technology, India
Fabien Feschet　　　　　　　　Université d'Auvergne, France
Leila De Floriani　　　　　　　University of Maryland, USA
Chiou-Shann Fuh　　　　　　　National Taiwan University, Taiwan
Atsushi Imiya　　　　　　　　 IMIT, Chiba University, Japan
Krassimira Ivanova　　　　　　Bulgarian Academy of Sciences, Bulgaria
Kamen Kanev　　　　　　　　 Shizuoka University, Japan
Kostadin Koroutchev　　　　　Universidad Autónoma de Madrid, Spain
Walter G. Kropatsch　　　　　 Vienna University of Technology, Austria
Jerome Liang　　　　　　　　　SUNY Stony Brook, USA
Tibor Lukić　　　　　　　　　 University of Novi Sad, Serbia
Benedek Nagy　　　　　　　　 Eastern Mediterranean University, North Cyprus

Kálmán Palágyi	University of Szeged, Hungary
Meenakshi Paramasivan	University of Trier, Gemany
Hemerson Pistori	Dom Bosco Catholic University, Brazil
Konrad Polthier	Freie Universitaet Berlin, Germany
Paolo Remagnino	Kingston University, UK
Nikolay Sirakov	Texas A&M University – Commerce, USA
Josef Slapal	Technical University of Brno, Czechia
Ivan Štajduhar	University of Rijeka, Croatia
K. G. Subramanian	Madras Christian College, India
João Manuel R. S. Tavares	University of Porto, Portugal
D. G. Thomas	Madras Christian College, India
László Varga	University of Szeged, Hungary
Petra Wiederhold	CINVESTAV-IPN, Mexico
Jinhui Xu	SUNY University at Buffalo, USA

Contents

A Model for Optimizing Recalculation Schedules to Minimize Regret   1
    Bethany Austhof and Lev Reyzin

A Theory of Learning with Competing Objectives and User Feedback   10
    Pranjal Awasthi, Corinna Cortes, Yishay Mansour, and Mehryar Mohri

Trick Costs for $\alpha\mu$ and New Relatives   50
    Samuel Bounan and Stefan Edelkamp

A Differential Approach for Several NP-hard Optimization Problems   68
    Sangram K. Jena, K. Subramani, and Alvaro Velasquez

On the Computational Complexities of Finding Selected Refutations
of Linear Programs   81
    K. Subramani and Piotr Wojciechowski

Extending the Tractability of the Clique Problem via Graph Classes
Generalizing Treewidth   94
    Philippe Jégou

Principled Approaches for Learning to Defer with Multiple Experts   107
    Anqi Mao, Mehryar Mohri, and Yutao Zhong

On Sample Reuse Methods for Answering k-wise Statistical Queries   136
    Lev Reyzin and Duan Tu

Neural Diffusion Graph Convolutional Network for Predicting Heat
Transfer in Selective Laser Melting   150
    Benjamin Uhrich, Tim Häntschel, Martin Schäfer, and Erhard Rahm

New Proportion Measures of Discrimination Based on Natural Direct
and Indirect Effects   165
    Ryusei Shingaki and Manabu Kuroki

Addressing Discretization Artifacts in Tomography by Accessing
and Balancing Pixel Coverage of Projections   182
    Csaba Olasz

Finding the Straight Skeleton for 3D Orthogonal Polyhedrons:
A Combinatorial Approach 206
　　Anukul Maity, Mousumi Dutt, and Arindam Biswas

Towards a Unifying View on Monotone Constructive Definitions 218
　　Linde Vanbesien, Samuele Pollaci, Bart Bogaerts, and Marc Denecker

Partial Boolean Functions for QBF Semantics 236
　　Allen van Gelder

Author Index 255

A Model for Optimizing Recalculation Schedules to Minimize Regret

Bethany Austhof[✉] and Lev Reyzin[✉]

Department of Mathematics, Statistics, and Computer Science,
University of Illinois at Chicago, Chicago, IL 60607, USA
{bausth2,lreyzin}@uic.edu

Abstract. In this paper we analyze online problems from the perspective of when to switch solutions when the cost is of doing so is high – we call such a solution change a "recalculation." We analyze this problem under the assumption we have algorithms that achieve per-round regret (which can be otherwise thought of as point-wise error, or other well-studied quantities) of the form $O(1/t^\varepsilon)$ after seeing t data-points. We study schedules with a constant number and an increasing number of recalculations in the total number of datapoints, and we examine when achieving optimal cumulative regret is possible.

Keywords: Regret minimization · Recalculation schedules · Online learning

1 Introduction

In many online settings, one faces the problem on when to recalculate a given solution based on new data. This phenomenon occurs in various settings by different names: in the bandit framework, "sticky" decisions prevent the learner from switching arms without paying a cost [6,10,12], in the online clustering setting every time one switchs solutions, one may need to pay for new centers [3,4,8,14], in facility location, adding a new facility adds cost [7,13].

The specific details of different problems require the analysis of their respective structures and often have their own involved solutions. Here, we attempt to take a broader view, and we adopt the learning language of additive regret (instead of "competitive ratios," etc. from the streaming literature, though the results apply just as well there).

We assume we have a black-box algorithm $\mathcal{A}$ that produces an estimator with expected **per-round regret** of $\mathcal{R}_t(x) \in [0, 1]$ on a new data-point x when given samples $S_t = \{x_1 \ldots x_t\}$ and when point x is drawn from the same distribution as $x_1, \ldots x_t$, which we will generically refer to as D. Given the assumption above, we will drop the argument from the function $\mathcal{R}$. We also define $\mathcal{R}_0 = 1$, which means that until samples are given to $\mathcal{A}$, full regret is suffered by its (lack of) solution.

We call giving samples $\{x_1, \ldots x_t\}$ to $\mathcal{A}$ a **recalculation** at round t. We work in the setting where calls to $\mathcal{A}$ are expensive and need to be traded off against the benefits to improving regret. A call to $\mathcal{A}$ involves recalculating, and possibly committing to, a new solution to a given problem. For example, in the case of facility location, it involves the costly opening of new facilities (or the moving of old ones).

Hence, we are interested in understanding the optimal recalculation strategy as minimize cumulative regret for various budgets on calls to $\mathcal{A}$. Given a strategy that recalculates at rounds $C = \{t_1, t_2, \ldots\}$, where each t_i represents a sequential point drawn from the distribution, the **expected cumulative regret** $\mathcal{R}^T$ would be

$$\mathcal{R}^T = \sum_{i=1}^{T} \max_{t \in C \text{ s.t. } t < i} \mathcal{R}_t \tag{1}$$

In this paper, we will consider per-round regret guarantees of the form $t^{-\varepsilon}$

$$\mathcal{R}_t = O(1/t^\varepsilon), \tag{2}$$

which is known to be the asymptotically optimal regret for many problems in bandit and online learning in the stochastic setting [2]. Rephrasing this in our terminology, we have that for the recalculation schedule $C = \{1, \ldots, T\}$, the expected cumulative regret for $0 \leq \varepsilon < 1$

$$\mathcal{R}^T = \sum_{t=1}^{T} \mathcal{R}_{t-1} = O\left(\sum_{t=1}^{T} \frac{1}{t^\varepsilon}\right) = O(T^{1-\varepsilon}),$$

So even when recalculating at every round, a regret guarantee smaller than $O(T^{1-\varepsilon})$ is not possible, and we therefore call this rate **optimal**. This is the quantity we will compare to while attempting to do many fewer recalculations.

To simplify notation further, we will henceforth use $\mathcal{R}$ for expected regret. Therefore, all of our results will be on bounding regret in expectation.

Before proceeding, we note that another interesting setting $\varepsilon = 1$ for $\mathcal{R}_t = O(1/t)$, which is, for example, relevant for problems of estimation of parameters to minimize mean squared error (MSE), where the squared error of the empirical average of t samples versus the true estimate scales as σ^2/t [5]. Here, we would have

$$\mathcal{R}^T = O\left(\sum_{t=1}^{T} \frac{1}{t}\right) = O(\log T).$$

2 Warm up

We begin by analyzing the special case of $\mathcal{R}_t = 1/\sqrt{t}$. The argument in the previous section shows that the optimum regret in this situation is $O(\sqrt{T})$. However, this was the setting in which we recalculated every time we drew a point, which in the online setting isn't always practicable. Thus, inducing a trade-off between the amount of times we recalculate and the regret formed in our calculation.

Proposition 1. *Let D be a probability distribution from which T data points are sampled online and let $\mathcal{R}_t = O(1/\sqrt{t})$. Using one recalculation, one can achieve an expected regret of $O(T^{2/3})$.*

Proof. We first begin by calculating the optimal expected regret of calculating the algorithm just once. Let c be a constant such that $0 \leq c \leq 1$. We have that the expected regret must be:

$$\sum_{t=1}^{T^c} \mathcal{R}_0 + \sum_{t=T^c+1}^{T} \mathcal{R}_{T^c} = O\left(T^c + T^{1-c/2}\right).$$

We note that this is just a simple optimization of the right hand side, so we equate the exponents and get $c = 2/3$. This recovers the well-known bound of ε-first, (ε-greedy, and epoch-greedy) sampling [11, 15, 16].

3 Uniform Recalculations

One naive idea is to create a sampling schedule that solves our problem is to sample uniformly, so after observing a set ℓ amount of points we sample again and continue in this manner. However, we find this to at best require $O(T^\varepsilon)$ recalculations to get asymptotically optimum regret.

Lemma 1. *Let D be a probability distribution from which T data points are sampled online and let $\mathcal{R}_t = O(1/t^\varepsilon)$ for $0 \leq \varepsilon < 1$. If we recalculate uniformly after observing every ℓ points (and therefore recalculate T/ℓ times), then we incur an expected regret of $O\left(\ell + T^{1-\varepsilon}\right)$.*

Proof. We calculate the expected regret for recalculating after every ℓ points, we carry out the calculation by using the general formula for a uniform schedule.

$$\sum_{i=0}^{T/\ell} \sum_{t=1}^{\ell} \mathcal{R}_{\ell i} = \ell + O\left(\sum_{i=1}^{T/\ell} \sum_{t=1}^{\ell} \frac{1}{(i\ell)^\varepsilon}\right)$$

$$= \ell + O\left(\sum_{i=1}^{T/\ell} \frac{\ell^{1-\varepsilon}}{i^\varepsilon}\right)$$

$$= O\left(\ell + T^{1-\varepsilon}\right),$$

which completes the proof. □

We note briefly, that this proof holds in the $\varepsilon = 1$ case, we just get the result $\mathcal{R} = O(\ell + \log(T))$.

As the lemma above shows, a constant number of recalculations would imply linearly many recalculations between rounds and therefore linear regret. Therefore, we need a smarter recalculation strategy if we want to recalculate only a constant number of times. This is presented in the following section.

We see in the previous section that uniform recalculation scheduling fails to guarantee small expected regret with a small number of recalculations. So we pivot to non-uniform schedules. Taking, first, a look at what we can do with a constant number of recalculations.

4 Constant Number of Recalculations

We now consider the case when we are allowed a constant number of recalculations, but they need not be uniformly spaced.

Proposition 2. *Let D be a probability distribution from which T data points are sampled online and let $\mathcal{R}_t = O(1/t^\varepsilon)$ for $0 \leq \varepsilon \leq 1$. Using one recalculation, one can achieve an expected cumulative regret of*

$$\mathcal{R}^T = O\left(T^{1/(1+\varepsilon)}\right).$$

Proof. Consider recalculating after T^c rounds for $c = \frac{1}{1+\varepsilon}$. We suffer $O\left(T^{1/(1+\varepsilon)}\right)$ cumulative regret on those rounds. For the remaining $T - T^c \leq T$ rounds, we suffer at most

$$\begin{aligned}O\left(T/T^{c\varepsilon}\right) &= O\left(T^{1-c\varepsilon}\right) \\ &= O\left(T^{1-\varepsilon/(1+\varepsilon)}\right) \\ &= O\left(T^{1/(1+\varepsilon)}\right)\end{aligned}$$

regret. □

Proposition 3. *Let D be a probability distribution from which T data points are sampled online and let $\mathcal{R}_t = O(1/t^\varepsilon)$ for $0 \leq \varepsilon \leq 1$. Using two recalculations, one can achieve an expected cumulative regret of*

$$\mathcal{R}^T = O\left(T^{1/(1+\varepsilon+\varepsilon^2)}\right).$$

Proof. Consider recalculating after T^{c_1} and T^{c_2} for $0 \leq c_1 \leq c_2 \leq 1$. Extending the argument in Lemma 2, we incur a total regret of $\mathcal{R}^T \leq T^{c_1} + T^{c_2 - c_1 \varepsilon} + T^{1 - c_2 \varepsilon}$. Equalizing the three terms we get the equations

$$c_1 = c_2 - c_1\varepsilon \text{ and } c_1 = 1 - c_2\varepsilon.$$

Solving the above yields $c_1 = \frac{1}{1+\varepsilon+\varepsilon^2}$ and yields an expected cumulative regret of $O(T^{c_1})$. □

Theorem 1. *Let D be a probability distribution from which T data points are sampled online and let $\mathcal{R}_t = O(1/t^\varepsilon)$ for $0 \leq \varepsilon < 1$. Using n recalculations, one can achieve an expected cumulative regret of*

$$\mathcal{R}^T = O\left(nT^{\frac{1-\varepsilon}{1-\varepsilon^{n+1}}}\right).$$

Proof. Generalizing from the above, we divide into n recalculations and bound $\mathcal{R}^T \leq \sum_{i=1}^{n+1} T^{c_i - c_{i-1}\varepsilon}$ setting $c_0 = 0$ and $c_{n+1} = 1$.

We argue that solving for the system of equations leads to the following recursive formula:
$$c_j = c_{j+1} \frac{\sum_{i=0}^{j-1} \varepsilon^i}{\sum_{i=0}^{j} \varepsilon^i}.$$

We proceed inductively, we first see that $c_1 = c_2 - \varepsilon c_1$, which gets us $c_1 = c_2(1+\varepsilon)^{-1}$, as desired. Moving on, we assume this holds for c_{j-1} and solve for c_j.

$$c_j - \varepsilon c_{j-1} = c_{j+1} - \varepsilon c_j$$
$$c_j(1+\varepsilon) = c_{j+1} + \varepsilon c_{j-1}$$
$$c_j(1+\varepsilon) = c_{j+1} + c_j \varepsilon \frac{\sum_{i=0}^{j-2} \varepsilon^i}{\sum_{i=0}^{j-1} \varepsilon^i}$$
$$c_j\left(1 + \varepsilon - \frac{\sum_{i=1}^{j-1} \varepsilon^i}{\sum_{i=0}^{j-1} \varepsilon^i}\right) = c_{j+1}$$
$$c_j\left(1 + \varepsilon - 1 + \frac{1}{\sum_{i=0}^{j-1} \varepsilon^i}\right) = c_{j+1}$$
$$c_j \frac{\sum_{i=0}^{j} \varepsilon^i}{\sum_{i=0}^{j-1} \varepsilon^i} = c_{j+1}$$
$$c_j = c_{j+1} \frac{\sum_{i=0}^{j-1} \varepsilon^i}{\sum_{i=0}^{j} \varepsilon^i}.$$

So we have a recursive formula and now we can see that if we terminate after n recalculations we have a telescoping product, and we have that
$$c_1 = \frac{1}{\sum_{i=0}^{n} \varepsilon^i} = \frac{1-\varepsilon}{1-\varepsilon^{n+1}}.$$

Since the $n+1$ phases have equal regret, the total regret is $O(nT_1^c)$, producing the bound of $\mathcal{R}^T = O\left(nT^{\frac{1-\varepsilon}{1-\varepsilon^{n+1}}}\right)$. This bound tends to T^ε for arbitrarily high constant n. □

We observe that this proof does hold for $\varepsilon = 1$, we just leave c_1 unsimplified and have that the expected cumulative regret in this case is $O(nT^{1/n})$.

5 Schedules with Increasing Recalculations in T

After seeing some extreme examples and results on constants, we wish to discover how to improve the amount of times needed to recalculate while still maintaining the optimum expected regret. By employing a "doubling trick" argument, we see that we can easily reduce to $\log(T)$ recalculations. We note that other settings have also been studied that achieve optimal regret using logarithmically many policy "switches," [1,9].

Lemma 2. *Let D be a probability distribution from which T points are sampled online and let $\mathcal{R}_t = O(1/t^\varepsilon)$ for $0 \leq \varepsilon < 1$. There exists a sequence of $\log(T)$ recalculations for which we can achieve optimal expected regret of $O(T^{1-\varepsilon})$.*

Proof. Let $T = 2^m$, with m sufficiently large, and let $\mathcal{L} = \{2^0, 2^1, \ldots 2^{m-1}\}$. If we recalculate the algorithm after seeing each point in $\mathcal{L}$, then we have the following expected cost:

$$\sum_{i=0}^{m} \sum_{t=1}^{2^i} \mathcal{R} \leq O\left(\sum_{i=0}^{m} \sum_{t=1}^{2^i} \left(\frac{1}{2^{(i-1)\varepsilon}}\right)\right)$$

$$= O\left(\sum_{i=0}^{m} \left(2^{i-(i-1)\varepsilon}\right)\right)$$

$$= O\left(\sum_{i=0}^{m} \left(2^{i(1-\varepsilon)}\right)\right)$$

$$= O\left(2^{(1-\varepsilon)m}\right)$$

$$= O\left(T^{(1-\varepsilon)}\right).$$

This establishes that we can recalculate $\log(T)$ times and achieve optimal expected regret of $T^{1-\varepsilon}$. □

We observe that if the regret is of the form $\mathcal{R}_t = O(1/t)$ (the $\varepsilon = 1$ case), then the regret will be bounded by $O(\log T)$ in the above schedule.

We've established that we can achieve optimal expected regret with only $\log(T)$ recalculations, we go on to establish that we cannot improve this to $\log \log(T)$, by showing that the expected regret when taking $\log \log(T)$ recalculations is $O(\log \log(T)\sqrt{T})$.

Lemma 3. *Let D be a probability distribution from which T are sampled online and let $\mathcal{R}_t = O(1/t^\varepsilon)$ for $0 < \varepsilon < 1$, where $c = 1/\varepsilon$. There exists a sequence of $\log \log(T)$ recalculations such that we can achieve an expected regret of $O(\log \log(T) T^{1-\varepsilon})$.*

Proof. Let $T = c^{c^m}$, with m sufficiently large. Let $\mathcal{L} = \{\ell_0, \ell_1, \ldots \ell_m\}$. We recalculate after seeing each point in $\mathcal{L}$.

We wish to find a schedule that will cover all T with $\log \log(T)$ recalculations. We do this with the following schedule: at epoch i, once we have seen a total of $T^{1-c^{-i}}$ points we recalculate. We show that if we cover half of T at epoch $\log \log(T)$, then in the next epoch we will have seen all of T. So we establish this recalculation schedule achieves this. Let y be the amount of times needed to see half of T under the above scheduling scheme, so:

A Model for Optimizing Recalculation Schedules to Minimize Regret

$$1/c = T^{-c^{-y}}$$
$$-\log_c(c) = (-c^{-y})\log_c(T)$$
$$\frac{1}{\log_c(T)} = c^{-y}$$
$$y = \log_c \log_c(T).$$

We note that we found the exact point at which we would have half, and that this makes $\log\log(T)$ the minimum amount of rounds needed to cover all of T with this particular schedule. We can be sure of this since there were no assumptions made on the size of T, so we couldn't reduce to $\log\log\log(T)$ rounds. So now, we can sum our per-round cost over the amount of rounds we must run:

$$\sum_{i=0}^{m} \sum_{t=1}^{T^{1-c^{-i}}} \mathcal{R}_{T^{1-c^{-(i-1)}}} \leq \sum_{i=1}^{m} \sum_{t=1}^{T^{1-c^{-i}}} T^{-\frac{c^{i-1}-1}{c^i}}$$
$$= \sum_{i=1}^{m} T^{1-c^{-i}-c^{-1}+c^{-i}}$$
$$= \log\log(T) T^{1-c^{-1}}$$
$$= \log\log(T) T^{1-\varepsilon}.$$

Computing the inner sum, we see that at each round, we will have $O(T^{1-\varepsilon})$ expected regret. □

We note this proof does not extend to the $\varepsilon = 1$ case, since the argument hinges on the expected regret being bounded by $O(T^{1-\varepsilon})$.

Now we establish a lower bound, in particular that we cannot achieve an expected regret of $O(T^{1-\varepsilon})$ with $\log\log(T)$ recalculations.

Lemma 4. *Any schedule that performs $O(\log\log T)$ recalculations using an algorithm that suffers per-round regret of $\mathcal{R} = O(1/t^\varepsilon)$, $0 < \varepsilon < 1$, must suffer $\omega(T^{1-\varepsilon})$ cumulative regret.*

Proof. Given $\log\log(T)$ recalculations, we are lower-bounded by a regret of $O(T^{1-\varepsilon})$. Specifically, we show that we cannot achieve a regret of $O(T^{1-\varepsilon})$ given $\log\log(T)$ recalculations. So assume to the contrary, that there exists a schedule that does this. Now, to achieve this clearly the regret incurred at each recalculation has to be $O(T^{1-\varepsilon})$. Based on this we craft an inductive argument on the size of each recalculation round. At round $i-1$, the size, $T^{c_{i-1}}$ must be $O(T^{1-\varepsilon^i})$. As a base case, we see that $T^{c_0} = O(T^{1-\varepsilon})$. We move onto the inductive step: We note that the regret for round i is

$$T^{c_i} T^{-\varepsilon c_{i-1}} = O(T^{1-\varepsilon})$$
$$T^{c_i} = O(T^{1-\varepsilon+\varepsilon c_{i-1}}).$$

This implies,
$$c_i \leq 1 - \varepsilon + \varepsilon(1 - \varepsilon^{i-2})$$
$$c_i \leq 1 - \varepsilon^{i-1}.$$

This, then results in the total portion of T witnessed as $\sum_{n=1}^{\log\log(T)} T^{1-\varepsilon^n}$. Now, since
$$\sum_{n=1}^{\log\log(T)} T^{1-\varepsilon^n} \leq \log\log(T) T^{1-\varepsilon^{\log\log(T)}} = o(T),$$
we have failed to witness all of T in $\log\log(T)$ rounds and have reached a contradiction. □

We note this proof does not extend to the $\varepsilon = 1$ case, since the argument hinges on the expected regret being bounded by $O(T^{1-\varepsilon})$.

Acknowledgments. This work was supported in part by National Science Foundation grant ECCS-2217023. We thank the anonymous reviewers of this paper for their helpful comments.

References

1. Abbasi-Yadkori, Y., Pál, D., Szepesvári, C.: Improved algorithms for linear stochastic bandits. In: Shawe-Taylor, J., Zemel, R.S., Bartlett, P.L., Pereira, F.C.N., Weinberger, K.Q. (eds.) Advances in Neural Information Processing Systems 24: 25th Annual Conference on Neural Information Processing Systems 2011. Proceedings of a meeting held Granada, Spain, 12–14 December 2011, pp. 2312–2320 (2011)
2. Auer, P., Cesa-Bianchi, N., Freund, Y., Schapire, R.E.: The nonstochastic multi-armed bandit problem. SIAM J. Comput. **32**(1), 48–77 (2002)
3. Bhattacharjee, R., Imola, J., Moshkovitz, M., Dasgupta, S.: Online k-means clustering on arbitrary data streams. In: Agrawal, S., Orabona, F. (eds.) International Conference on Algorithmic Learning Theory, Singapore, 20–23 February 2023. Proceedings of Machine Learning Research, vol. 201, pp. 204–236. PMLR (2023)
4. Bhattacharjee, R., Moshkovitz, M.: No-substitution k-means clustering with adversarial order. In: Feldman, V., Ligett, K., Sabato, S. (eds.) Algorithmic Learning Theory, Virtual Conference, Worldwide. Proceedings of Machine Learning Research, 16–19 March 2021, vol. 132, pp. 345–366. PMLR (2021)
5. DeGroot, M.H.: Probability and Statistics (1986)
6. Dekel, O., Ding, J., Koren, T., Peres, Y.: Bandits with switching costs: $T^{2/3}$ regret. In: Shmoys, D.B. (ed.) Symposium on Theory of Computing, STOC 2014, New York, NY, USA, 31 May–03 June 2014, pp. 459–467. ACM (2014)
7. Fotakis, D.: On the competitive ratio for online facility location. Algorithmica **50**(1), 1–57 (2008)
8. Hess, T., Moshkovitz, M., Sabato, S.: A constant approximation algorithm for sequential random-order no-substitution k-median clustering. In: Ranzato, M., Beygelzimer, A., Dauphin, Y.N., Liang, P., Vaughan, J.W. (eds.) Advances in Neural Information Processing Systems 34: Annual Conference on Neural Information Processing Systems 2021, NeurIPS 2021, 6–14 December 2021, Virtual, pp. 3298–3308 (2021)

9. Jaksch, T., Ortner, R., Auer, P.: Near-optimal regret bounds for reinforcement learning. J. Mach. Learn. Res. **11**, 1563–1600 (2010)
10. Kash, I.A., Reyzin, L., Yu, Z.: Slowly changing adversarial bandit algorithms are provably efficient for discounted mdps. CoRR arxiv:2205.09056 (2022)
11. Langford, J., Zhang, T.: The epoch-greedy algorithm for contextual multi-armed bandits. Adv. Neural. Inf. Process. Syst. **20**(1), 96–1 (2007)
12. Machado, M.C., Bellemare, M.G., Talvitie, E., Veness, J., Hausknecht, M.J., Bowling, M.: Revisiting the arcade learning environment: evaluation protocols and open problems for general agents (extended abstract). In: Lang, J. (ed.) Proceedings of the Twenty-Seventh International Joint Conference on Artificial Intelligence, IJCAI 2018, Stockholm, Sweden, 13–19 July 2018, pp. 5573–5577. ijcai.org (2018)
13. Meyerson, A.: Online facility location. In: 42nd Annual Symposium on Foundations of Computer Science, FOCS 2001, Las Vegas, Nevada, USA, 14–17 October 2001, pp. 426–431. IEEE Computer Society (2001)
14. Moshkovitz, M.: Unexpected effects of online no-substitution k-means clustering. In: Feldman, V., Ligett, K., Sabato, S. (eds.) Algorithmic Learning Theory, Virtual Conference, Worldwide. Proceedings of Machine Learning Research, 16–19 March 2021, vol. 132, pp. 892–930. PMLR (2021)
15. Sutton, R.S., Barto, A.G.: Reinforcement Learning: An Introduction. MIT press, Cambridge (2018)
16. Tran-Thanh, L., Chapman, A., De Cote, E.M., Rogers, A., Jennings, N.R.: Epsilon–first policies for budget–limited multi-armed bandits. In: Proceedings of the AAAI Conference on Artificial Intelligence, vol. 24, pp. 1211–1216 (2010)

A Theory of Learning with Competing Objectives and User Feedback

Pranjal Awasthi[1(✉)], Corinna Cortes[1], Yishay Mansour[1,2], and Mehryar Mohri[1,3]

[1] Google Research, Mountain View, USA
pranjalawasthi@google.com
[2] Tel Aviv University, Tel Aviv, Israel
[3] Courant Institute of Mathematicl Sciences, New York, USA

Abstract. Large-scale deployed learning systems are often evaluated along multiple objectives or criteria. But, how can we learn or optimize such complex systems, with potentially conflicting or even incompatible objectives? How can we improve the system when user feedback becomes available, feedback possibly alerting to issues not previously optimized for by the system? We present a new theoretical model for learning and optimizing such complex systems. Rather than committing to a static or pre-defined tradeoff for the multiple objectives, our model is guided by the feedback received, which is used to update its internal state. Our model supports multiple objectives that can be of very general form and takes into account their potential incompatibilities. We consider both a stochastic and an adversarial setting. In the stochastic setting, we show that our framework can be naturally cast as a Markov Decision Process with stochastic losses, for which we give efficient vanishing regret algorithmic solutions. In the adversarial setting, we design efficient algorithms with competitive ratio guarantees. We also report the results of experiments with our stochastic algorithms validating their effectiveness.

Keywords: Multiobjective optimization · Online learning · ML fairness

1 Introduction

Learning algorithms trained on large amounts of data are increasingly adopted in a variety of applications and form the engine driving complex large-scale systems such as e-commerce platforms, online advertising auctions and recommender systems. Their system designer must take into account multiple metrics when optimizing them [26, 33, 36]. As an example, consider the case of a recommendation system for recipes, videos or fashion. There is no single metric that defines what a good recommendation engine should do. One needs to carefully take into consideration metrics measuring the quality of recommendations provided to end-users, their relevance and utility, the long-term growth of the content creators, and the overall revenue generated for the hosting platform. Furthermore, it is crucial to consider the risk of bias in these systems [21,43, 48]. Hence, additional metrics may need to be incorporated, such as performance across demographic groups, geographical locations or other identity terms. This can easily lead to hundreds of metrics that need to be simultaneously optimized for user satisfaction.

Further complicating the above scenario is the fact that often the multiple metrics considered are incompatible and inherently in conflict with each other [24,30,40]. For instance, in the context of a recommendation system, there is a tension between maximizing revenue via ad placements and maximizing end-user "happiness". Another tension may be between maximizing quality versus diversity of recommendations. In many cases, resolving such conflicts may force the designer to make hard choices among notions that seem perfectly reasonable in isolation, weighing in current use-patterns, wins and losses. An illuminating example is the analysis of the COMPAS tool for predicting recidivism by [2]. The authors showed that, among black defendants who do not recidivate, the tool predicted incorrectly at twice the rate than it did for white defendants who did not recidivate, i.e., the tool was unfair according to the *false positive rate* metric. The creator of the tool, Northpointe, responded by demonstrating that the tool was fair according to other natural measures such as AUC (Area Under the ROC Curve), for which each group had similar values. Later work showed that this tension is inherent and that it is often impossible to simultaneously satisfy multiple seemingly natural criteria [30] (see also [16]).

The above discussion raises the question of how one should define the optimal trade-off among multiple conflicting metrics to optimize for user satisfaction. A natural approach is to define the trade-offs in a static manner, either by using domain knowledge and human expertise, or by analyzing past historical data. Another line of work studies optimization in the presence of multiple objectives by designing algorithms that compete with *any* linear combination of the objectives [10,38] or by designing pareto-optimal solutions [40,41]. However, these solutions may be sub-optimal for the richer situation where user feedback is available. While algorithms tailored to a specific metric or a combination of metrics would be effective at first, experience shows that they become non pertinent over time: once a system is deployed and it interacts with its end-users, inefficiencies in the system design emerge, as evident via the user feedback, which in turn could lead one to prefer metrics originally not accounted for [34]. In the context of the COMPAS tool discussed above, this would correspond to the situation where user complaints make the system designers change loss function to ensure equal false-positive rates. The important aspect is that the underlying data distribution on which the tool has been trained doesn't change, new user feedback simply alerts the designers to short-comings of the system. Motivated by the above, in this work, we present a theoretical data-driven model for optimizing multiple conflicting metrics by taking into account the user feedback. Our proposed framework allows for the design of algorithmic solutions with strong theoretical guarantees.

In the context of a recommender system, user initiated feedback may be a"dislike", "too spicy", or "age inappropriate" [9], but feedback may also be indirectly observed by, e.g., high abandonment rates or low click-through rates. Going from complaints to actionable solutions involves many steps. First, the complaints are analyzed, typically by human specialists, and attributed to a set of predefined criteria, such as low accuracy of classifiers, false positive rates or AUC scores. Each complaint could trigger several criteria and a human specialist can monitor the aggregate performance on each criterion. Since criteria are often incompatible, based on the analysis of the complaints and their effect on the criteria, a decision is made to allocate resources to improve a subset of them and this process repeats [49]. While human involvement is crucial in the above process, a large portion of the above process could be made algorithmic and automated.

Our model assumes predetermined costs for user complaints along the multiple metrics. The difficulty in optimizing for user happiness arises from the fact that the nature and volume of the complaints depend on the state of the model. Of course if no complaints is received, an optimal state has been reached, but most often complaints will arise. Fixing the model to optimize for this set of complaints will most likely spur a different set of complaints, etc. Only by visiting all incompatible states of the model and observing the associated complaint set would one be able to fully optimize the model. Such an exhaustive search is prohibitive from both a time and development perspective. This paper presents a model that effectively reaches a beneficial state and provides performance guarantees.

The rest of the paper is organized as follows. In Sect. 2, we define our model. In the stochastic setting (Sect. 3), we show that our framework can be cast as a Markov Decision Process with stochastic losses, for which we give efficient vanishing regret algorithmic solutions. We also further discuss our modeling assumptions and extensions. The adversarial setting is discussed in Appendix D. Here, we give algorithms with competitive ratio guarantees. Section 4 demonstrates how our framework can be realized in practice and reports the results of experiments with our algorithms in the stochastic setting that demonstrate their effectiveness and the applicability of our model. We also defer the related work discussion as well as many of the proofs to the appendix.

2 Conflict Resolution Model

We consider optimization in the presence of multiple criteria, where not all criteria can be satisfied simultaneously. The constraints are specified by an undirected graph $\mathcal{G} = (V, E)$, where each vertex represents a criterion and where an edge between vertices v_i and v_j indicates that criteria v_i and v_j cannot be simultaneously satisfied. We denote by $V = \{v_1, \ldots, v_k\}$ the set of k criteria considered. Figure 1 illustrates these definitions. Note that vertices may represent joint criteria as in Fig. 1(b).

We consider a machine learning system that evolves over a sequence of time steps in the presence of the criteria represented by the graph $\mathcal{G}$. At each time step t, the system is in some state s_t characterized by its performance on all criteria in V. Note, a state is distinct from a vertex of $\mathcal{G}$. The system then receives a new batch of feedback that depend on its current state and incurs a loss. The objective of the algorithm is to minimize the total cost incurred over a period of time, which includes the total loss accrued, as well as the total cost of fixing various criteria over that period. We envision that the algorithm is solving a constraint optimization problem with the criteria as constraints.

The assignment of a complaint to criteria can be achieved by human analysis or via a multi-class multi-label classifier trained on past data and making use of known classifiers for specific criteria. Even when a complaint is related to a single criterion, we do not simply advocate taking that raw feedback as the ground truth. We discuss the risks associated with doing so in Sect. 3 and Appendix F, in the context of the COMPAS example. To further improve and maintain the accuracy of this multi-class multi-label classifier, in practice, there may be ongoing data labeling and assistance by expert auditors analyzing complaints. Note that not all complaints received by the system are relevant and the classifier, or a human in the loop, may decide to not assign

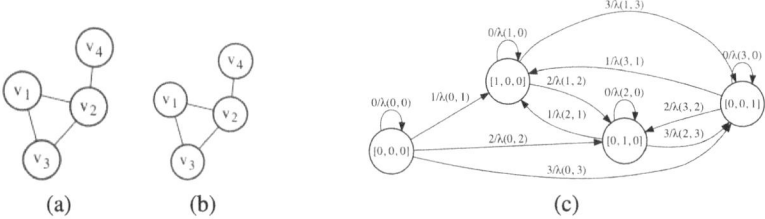

Fig. 1. (a) Illustration of constraints graph $\mathcal{G}$. Vertices v_1, v_2, v_3, v_4 represent 4 different criteria. (b) More generally, each vertex can represent a joint criterion, for example $v_1 \wedge v_2$. This helps specify joint constraints such as the following: v_1, v_2, and v_3 cannot be simultaneously satisfied. (c) Illustration of the MDP for a fully connected incompatibility graph $\mathcal{G}$ over three criteria. The state set is $\mathcal{S} = \{\mathbf{0} = [0,0,0], \mathbf{1} = [1,0,0], \mathbf{2} = [0,1,0], \mathbf{3} = [0,0,1]\}$, the action set $\mathcal{A} = \{0, 1, 2, 3\}$. Each transition is labeled with $a/\lambda(s, a)$, where a is the action taken from s and where $\lambda(s, a)$ is the total loss incurred as a result.

a complaint to any criterion. This also helps protect the system against potential attacks by coordinated users. Recent work on interactive models for ML fairness has studied this for specific metrics and auditor behavior [5].

Loss. As a result of the complaints, the system incurs a loss and responds by changing its state. The definition of the loss, which depends on the criteria affected by the complaints is critical, a poor choice can yield a so-called *loudest voice* effect (see discussion in Sect. 3). The notion of complaints and the associated loss may seem abstract at this point. In Sect. 4, we demonstrate how our model can be applied in practice.

Graph and Criteria. The assumption that the graph $\mathcal{G}$ is known a priori may seem restrictive. However, in many settings, graph $\mathcal{G}$ can be derived from analyzing past complaints and by measuring how fixing one criterion affects the performance on others. For instance, in the recommendation system example discussed above, where each metric corresponds to the false positive rate on a different slice of the data, one can easily use past data to see how optimizing the false positive rate on one slice affects the other and get the graph of incompatibilities. See the experiments in Sect. 4 for a more concrete example. Our model also provides the flexibility of accounting for incompatibilities among criteria such as those discussed by [30] and [16]. This can be achieved by augmenting the graph with vertices representing joint criteria as in Fig. 1(b). The graph stipulates in particular that v_1, v_2 and v_3 cannot be all simultaneously satisfied.

States. We will adopt the following simplifying assumptions and will later discuss their extensions or relaxation in Sect. 3. We assume that each criterion can only be in one of two states: *fixed*, meaning that criterion v_i is met or is not violated, or *unfixed*, meaning the opposite. Hence, the overall state of the system can be described by a k-dimensional Boolean vector.[1] An action corresponds to fixing a particular criterion,

---

[1] Note, that this implies that the number of states is exponential, hence, any algorithm which depends polynomially on the number of states would have an exponential overhead.

or set of criteria, and moving to a different state. *Fixing* the criteria associated to v_i entails an algorithmic and resource allocation cost that we denote by c_i. Initially, all criteria are unfixed. At each time step, a conflict resolution system or algorithm selects some action, which may be to fix an unfixed vertex v_i, thereby incurring the cost c_i and *unfixing* any vertex adjacent to v_i, or the algorithm may select the null action, not to fix or unfix any vertex and wait to collect more data. Note that the incompatibilities in our model defined via edges in the graph are data agnostic. In practice, it is possible that two generally incompatible criteria can be simultaneously satisfied for a given dataset, say via incorporating a slack. This is a direction for future work.

Fixing Costs. This can be estimated from past experience. In the absence of any prior information, one could assume a unit fixing cost. We deliberately avoid making specific choices. This gives us flexibility in dealing with different types of metrics in a unified manner. While the focus of our study is theoretical, let us emphasize that our model is easily applicable and implementable in practice. We further discuss this in the end of Sect. 3 and illustrate it in Sect. 4.

3 Stochastic Setting

We first detail a stochastic setting of our model that can be described in terms of a Markov Decision Process (MDP). Next, we present algorithms with strong regret guarantees.

Description. The distribution of complaints received by the system is a function of its current *state*, i.e., the current set of fixed or unfixed criteria v_i. Thus, we consider an MDP with a state space $\mathcal{S} \subseteq \{0,1\}^k$ representing the set of bit vectors for criteria: a state $s \in \{0,1\}^k$ is defined by $s(i) = 0$ when criterion v_i is unfixed and $s(i) = 1$ when it is fixed. By definition of the incompatibility graph $\mathcal{G}$, s is a valid state iff the set of fixed criteria at s is an independent set of $\mathcal{G}$.

When in state $s \in \mathcal{S}$, the system incurs a loss ℓ_i^s due to complaints related to criterion $i \in [k]$. Loss ℓ_i^s is a random variable assumed to take values in $[0, B]$ with mean μ_i^s. We do not assume independence across criteria, i.e., ℓ_i^s and ℓ_j^s may be dependent for a given state s. The action set is $\mathcal{A} = \{0, 1, \ldots, k\}$. A non-zero action i corresponds to fixing criterion i. Action 0 is the null action, i.e., no criterion is fixed. Transitions are deterministic: given state s and action $i \in \mathcal{A}$, the next state is s if $i = 0$, otherwise, for $i \neq 0$ the next state is the state s' that only differs from s by $s'(i) = 1$ and (possibly) $s'(j) = 0$ for all $j \in N(i)$, where $N(i)$ is the neighbors of v_i in $\mathcal{G}$, since neighbors of i must be unfixed once i is fixed.

Each action $a = i$ admits a fixing cost c_i. The cost for unfixing, as well as the null action, is zero. The loss incurred when taking action a at state s is the sum of the fixing cost c_a and the complaint losses at the (possibly) next state s': $\lambda(s,a) = c_a + \sum_{i=1}^{k} \ell_i^{s'}$. The expected loss of transition (s, a, s') is:

$$\mathbb{E}\left[c_a + \sum_{i=1}^{k} \ell_i^{s'}\right] = c_a + \sum_{i=1}^{k} \mu_i^{s'}. \tag{1}$$

Note, c_a and the losses $\ell_i^{s'}$ are observed by the algorithm, but the mean values $\mu_i^{s'}$ are unknown. To keep the formalism simple we assume that the cost c_a of taking an action a is independent of the current state s.

Figure 1(c) illustrates our stochastic model for three mutually incompatible criteria. The notion of each metric in a binary state is a simplifying modeling assumption for our theory. We discuss this more at the end of this section.

Correlation Sets. In practice, the distribution of complaints related to a criterion v_i at two different states may be related. To capture these correlations in a general way, we assume that a collection $\mathcal{C} = \{\mathcal{C}_1, \mathcal{C}_2, \ldots, \mathcal{C}_n\}$ of *correlation sets* is given, where each $\mathcal{C}_j$ is a subset of the k criteria and has size at most m. By allowing correlation sets of varying sizes, we can capture a range of dependencies that may exist between different criteria. These dependencies affect the loss observed by the algorithm at each time.

$$\begin{array}{l} \mathcal{C}_1 \boxed{V_1} \boxed{V_2} \\ \mathcal{C}_2 \boxed{V_4} \boxed{V_3} \end{array} \quad \mathcal{C}_3 \qquad \begin{array}{l} \mu_1 = \theta_1^s \\ \mu_2 = \theta_1^s + \theta_3^s \\ \mu_3 = \theta_3^s \\ \mu_4 = \theta_2^s \end{array}$$

Fig. 2. Example of correlation sets and associated losses for a graph with four criteria.

We assume that at a given state s, each set $\mathcal{C}_j$ generates losses with mean value θ_j^s per vertex, and that if two states s and s' admit the same configuration for the vertices in $\mathcal{C}_j$, then they share the same parameter $\theta_j^s = \theta_j^{s'}$. Given a criterion i and a state s, we assume that the loss incurred by criterion i equals the sum of the individual losses due to each correlation set $\mathcal{C}_j$ that contains i. Thus, μ_i^s can be expressed as follows: $\mu_i^s = \sum_{j=1}^n \theta_j^s \mathbb{I}(i \in \mathcal{C}_j)$. If a criteria is not correlated with any other vertex, we add to $\mathcal{C}$ a correlation set of size one for that criterion. See Fig. 2 for an illustration. For each $j \in [n]$, there are at most 2^m configurations for the vertices of $\mathcal{C}_j$ in a state s, hence there are at most $2^m n$ distinct parameters θ_j^s. Let $\boldsymbol{\theta}$ denote the vector of all distinct parameters θ_j^s. Our MDP model can then be denoted MDP$(\mathcal{S}, \mathcal{A}, \mathcal{C}, \boldsymbol{\theta})$.

Algorithm. We consider an online algorithm that at time t takes action a_t from state s_t and reaches state s_{t+1}, starting from the initial state $(0, \ldots, 0)$. settings, The objective of an algorithm can be formulated as that of learning a policy, that is a mapping $\pi: \mathcal{S} \to \mathcal{A}$, with a value close to that of the optimal. We are mainly interested in the cumulative loss of the algorithm over the course of T interactions with the environment. The goal is to minimize the pseudo-regret:

$$\text{Reg}(\mathcal{A}) = \sum_{t=1}^T \mathbb{E}\left[\lambda_t(s_t, a_t)\right] - \sum_{t=1}^T \mathbb{E}\left[\lambda_t(s_t^{\pi^*}, \pi^*(s_t^{\pi^*}))\right],$$

where $\lambda_t(s, a)$ is the total loss incurred by taking action a at state s at time t, $s_1 = (0, \ldots, 0)$ and π^* is the optimal policy. Note, λ_t is only a function of the current state and the action taken. The expectation is over the random generation of the complaint losses. Given the correlation sets and the parameter θ, the optimal policy π^* corresponds to moving from the initial state $(0,\ldots,0)$ to the state $s^* \in \mathcal{S}$ with the most favorable distribution and remaining at s^* forever. We define by $g(s)$ the expected (per time step) loss incurred by staying in state s, that is, $g(s) \doteq \sum_{i=1}^{k} \mu_i^s$. The optimal state s^* is then defined as follows:

$$s^* = \operatorname*{argmin}_{s \in \mathcal{S}} g(s). \qquad (2)$$

Note, in this definition we disregard the one-time cost of moving to a state from the initial state, since in the long run the expected cost incurred by staying at a given state governs the choice of the optimal state. Since our problem can be seen as that of learning with a deterministic MDP with stochastic losses, we could adopt an existing algorithm for that problem [23]. However, the running-time of such algorithms would directly depend on the size of the state space $\mathcal{S}$, which here is exponential in k, and that of the action set $\mathcal{A}$. Furthermore, the regret guarantees of these algorithms would also depend on $|\mathcal{S}||\mathcal{A}|$. Instead, by exploiting the structure of the MDP, we can design vanishing regret algorithms with a computational complexity that is only polynomial in k and the number of parameters. We will assume access to an oracle that, given θ, can optimize (2). In Appendix B, we show how to approximately solve (2) for the case of singleton correlation sets, where the true parameters θ can also be estimated efficiently (see Theorem 4).

There are important distinctions between our proposed model and the standard online learning setting. We have a notion of a state that evolves as a result of the stochastic losses suffered. These losses in turn depend on the distribution of complaints received. This distribution should not be confused with the distribution of data over which classifiers may be trained to fix a criterion. The latter is assumed constant in time. Via correlation sets we can model complex correlations among the criteria when defining the stochastic losses.

Case $m = 2$. To illustrate the ideas behind our general algorithm, we first consider a simpler setting where correlation sets are defined on subsets of size at most two. This setting also captures an important case where fixing a particular criterion affects the complaints of its neighbors. The algorithmic challenge we face here is to avoid exploring the exponentially many states in the MDP. Instead, we will design an algorithm that spends an initial exploration phase by visiting a specific subset of states of size at most $4n$. This subset denoted by $\mathcal{K}$, that we call a *cover* of $\mathcal{C}$ will help the algorithm estimate the expected loss of any state in the MDP given the estimates of losses for states in the cover. After the exploration phase, the algorithm creates an estimate $\hat{\theta}$ of the true parameter vector θ, uses the optimization oracle for solving Eq. (2) to find a near optimal state $\hat{s}$ and selects to stay at state $\hat{s}$ for the remaining time steps. We next define the cover.

For two criteria i, j and $b \in \{0, 1\}$, we say that (i, j, b) is a *dichotomy* if there exist two states $s, s' \in \mathcal{S}$ such that: (1) $s(j) = 0$ and $s'(j) = 1$, and (2) $s(i) = s'(i) = b$. We call the two states s, s' an (i, j, b)-pair. Note that if an edge (v_i, v_j) is present in $\mathcal{G}$, then $(i, j, 1)$ cannot be a dichotomy, since criteria i and j cannot be fixed simultaneously. A cover $\mathcal{K}$ of $\mathcal{C}$ is simply a subset of the states in the MDP that contains an (i, j, b)-pair for every $\{i, j\} \in \mathcal{C}$ and valid dichotomy (i, j, b).

> **Input:** The graph $\mathcal{G}$, correlation sets $\mathcal{C}$, fixing costs c_i.
> 1. Pick a cover $\mathcal{K} = \{s_1, s_2, \ldots, s_r\}$ of $\mathcal{C}$.
> 2. Let $N = 10 \frac{T^{2/3}(\log rkT)^{1/3}}{r^{2/3}}$.
> 3. For each state $s \in \mathcal{K}$ do:
>   – Move from current state to s in at most k time steps.
>   – Play action $a = 0$ in state s for the next N time steps to obtain an estimate $\widehat{\mu}_i^s$ for all $i \in [k]$.
> 4. Using the estimated losses for the states in $\mathcal{K}$ and Equation (3), run the oracle for the optimization (2) to obtain an approximately optimal state $\hat{s}$.
> 5. Move from current state to $\hat{s}$ and play action $a = 0$ from $\hat{s}$ for the remaining time steps.

Fig. 3. Algorithm for $m = 2$ achieving $\tilde{O}(T^{2/3})$ pseudo-regret.

Furthermore, for every singleton set $\{i\}$ in $\mathcal{C}$, $\mathcal{K}$ contains states s, s' such that $s(i) = 0$, $s'(i) = 1$ and $s(j) = s'(j)$ for all $j \neq i$. Note that we only need the cover to contain an (i, j, b)-pair if $\{i, j\}$ is a correlation set. Hence, it is easy to see that when $m=2$, there is always a cover of size at most $4n$. Next, we state our key result that estimating the loss values for the states in a cover is sufficient.

Theorem 1. *Let $\mathcal{K}$ be a cover for $\mathcal{C}$. For any state $s \in \mathcal{S}$ and any $i \in [k]$ with $s(i) = b$, we have:*

$$\mu_i^s = \mu_i^{s'} + \sum_{j=1}^{k} X_b^{i,j} \left[\mathbb{I}(s(j) = 1) \mathbb{I}(s'(j) = 0) \right]$$
$$- \sum_{j=1}^{k} X_b^{i,j} \left[\mathbb{I}(s(j) = 0) \mathbb{I}(s'(j) = 1) \right], \quad (3)$$

where s' is any state in $\mathcal{K}$ with $s'(i) = b$, and for $\{i, j\} \in \mathcal{C}$, $X_b^{i,j} := \mu_i^{s_1} - \mu_i^{s_2}$ where (s_1, s_2) is some (i, j, b) pair. If $\{i, j\} \notin \mathcal{C}$, we define $X_b^{i,j}$ to be zero.

From the above theorem we have the following guarantee.

Theorem 2. *Consider an MDP$(\mathcal{S}, \mathcal{A}, \mathcal{C}, \theta)$ with losses in $[0, B]$, maximum fixing cost c, and correlations sets of size at most $m = 2$. Let $\mathcal{K}$ be a cover of $\mathcal{C}$ of size $r \leq 4n$, then, the algorithm of Fig. 3 achieves a pseudo-regret bounded by $O(kr^{1/3}(c + B)(\log rkT)^{1/3}T^{2/3})$. Furthermore, given access to an oracle for (2), the algorithm runs in time polynomial in k and $n = |\mathcal{C}|$.*

There is a natural extension to arbitrary correlation sets via extending the notion of a dichotomy and a cover (Algorithm in Fig. 7, Appendix B). Our algorithms are also scalable. During step 1 they only explore the states in the cover $\mathcal{K}$ that could be much smaller than the full state space.

Beyond $T^{\frac{2}{3}}$ Regret. Next, we present algorithms that achieve $\tilde{O}(\sqrt{T})$ regret, first in the case $m = 1$, next for any m, under the assumption that each criterion does not participate in too many correlations sets. Although our problem can be cast as an instance of the

---

Input: graph $\mathcal{G}$, correlation sets $\mathcal{C}$, fixing costs c_i.

1. Let $\mathcal{K}$ be the cover of size $k + 1$ that includes the all zeros state and the states corresponding to indicator vectors of the k vertices.
2. Move to each state in the cover once and update the optimistic estimates according to (4).
3. For episodes $h = 1, 2, \ldots$ do:
   - Run the optimization oracle for solving Eq. (2) with the optimistic estimates as in (4) to get a state s.
   - Move to state s. Stay in state s for $t(h)$ time steps and update corresponding estimates using (4). Here $t(h) = \min_i \tau_{i,t_h}^{s(i)}$ and t_h is the total number of time steps before episode h starts.

Fig. 4. Online algorithm for $m = 1$ with $\tilde{O}(\sqrt{T})$ regret.

stochastic multi-armed bandit problem with switching costs, and arms corresponding to the states in the MDP, existing algorithms achieving $\tilde{O}(\sqrt{T})$ have time complexity that depends on the number of arms which in our case is exponential (2^k) [8,42]. We will show here that, in most realistic instances of our model, we can achieve $\tilde{O}(\sqrt{T})$ regret efficiently. When correlation sets are of size one, the parameter vector $\boldsymbol{\theta}$ can be described using the following $2k$ parameters: for each $i \in [k]$, let γ_i^0 denote the expected loss incurred by criterion i when it is unfixed and γ_i^1 its expected loss when it is fixed. Our proposed algorithm is similar to the UCB algorithm [3]. For every vertex i, let $\tau_{i,t}^0$ be the total number of time steps up to t (including t) during which v_i is in an unfixed position and let $\tau_{i,t}^1$ be the number of times steps up to t during which v_i is in a fixed position. Fix $\delta \in (0, 1)$ and let $\hat{\gamma}_{i,t}^b$ be the empirical average loss observed when vertex v_i is in state b, for $b \in \{0, 1\}$. Our algorithm maintains optimistic estimates

$$\tilde{\gamma}_{i,t}^b = \hat{\gamma}_{i,t}^b - 10B\sqrt{\frac{\log(kT/\delta)}{\tau_{i,t}^b}}. \qquad (4)$$

The algorithm divides the T time steps into consecutive intervals that we call as *episodes*. In episode h, the algorithm moves to and stays at a fixed state for $t(h)$ time steps. in a fixed state. At the end of the episode it makes a query to the optimization oracle (using the current optimistic estimates) to decide on the state to go to for the next episode. The algorithm carefully chooses $t(h)$ to maintain low regret. The algorithm is described in Fig. 4. We will prove that it benefits from the following regret guarantee.

Theorem 3. *Consider MDP*$(\mathcal{S}, \mathcal{A}, \mathcal{C}, \boldsymbol{\theta})$ *with losses in* $[0, B]$ *and maximum fixing cost c. Given correlations sets $\mathcal{C}$ of size one, the algorithm of Fig. 4 achieves a pseudo-regret bounded by $O(k^2(c + B)^2\sqrt{T}\log T)$. Furthermore, given access to an oracle for (2), the algorithm runs in time polynomial in k.*

The algorithm of Fig. 4 can be extended to higher m values (see Fig. 9 in Appendix C).

Modeling Assumptions and Extensions. Here we briefly discuss assumptions and extensions.

Scalability. The running time of our algorithms depends linearly on the size of the cover $\mathcal{K}$. While in the worst case the size of the cover could be exponential in n, m, in practice, we expect it to be rather small.

Loss Function. The choice of the loss function is critical. We made the simplifying assumption that the loss at each time step is additive in the losses incurred by correlation sets. A careless choice of what the additive losses correspond to may result in a sub-optimal overall. For example, a poor choice is one that uses the volume of complaints, i.e., how many complaints have triggered a criterion. This will make us vulnerable to the loudest voices in the system. In Sect. 4, we discuss how our framework can be implemented in practice and present reasonable choices for the loss function. We further discuss the choice of the loss function in the case of the COMPAS example in Appendix F.

Adversarial Manipulation. Our model may be vulnerable to strategic coordination. A malicious group of users can form a sub-community generating a large number of complaints to press the system to include a new criterion in the graph. The presence of such poor criteria may result in an overall suboptimal system. Modeling this scenario is beyond the scope of the current work.

Continuous States. While this is a direction for future work, our method offers a simple way to achieve this by adding, for each criterion i, new criteria to the graph with different levels or thresholds $\tau_1, \tau_2, \ldots$ for satisfying i.

4 Experiments

Real-World Dataset. We next illustrate how our framework can be applied to real-world data. Due to space constraints, we provide a brief description here and refer the reader to Appendix E for details and more results. We studied the UCI Adult dataset [31] which includes $48{,}852$ examples, each represented by 124 features, after processing. Each data point corresponds to a person and the label is a $0/1$-value representing whether the income of the person is more or less than \$50,000. The dataset contains information about sensitive attributes such as race and gender. We simulated an online scenario where a classifier is making predictions on the income of individuals. At each time step, a batch of complaints arrive, the system incurs a loss and responds by transitioning to a different state (and updating the classifier). We now describe the various components.

Graph $\mathcal{G}$: We used race in {black, white} as an attribute to obtain two sub-populations and considered two natural criteria, namely the true positive rate and the AUC score, equivalent to the criteria of the COMPAS tool. This leads to four vertices $tpr_w, tpr_b, auc_w, auc_b$. Furthermore, we added the overall classifier accuracy as a fifth vertex. We consider a unit fixing cost for all criteria.

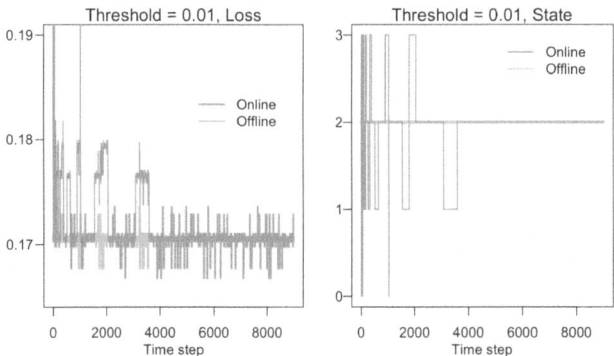

Fig. 5. Performance of the Algorithm of Fig. 4. Loss of the offline and online algorithms and states chosen by the online algorithm, as a function of the time steps.

Losses and Correlation Sets: We consider correlation sets of size one, and the total loss of a state is the sum of the losses of each criterion. For the accuracy vertex, we define the loss to be the error of the classifier. For a vertex, say tpr_w, if the overall tpr of the classifier and the tpr on the white population deviated by more than τ, we penalized the classifier linearly: the loss for tpr_w was defined as: $\max(0, |tpr_{overall} - tpr_w| - \tau)$. Other losses are defined similarly. We set $\tau = 0.005$.

Incompatibilities and State Transitions: We solved, for each state $s \in \{0,1\}^5$, an optimization via the tensorflow constrained optimization toolkit [12, 13] to get a classifier. We evaluated the classifier on a test set and if the loss of any criterion was more than a specific threshold (0.01), we considered the state invalid. As an example, in the instance corresponding to Fig. 5, we obtained four valid states. We obtained the state transitions as follows. If the algorithm asked to fix v_i in state s, we set $s(i) = 1$ to go to the next state s'. While s' is invalid, we unfixed the criterion (not including v_i) with the highest loss in the state s' to reach another state.

Simulating Complaints: We divided the dataset into 16,000 examples that we used to update our classifier at each time step and the remaining *test* set to simulate the arrival of complaints. At each step, we randomly selected a batch of examples from the test set to generate complaints. This batch was used to compute the loss at the given time step.

Benchmark and Results: We compared our Algorithm of Figure 4 with an offline optimal solution computed by finding the state with the minimum average loss over the sequence of complaints. The results are in Fig. 5. We plot the loss of the algorithm as compared to the benchmark, as well as the states chosen by the algorithm, as a function of the time steps. Our algorithm quickly converges to the offline optimal solution after an initial exploration phase. Note that the choice of the loss functions was important in this case and that we did not weight each criterion by the volume of the complaints. This demonstrates that our algorithms, when combined with a good choice of the loss function, can be useful in practice. See Appendix E for details and additional experiments.

5 Conclusion

We presented a new data-driven model of online optimization from user feedback in the presence of multiple criteria, with algorithms benefiting from theoretical guarantees both in the stochastic and the adversarial setting. We provided empirical evidence that our model can be effectively realized in practice. Several extensions are worth exploring in future work. These include fixing costs that can vary with time to capture varying algorithmic price and human effort cost. Similarly, the losses in our stochastic model could be time-dependent to express the growing cost of a criterion not being addressed.

Acknowledgements. YM has received funding from the European Research Council (ERC) under the European Union's Horizon 2020 research and innovation program (grant agreement No. 882396), by the Israel Science Foundation (grant number 993/17), the Yandex Initiative for Machine Learning at Tel Aviv University and a grant from the Tel Aviv University Center for AI and Data Science (TAD).

A Related Work

There is extensive literature on optimizing multiple metrics or objectives under specific criteria. The recent works of [10,38] consider optimizing in the presence of multiple base objectives. Given objectives $L_1,\ldots,L_i$ these works aim to design "agnostic" algorithms that can simultaneously compete with any linear or convex combination of the objectives. Another line of work considers design algorithms that can achieve the Pareto optimal solution [25,35,40,41].

Another line of work considers optimizing multiple constraints (inspired by group fairness metrics) via constrained non-convex optimization [1,12,44]. These publications either reduce the problem to that of cost-sensitive classification [1,15] or replace the non-convex constraints by convex proxies and next optimize them via external or swap regret minimization algorithms [12,13].

There have also been studies of the inherent tension between satisfying multiple metrics. [30] and [16] demonstrate that it is impossible to satisfy equal opportunity and calibration at the same time. Inspired from fairness applications the work of [37] studies the tradeoff between accuracy and other metrics of interest such as false positive and false negative rates.

Recent works have also studied the long-term impact of optimizing multiple conflicting criteria in settings with feedback mechanisms [20,27,34,39]. [34] show that, in certain situations, constrained loss minimization to equalize certain criteria could lead to further disparate impact on the end users in the long run. [20] proposed algorithms for minimizing such disparate impact in settings involving repeated loss minimization. More recently, [22,47] study the problem of satisfying multiple constraints in reinforcement learning settings involving a Markov Decision Process. The authors in [22] consider learning in an MDP where the criteria to be optimized require that the algorithm never takes an action a over action a' if the long-term reward is higher. It is clear to see that the optimal policy for the MDP indeed satisfies this property. Hence, there does exist a policy that satisfies the required criterion. However, the authors show that finding a near optimal policy while satisfying the criterion requires time exponential in the size of the state space.

[47] consider other metrics such as demographic parity in the context of learning in MDPs. [14] show that existing importance sampling methods for off-policy policy selection in reinforcement learning can lead to bad outcomes according to other natural criteria and present algorithms to mitigate this effect.

While our work also involves learning in a Markov Decision Process (MDP) and optimizing multiple criteria in the long term, the setup and the motivation are different. Unlike all the previous work mentioned, we do not commit to a fixed definition of quality or a metric, and allow for arbitrary criteria. Hence, states in our MDP correspond to the current configurations of different criteria. Rather than studying each metric in isolation, the objective of our work is to propose a data-driven model that can learn from feedback, a near-optimal configuration of the metrics to impose on the system. To the best of our knowledge, ours is the first work to incorporate optimizing metrics of arbitrary types in an online setting. In this context, inspired by fairness applications, the recent work of [29] studies a specific combination of group and individual fairness metrics. The authors consider a setting where there is a distribution over individuals as well as a distribution over classification tasks. They consider algorithms for achieving *average* individual fairness, that is in expectation over classification tasks, the performance of the algorithm on a group fairness metric such as demographic parity should be the same for each individual.

An important aspect of our stochastic MDP-based model requires the ability to observe the losses associated with different criteria at each time. This relates to the problem of evaluating and monitoring the performance of the system according to different metrics from data. There has been work in recent years on developing auditing and monitoring approaches [4,6,17]. Furthermore, many metrics require access to both labeled data and to certain sensitive attribute information such as race or gender, for accurate evaluation. A recent line of work has studied this estimation problem when one has limited and/or noisy access to sensitive attribute information [11,18,32,46]. Finally, we note that our model learns from feedback received as a form of complaints. These complaints are a result of a (potentially incorrect) decision made by an ML system. There has been recent work in developing counterfactual based explanations [45] for such decisions and exploring recourse strategies [19].

B Stochastic Setting

We first show that in the stochastic model, if correlation sets are of size one then one can efficiently approximate the cost of the optimal state up to a factor of two.

Theorem 4. *If correlations sets are of size one ($m = 1$), then, for any $\epsilon, \delta > 0$, the true parameter vector for MDP($\mathcal{S}, \mathcal{A}, \mathcal{C}, \boldsymbol{\theta}$) can be approximated to ϵ-accuracy in ℓ_∞-norm with probability at least $1 - \delta$, in at most $O(\frac{B^2 k}{\epsilon^2} \log(\frac{k}{\delta}))$ time steps and exploring at most $k + 1$ specific states in $\mathcal{S}$. Furthermore, given a parameter vector $\boldsymbol{\theta}$, there is an algorithm that runs in time polynomial in k and finds an approximately optimal state s' such that $g(s') \leq 2 \min_{s \in \mathcal{S}} g(s)$.*

Proof. Notice that when correlation sets are of size one, the expected loss incurred for criterion v_i at any given state s solely depends on whether $s(i) = 0$ or $s(i) = 1$. Hence in this case the MDP consists of $2k$ parameters where we use γ_i^1 and γ_i^0 to denote the expected losses incurred by vertex i when it is in fixed and unfixed position respectively. For any $\delta > 0$, by Hoeffding's inequality, we have that if we stay in state $s = (0, 0, \ldots, 0)$ for $N = \frac{B^2}{\epsilon^2} \log(2k/\delta)$ time steps then with probability at least $1 - \frac{\delta}{2}$, we have each γ_i^0 estimated up to ϵ accuracy. Let $e_i \in \{0, 1\}^k$ denote the indicator vector for i. If we stay in state $s = e_i$ for $\frac{B^2}{\epsilon^2} \log(2k/\delta)$ time steps, then with probability at least $1 - \frac{\delta}{2}$ we have γ_i^1 estimated up to ϵ accuracy. Hence, overall after $O(\frac{B^2 k}{\epsilon^2} \log(\frac{k}{\delta}))$ time steps, we have each parameter estimated up to ϵ accuracy. Notice that in total we observe at most $k+1$ states.

Next, we show how to efficiently approximate the loss of the best state. Given the parameters of the MDP each vertex has two costs $\Lambda_i^{(1)} = \gamma_i^0$, denoting the cost incurred if the vertex is unfixed and $\Lambda_i^{(2)} = c_i + \gamma_i^1$, denoting the cost incurred if the vertex is fixed. Without loss of generality assume that $\Lambda_i^{(1)} > \Lambda_i^{(2)}$ (any vertex that does not satisfy this can be safely left unfixed). For each i, define $y_i = 1$ if vertex i is unfixed otherwise define $y_i = 0$. Then the offline problem of finding the best state can be written as

$$\min \sum_{i=1}^k (1 - y_i)\Lambda_i^2 + y_i \Lambda_i^1 = \sum_{i=1}^k y_i \gamma_i + \sum_{i=1}^k \Lambda_i^{(2)}$$

$$\text{s.t. } y_i \in \{0, 1\}$$

$$y_i + y_j \geq 1, \; \forall (v_i, v_j) \in E.$$

Here $\gamma_i = \Lambda_i^{(1)} - \Lambda_i^{(2)} > 0$. By relaxing y_i to be in $[0, 1]$ and solving the corresponding linear programming relaxation, we get a solution $y_1^*, y_2^*, \ldots, y_k^*$. Let LPval denote the linear programming objective value achieved by $y_1^*, y_2^*, \ldots, y_k^*$. Since the linear programming formulation is a valid relaxation of the problem of finding the best state, we have LPval $\leq \min_{s \in S} g(s)$.

We output the state s' in which a vertex i if and only if $y_i^* < 1/2$. Let S be the set of fixed vertices. We have

$$g(s') = \sum_{i \in S} \Lambda_i^{(2)} + \sum_{i \notin S} \Lambda_i^{(1)}$$

$$= \sum_{i=1}^k \Lambda_i^{(2)} + \sum_{i \notin S} (\Lambda_i^{(1)} - \Lambda_i^{(2)})$$

$$= \sum_{i=1}^k \Lambda_i^{(2)} + \sum_{i \notin S} \gamma_i$$

$$< \sum_{i=1}^k \Lambda_i^{(2)} + 2 \sum_{i \notin S} y_i^* \gamma_i$$

$$< 2 \left(\sum_{i=1}^k \Lambda_i^{(2)} + \sum_{i=1}^k y_i^* \gamma_i \right)$$

$$< 2 \cdot \text{LPval}$$

$$\leq \min_{s \in S} 2 g_{\mathbf{p}}(s).$$

B.1 Case $m = 2$

To illustrate the ideas behind our general algorithm, we first consider a simpler setting where correlation sets are defined on subsets of size at most two. This setting also captures an important case where fixing a particular criterion affects the rate of complaints of its neighbors.

Our algorithm consists of an exploration phase where it observes the losses for a specific subset of at most $4n$ states. We will show that after the exploration phase, the algorithm can accurately estimate the expected loss for any other state $s \in \mathcal{S}$. Notice that the number of states in $\mathcal{S}$ is in general exponential in k. Thus, the subset of states to observe must be carefully chosen and must take into account the structure of the graph $\mathcal{G}$. After the exploration phase, the algorithm creates an estimate $\hat{\theta}$ of the true parameter vector θ, uses the optimization oracle for solving Eq. (2) to find a near optimal state $\hat{s}$ and selects to stay at state $\hat{s}$ for the remaining time steps.

Let c denote the maximum fixing cost: $c = \max_{i \in [k]} c_i$. We will show that the pseudo-regret of our algorithm is bounded by $O(k \log k (c + B)^{1/3} T^{2/3} \log(kT))$. We first describe how we select the subset of states to observe in the exploration phase.

We say that (i, j, b) is a *dichotomy* if for two criteria i and j and for $b \in \{0, 1\}$, there exist two states $s, s' \in \mathcal{S}$ such that: (1) $s(j) = 0$ and $s'(j) = 1$, and (2) $s(i) = s'(i) = b$. Note that if an edge (v_i, v_j) is present in $\mathcal{G}$, then $(i, j, 1)$ cannot be a dichotomy, since criteria i and j cannot be fixed simultaneously.

Definition 1. *Consider a subset $\mathcal{K} \subset \mathcal{S}$. We will say that $\mathcal{K}$ is a cover for $\mathcal{C}$ if for any dichotomy (i, j, b), where $\{i, j\}$ is a correlation set ($\{i, j\} \in \mathcal{C}$) there exist two states $s, s' \in \mathcal{K}$ such that:*

(1) they agree in all criteria except criterion j: $s(l) = s'(l)$ for all $l \neq j$;
(2) criteria i is in state b in both: $s(i) = s'(i) = b$;
(3) we have that $s(j) = 0$ and $s'(j) = 1$.

We call such a pair (s, s') an (i, j, b)-pair.

Furthermore, for every singleton set $\{i\}$ in $\mathcal{C}$, the cover $\mathcal{K}$ contains states s, s' such that $s(i) = 0, s'(i) = 1$ and $s(j) = s'(j)$ for all $j \neq i$. We can always find a cover $\mathcal{K}$ of size at most $4n$ by picking for each $\{i, j\} \in \mathcal{C}$, at most four states corresponding to different bit configurations for i and j, with all other bits set to zero. For any valid dichotomy (i, j, b) we define $X_b^{i,j}$ as

$$X_b^{i,j} \doteq \mu_i^s - \mu_i^{s'}, \qquad (5)$$

where $s, s' \in \mathcal{K}$ is an (i, j, b) pair. If $\{i, j\} \notin \mathcal{C}$ we define $X_b^{i,j}$ to be zero. Notice that the values $X_b^{i,j}$ can be approximated from estimating the loss values of states in the cover. Next, we state our key result showing that, given the loss values for the states in a cover, we can accurately estimate the loss values for any vertex in any other state.

Theorem 5 (Theorem 1). *Let $\mathcal{K}$ be a cover for $\mathcal{C}$. Then, for any state $s \in \mathcal{S}$ and any $i \in [k]$ with $s(i) = b$, we have:*

$$\mu_i^s = \mu_i^{s''} + \sum_{j=1}^{k} X_b^{i,j} \left[\mathbb{I}(s(j) = 1)\, \mathbb{I}(s''(j) = 0) - \mathbb{I}(s(j) = 0)\, \mathbb{I}(s''(j) = 1) \right], \quad (6)$$

where s'' is any state in $\mathcal{K}$ with $s''(i) = b$.

Proof. Consider a correlation set $\{i, j\}$. The expected loss incurred by vertex v_i or v_j due to this set in any given state depends solely on the configuration of v_i and v_j in that state. Hence there are four parameters in the $\boldsymbol{\theta}$ vector corresponding to the correlation set $\{i, j\}$ and we denote them using $\gamma_{i,j}^{a,b}$, where $a, b \in \{0, 1\}$. Let s, s' be an (i, j, b) pair. When we switch from s to s' the only difference in the expected losses for vertex i comes from the pair (i, j). Hence we have

$$\mu_i^{s'} - \mu_i^s = \gamma_{i,j}^{b,1} - \gamma_{i,j}^{b,0} =: X_b^{i,j}.$$

Hence, given the loss estimates for states in $\mathcal{K}$ we can estimate $X_b^{i,j}$ for each $i, j \in [k]$ and $b \in \{0, 1\}$. Next, given an arbitrary state s with $s(i) = b$ let $s'' \in \mathcal{K}$ such that $s''(i) = b$. We have

$$\mu_i^s = \mu_i^{s''} + \sum_{\substack{j:s(j)=0 \\ s''(j)=1}} (\gamma_{i,j}^{b,0} - \gamma_{i,j}^{b,1}) + \sum_{\substack{j:s(j)=1 \\ s''(j)=0}} (\gamma_{i,j}^{b,1} - \gamma_{i,j}^{b,0})$$

$$= \mu_i^{s''} + \sum_{\substack{j:s(j)=1, \\ s''(j)=0}} X_b^{i,j} - \sum_{\substack{j:s(j)=0, \\ s''(j)=1}} X_b^{i,j}$$

$$= \mu_i^{s''} + \sum_{j=1}^{k} X_b^{i,j} \left[\mathbb{I}(s(j) = 1)\, \mathbb{I}(s''(j) = 0) - \mathbb{I}(s(j) = 0)\, \mathbb{I}(s''(j) = 1) \right].$$

From the above theorem we have the following guarantee.

Input: The graph $\mathcal{G}$, correlation sets $\mathcal{C}$, fixing costs c_i.

1. Pick a cover $\mathcal{K} = \{s_1, s_2, \ldots, s_r\}$ of $\mathcal{C}$.
2. Let $N = 10 \frac{T^{2/3}(\log rkT)^{1/3}}{r^{2/3}}$.
3. For each state $s \in \mathcal{K}$ do:
   - Move from current state to s in at most k time steps.
   - Play action $a = 0$ in state s for the next N time steps to obtain an estimate $\widehat{\mu}_i^s$ for all $i \in [k]$.
4. Using the estimated losses for the states in $\mathcal{K}$ and Equation (6), run the oracle for the optimization (2) to obtain an approximately optimal state $\hat{s}$.
5. Move from current state to $\hat{s}$ and play action $a = 0$ from $\hat{s}$ for the remaining time steps.

Fig. 6. Online algorithm for $m = 2$ achieving $\tilde{O}(T^{2/3})$ pseudo-regret.

Theorem 6 (Theorem 2). *Consider an MDP$(\mathcal{S}, \mathcal{A}, \mathcal{C}, \boldsymbol{\theta})$ with losses in $[0, B]$, a maximum fixing cost c, and correlations sets of size at most $m = 2$. Let $\mathcal{K}$ be a cover of $\mathcal{C}$ of size $r \leq 4n$, then, the algorithm of Fig. 6 (same as Fig. 3) achieves a pseudo-regret bounded by $O(kr^{1/3}(c+B)(\log rkT)^{1/3}T^{2/3})$. Furthermore, given access to the optimization oracle for Eq. (2), the algorithm runs in time polynomial in k and $n = |\mathcal{C}|$.*

Proof. In each time step the maximum loss incurred by any criterion is bounded by $c+B$. Let $\{s_1, s_2, \ldots, s_r\}$ be the states in $\mathcal{K}$. During the exploration phase the algorithm stays in each state for N time steps and incurs a total loss bounded by $kNr(c+B)$. During the exploration phase the algorithm moves from one state to another in at most k steps and incurs a total additional loss of at most $rk^2(c+B)$. At any given state $s \in \mathcal{K}$ and vertex v_i, after N time steps we will, with probability at least $1 - \delta$, an estimate of μ_i^s up to an accuracy of $2B\sqrt{\frac{\log 1/\delta}{N}}$. Setting $\delta = 1/(rkT^4)$ and using union bound, we have that at the end of the exploration phase, with probability at least $1 - \frac{1}{T^4}$, the algorithm will have estimate $\hat{\mu}_i^s$ for all $s \in \mathcal{K}$ and $i \in [k]$ such that

$$\hat{\mu}_i^s - \mu_i^s \leq 4B\sqrt{\frac{\log rkT}{N}}. \tag{7}$$

Hence during the exploitation phase, with high probability, the algorithm will have the estimate for the expected loss of each state in $\mathcal{S}$, i.e., $\sum_i \mu_i^s$ up to an error of $4kB\sqrt{\frac{\log rkT}{N}}$. Combining the above we get that the total pseudo-regret of the algorithm is bounded by

$$\text{Reg}(\mathcal{A}) \leq kNr(c+B) + rk^2(c+B) + \left(1 - \frac{1}{T^4}\right)4kBT\sqrt{\frac{\log rkT}{N}} + \frac{1}{T^4}k(c+B)T.$$

Setting $N = 10\frac{T^{2/3}(\log rkT)^{1/3}}{r^{2/3}}$ we get that

$$\text{Reg}(\mathcal{A}) \leq O(kr^{1/3}(c+B)(\log rkT)^{1/3}T^{2/3}).$$

B.2 General cCse

The algorithm for the case of $m = 2$ naturally extends to arbitrary correlation set sizes. Overall the structure of the algorithm remains the same where we pick a cover of $\mathcal{C}$ and estimate the losses incurred in states that belong to the cover. Using the estimated losses we are able to approximately estimate the loss of any vertex at any other state. In order to do this we extend the definition of the cover as follows. Given correlation sets of arbitrary size in $\mathcal{C}$, a vertex v_i may participate in many of them. We say that vertices v_i and v_j share a correlation set, if they appear together in a set in $\mathcal{C}$. Consider the set of indices of all the vertices that v_i shares a correlation set with. We partition this set into disjoint subsets such that no two vertices in different subsets share a correlation set. For a given vertex v_i, we denote this collection of disjoint subsets by I_i. For example, if $\mathcal{C}$ contains sets $\{1, 2\}, \{1, 3\}$, and $\{1, 4\}$, then, I_1 consists of the set $\{2, 3, 4\}$. On the other hand if $\mathcal{C}$ contains sets $\{1, 2, 3\}, \{1, 3, 4\}$, and $\{1, 6, 7\}$ then, I_1 consists of sets $\{2,3,4\}$ and $\{6,7\}$. For a given state s and $J \in I_i$ we denote by $s(J)$ the vector s restricted to indices in J. Notice that, in the worst case, I_i will consist of a single set of size at most $\min(k-1, nm)$. However, for more structured cases (e.g., $m = 2$) we expect I_i to consist of subsets of small sizes.

Given $i \in [k]$, $J \in I_i$, $b \in \{0,1\}$ and vectors u_1, u_2, we say that (i, b, J, u_1, u_2) is a dichotomy, if there exist two states $s, s' \in S$ such that: (1) $s(J) = u_1$, $s'(J) = u_2$, (2) $s(i) = b = s'(i)$, and (3) s, s' agree in all other criteria. We call such a pair of states s, s' an (i, b, J, u_1, u_2) pair. We next extend the definition of a cover as follows. A subset $\mathcal{K} \subseteq S$ is called a cover of $\mathcal{C}$ if for any valid dichotomy (i, b, J, u_1, u_2), there exists an (i, b, J, u_1, u_2) pair $s, s' \in \mathcal{K}$. In general, we will always have a cover of size at most $n2^{mn}$. Similar to (5), for a valid dichotomy (i, b, J, u_1, u_2), we define $X_{b,J}^{i,u_1,u_2}$ as

$$X_{b,J}^{i,u_1,u_2} := \mu_i^s - \mu_i^{s'}, \tag{8}$$

where $s, s' \in \mathcal{K}$ is an (i, b, J, u_1, u_2) pair. Given the loss values in the states present in $\mathcal{K}$, we can estimate the loss of any other state using Theorem 7 stated below.

Theorem 7. *Let $\mathcal{K}$ be a cover for $\mathcal{C}$. Then, for any state $s \in S$ and any $i \in [k]$ with $s(i) = b$, we have:*

$$\mu_i^s = \mu_i^{s''} + \sum_{J \in I_i} X_{b,J}^{i,s(J),s''(J)} \tag{9}$$

Here s'' is any state in $\mathcal{K}$ with $s''(i) = b$.

Proof. Let $s, s' \in \mathcal{K}$ be an (i, b, J, u_1, u_2) pair. When we move from state s to s', the only difference between the expected losses incurred by vertex v_i comes from the configuration of the vertices in J. Hence there at at most $2^{|J|+1}$ distinct parameters governing the expected loss incurred by vertex i in a given state s due to the configuration of the vertices in J. Denoting these parameters by $\gamma_{i,J}^{b,s(J)}$ we have

$$\mu_i^{s'} - \mu_i^s = \gamma_{i,J}^{b,s'(J)} - \gamma_{i,J}^{b,s(J)} := X_{b,J}^{i,s'(J),s(J)}.$$

Given the loss values for the states in the cover $\mathcal{K}$, we can estimate the quantities $X_{b,J}^{i,s(J),s''(J)}$.

Next, for an arbitrary state s such that $s(i) = b$, let $s'' \in \mathcal{K}$ be such that $s''(i) = b$. We have

$$\mu_i^s = \mu_i^{s''} + \sum_{J \in I_i} \gamma_{i,J}^{b,s(J)} - \gamma_{i,J}^{b,s''(J)}$$

$$= \sum_{J \in I_i} X_{b,J}^{i,s(J),s''(J)}.$$

For general correlation sets with each vertex participating in at most n sets, we use (9) instead of (6) to estimate losses in step 4 of the algorithm in Fig. 6. The algorithm for general m is described in Fig. 7 and has the following associated regret guarantee. The proof is identical to the proof of Theorem 2.

Input: The graph $\mathcal{G}$, correlation sets $\mathcal{C}$, fixing costs c_i.

1. Pick a cover $\mathcal{K} = \{s_1, s_2, \ldots, s_r\}$ of $\mathcal{C}$.
2. Let $N = 10 \frac{T^{2/3}(\log rkT)^{1/3}}{r^{2/3}}$.
3. For each state $s \in \mathcal{K}$ do:
   - Move from current state to s in at most k time steps.
   - Play action $a = 0$ in state s for the next N time steps to obtain an estimate $\widehat{\mu}_i^s$ for all $i \in [k]$.
4. Using the estimated losses for the states in $\mathcal{K}$ and Equation (9), run the oracle for the optimization (2) to obtain an approximately optimal state $\hat{s}$.
5. Move from current state to $\hat{s}$ and play action $a = 0$ from $\hat{s}$ for the remaining time steps.

Fig. 7. Online algorithm for general m achieving $\tilde{O}(T^{2/3})$ pseudo-regret.

Theorem 8. *Consider an MDP$(\mathcal{S}, \mathcal{A}, \mathcal{C}, \boldsymbol{\theta})$ with losses bounded in $[0, B]$ and maximum cost of fixing a vertex being c. Given correlations sets $\mathcal{C}$ of size at most m, and a cover $\mathcal{K}$ of $\mathcal{C}$ of size $r \leq n2^{mn}$, the algorithm in Fig. 7 achieves a pseudo-regret bounded by $O(kr^{1/3}(c + B)(\log rkT)^{1/3}T^{2/3})$. Furthermore, given access to the optimization oracle for Eq. (2) the algorithm runs in time polynomial in k, $n = |\mathcal{C}|$ and $r = |\mathcal{K}|$.*

C Beyond $T^{\frac{2}{3}}$ Regret

In this section, we present algorithms for our problem that achieve $\tilde{O}(\sqrt{T})$ regret, first in the case $m = 1$, next for any m, under the natural assumption that each criterion does not participate in too many correlations sets.

Let us first point out that our problem can be cast as an instance of the stochastic multi-armed bandit problem with switching costs, where each state s is viewed as an arm and where the cost of transitions from state s to state s' is the switching cost between s and s'. For the instance of this problem with identical switching costs, [8][Appendix A] gave an algorithm achieving expected regret $\tilde{O}(\sqrt{T})$, via an arm-elimination technique with at most $O(\log \log T)$ switches. However, naturally, the regret guarantee and the time complexity of that algorithm depend on the number of arms, which in our case is exponential (2^k). We will show here that, in most realistic instances of our model, we can achieve $\tilde{O}(\sqrt{T})$ regret efficiently.

We first consider the case where the correlations sets in $\mathcal{C}$ are of size one ($m = 1$). In this case, the parameter vector $\boldsymbol{\theta}$ can be described using the following $2k$ parameters: for each $i \in [k]$, let γ_i^0 denote the expected loss incurred by criterion i when it is unfixed and γ_i^1 its expected loss when it is fixed. In this case, the cover $\mathcal{K}$ is of size $k + 1$ and includes the all-zero state, as well as k states corresponding to the indicator vectors of the k vertices. Our algorithm is similar to the UCB algorithm for multi-armed bandits [3] and maintains optimistic estimates of the parameters. For every vertex i, we denote by $\tau_{i,t}^0$ the total number of time steps up to t (including t) during which the vertex v_i is in an unfixed position and by $\tau_{i,t}^1$ the total number of times steps up to t during which vertex v_i is in a fixed position. Fix $\delta \in (0, 1)$ and let $\hat{\gamma}_{i,t}^b$ be the empirical expected

---

Input: graph $\mathcal{G}$, correlation sets $\mathcal{C}$, fixing costs c_i.

1. Let $\mathcal{K}$ be the cover of size $k + 1$ that includes the all zeros state and the states corresponding to indicator vectors of the k vertices.
2. Move to each state in the cover once and update the optimistic estimates according to (10).
3. For episodes $h = 1, 2, \ldots$ do:
    - Run the optimization oracle for solving Eq. (2) with the optimistic estimates as in (10) to get a state s.
    - Move from current state to state s. Stay in state s for $t(h)$ time steps and update the corresponding estimates using (10). Here $t(h) = \min_i \tau_{i,t_h}^{s(i)}$ and t_h is the total number of time steps before episode h starts.

Fig. 8. Online algorithm for $m = 1$ with $\tilde{O}(\sqrt{T})$ regret.

loss observed when vertex v_i is in state b, for $b \in \{0, 1\}$. Our algorithm maintains the following optimistic estimates at each time step t,

$$\tilde{\gamma}_{i,t}^b = \hat{\gamma}_{i,t}^b - 10B\sqrt{\frac{\log(kT/\delta)}{\tau_{i,t}^b}}. \tag{10}$$

To minimize the fixing cost incurred when transitioning from one state to another, our algorithm works in episodes. In each episode h, the algorithm first uses the current optimistic estimates to query the optimization oracle and determine the current best state s. Next, it remains at state s for $t(h)$ time steps before querying the oracle again. The number of time steps $t(h)$ will be chosen carefully to avoid incurring the fixing costs too often. The algorithm is described in Fig. 8 (same as Fig. 4 in main body). We will prove that it benefits from the following regret guarantee.

Theorem 9 (Theorem 3). *Consider an MDP$(\mathcal{S}, \mathcal{A}, \mathcal{C}, \boldsymbol{\theta})$ with losses bounded in $[0, B]$ and maximum cost of fixing a vertex being c. Given correlations sets $\mathcal{C}$ of size one, the algorithm of Fig. 8 (same as Fig. 4) achieves a pseudo-regret bounded by $O(k^2(c + B)^2\sqrt{T}\log T)$. Furthermore, given access to an oracle for (2), the algorithm runs in time polynomial in k.*

Proof. We first bound the total number of different states visited by the algorithm. Initially the algorithm visits $k+1$ states in the cover. After that, each time the optimization oracle returns a new state s, by the definition of $t(h)$, the number of time steps where some vertex is in a 0 or 1 position is doubled. Hence, at most $O(k \log T)$ calls are made to the optimization oracle. Noticing that one can move from one state to another in at most k time steps, the total loss incurred during the switching of the states is bounded by $O(k^2(c + B) \log T)$.

For $\epsilon > 0$ to be chosen later, we consider the episodes where the algorithm plays a state s with expected loss at most ϵ more than that of the best state s^*. The total expected regret accumulated in these *good* episodes is at most ϵT. We next bound the expected regret accumulated during the bad episodes.

From Hoeffding's inequality we have that for any time t, with probability at least $1 - \frac{\delta}{T^3}$, for all $i \in [k], b \in \{0,1\}$,

$$\tilde{\gamma}_{i,t}^b + 20B\sqrt{\frac{\log(kT/\delta)}{\tau_{i,t}^b}} \geq \gamma_i^b \geq \tilde{\gamma}_{i,t}^b. \qquad (11)$$

Let G be the good event that (11) holds for all $t \in [1, T]$. Conditioned on G we also have that for any state s and vertex i

$$\mu_i^s \geq \tilde{\mu}_i^s, \qquad (12)$$

where $\tilde{\mu}_i^s$ is the estimated loss using the optimistic estimates. We will bound the expected regret accumulated in the bad episodes conditioned on the event G above.

In order to do this we define certain key quantities. Consider a particular trajectory $\mathcal{T}$ of T time steps executed by the algorithm. Furthermore, let $\mathcal{T}$ be such that the good event in (11) holds during the T time steps. We associate the following random variables with the trajectory. Let N_ϵ be the total number of time steps spent in bad episodes. Furthermore, let Reg_ϵ be the total accumulated regret during these time steps. Then it is easy to see that $\mathbb{E}[\text{Reg}_\epsilon|G] > \epsilon N_\epsilon$. For each vertex v_i and $b \in \{0,1\}$ we define $\tau_\epsilon(i,b)$ to be the total number of time steps that vertex v_i spends in bad episodes in position b and $\tau_\epsilon(i,b,t)$ to be the total number of time steps spent in bad episodes up to time step t. Notice that

$$\sum_b \sum_i \tau_\epsilon(i,b) \leq 2kN_\epsilon. \qquad (13)$$

Consider a particular bad episode h and let s be the state returned by the optimization oracle during that episode. Then conditioned on the good event G, the total regret Reg_h accumulated during episode h satisfies

$$\mathbb{E}[\text{Reg}_h|\mathcal{T}] = \sum_i \left(\mu_i^s - \mu_i^{s^*}\right)t(h)$$

$$\leq \sum_i \left(\mu_i^s - \tilde{\mu}_i^{s^*}\right)t(h) \qquad \text{(from (12))}$$

$$\leq \sum_i \left(\mu_i^s - \tilde{\mu}_i^s\right)t(h) \quad \text{(since } s \text{ is best state according to the optimistic losses)}$$

$$\leq \sum_i \left(\gamma_i^{s(i)} - \tilde{\gamma}_{i,t_h}^{s(i)}\right)t(h)$$

$$\leq \sum_i 20B\sqrt{\frac{\log(kT/\delta)}{\tau_{i,t_h}^b}}t(h). \quad \text{(from (10))}$$

In the above inequality, the expectation is taken over the loss distribution for each vertex during states visited in the trajectory $\mathcal{T}$. we have that

$$\mathbb{E}[\text{Reg}_h|\mathcal{T}] \leq \sum_i 20B\sqrt{\frac{\log(kT/\delta)}{\tau_\epsilon(i,b,t_h)}}t(h).$$

Summing over bad episodes, the total expected regret in bad episodes can be bounded by

$$\mathbb{E}[\text{Reg}_\epsilon | \mathcal{T}] \leq \sum_i \sum_b \sum_{h:h \text{ is bad}} 20B \sqrt{\frac{\log(kT/\delta)}{\tau_\epsilon(i,b,t_h)}} t(h). \tag{14}$$

Notice that $\tau_\epsilon(i,b,t_h) = \sum_{h'<h:h' \text{ is bad}} t(h')$. Furthermore, we know that ([23]) for any sequence $z_1, z_2, \ldots, z_h$ of non-negative numbers such that $z_i \geq 1$,

$$\sum_{i=1}^h \frac{z_i}{\sqrt{\sum_{j=1}^{i-1} z_j}} \leq (1+\sqrt{2}) \sqrt{\sum_{i=1}^h z_i}. \tag{15}$$

From (15) we get:

$$\sum_{h:h \text{ is bad}} \frac{t(h)}{\sqrt{\tau_\epsilon(i,b,t_h)}} \leq \sqrt{\tau_\epsilon(i,b)}.$$

Substituting into (14) we get that

$$\mathbb{E}[\text{Reg}_\epsilon | \mathcal{T}] \leq \sum_i \sum_b 20B \sqrt{\log(kT/\delta)} \sqrt{\tau_\epsilon(i,b)}.$$

Using (13) we have that the above expected regret is maximized when $\tau_\epsilon(i,b)$ are equal, thereby implying

$$\mathbb{E}[\text{Reg}_\epsilon | \mathcal{T}] \leq 20Bk \sqrt{\log(kT/\delta)} \sqrt{N_\epsilon}.$$

Using the fact that $\mathbb{E}[\text{Reg}_\epsilon | G] > \epsilon N_\epsilon$ we get that conditioned on G,

$$N_\epsilon \leq \frac{400 B^2 k^2 \log(kT/\delta)}{\epsilon^2}.$$

Combining trajectories $\mathcal{T}$ where the good event G holds, we get that the total expected regret accumulated in the bad episodes satisfies

$$\mathbb{E}[\text{Reg}_\epsilon | G] \leq 20Bk \sqrt{\log(kT/\delta)} \sqrt{N_\epsilon}$$

$$\leq 400 B^2 k^2 \frac{\log(kT/\delta)}{\epsilon}.$$

Combining the above with the total expected regret accumulated in the good episodes, the loss of moving to different states, and the probability of good event G not holding, we get

$$\text{Reg}(\mathcal{A}) \leq 400 B^2 k^2 \frac{\log(kT/\delta)}{\epsilon} + \epsilon T + \frac{k(c+B)}{T^3} + O(k^2(c+B)\log T).$$

Setting $\epsilon = \frac{1}{\sqrt{T}}$ and $\delta = \frac{1}{T^4}$, we have the final bound

$$\text{Reg}(\mathcal{A}) \leq O\big((c+B)^2 k^2 \sqrt{T} \log(T)\big).$$

Input: The graph $\mathcal{G}$, correlation sets $\mathcal{C}$, fixing costs c_i.

1. Let $\mathcal{K}$ be the cover of size $O(k^{m+1})$.
2. Move to each state in the cover once and update the optimistic estimates according to (16).
3. For episodes $h = 1, 2, \ldots$ do:
   - Run the optimization oracle (2) with the optimistic estimates as in (16) to get a state s.
   - Move from current state to state s. Stay in state s for $t(h)$ time steps and update the corresponding estimates using (16). Here $t(h) = \min_i \tau_{i,t_h}^{\mathbf{s}(i)}$ and t_h is the total number of time steps before episode h starts.

Fig. 9. Online algorithm for higher m.

The above result extends to higher m values, assuming that each vertex does not participate in too many correlation sets. If a vertex v_i appears in at most $O(\log k)$ correlation sets, then the total loss incurred by vertex v_i in any state depends on the position of v_i and every other vertex that it is correlated with. Hence the total loss incurred by vertex v_i depends on an $O(m \log k)$-dimensional vector. For every configuration **b** of this vector, we associate with each vertex v_i, parameters $\gamma_i^{\mathbf{b}}$. Notice that there are at most $O(k^m)$ such parameters. Each parameter is in turn a sum of a subset of the parameters in $\boldsymbol{\theta}$. Notice that in this case the size of the cover $\mathcal{K}$ is upper bounded by $O(k^{m+1})$. Our algorithm for higher m values is similar to the one for $m = 1$, but instead maintains optimistic estimates of the parameters $\gamma_i^{\mathbf{b}}$ via

$$\tilde{\gamma}_{i,t}^{\mathbf{b}} = \hat{\gamma}_{i,t}^{\mathbf{b}} - 10B\sqrt{m \frac{\log(kT/\delta)}{\tau_{i,t}^{\mathbf{b}}}}. \tag{16}$$

Here $\tau_{i,t}^{\mathbf{b}}$ is the total time spent up to and including t where the vertex i and the vertices that it is correlated with are in configuration **b**. Similarly, for a given state s, we will denote by $\mathbf{s}(i)$, the configuration of the vertex i and the vertices that it is correlated with. The algorithm is sketched below

For $m \geq 1$, we obtain the following guarantee.

Theorem 10. *Consider an MDP$(\mathcal{S}, \mathcal{A}, \mathcal{C}, \boldsymbol{\theta})$ with losses bounded in $[0, B]$ and maximum cost of fixing a vertex being c. Given correlations sets $\mathcal{C}$ of size at most m such that each vertex participates in at most $O(\log k)$ sets, the the algorithm in Fig. 9 achieves a pseudo-regret bounded by $O(mk^{2m+2}(c+B)^2\sqrt{T} \log T)$. Furthermore, given access to an oracle for (2), the algorithm runs in time polynomial in $O(k^{m+1})$.*

Proof. The proof is very similar to the proof of Theorem 3. Since each time the optimization oracle is called the time spent in some configuration $\mathbf{s}(i)$ is doubled, we get that the total number of calls to the optimization oracle are bounded by $O(k^m \log T)$. Hence the total loss incurred during the exploration phase can be bounded by $O(k^m(c+B) \log T)$. Let G be the good event that (16) holds for all $t \in [1, T]$.

As before, the loss incurred during good episodes is bounded by ϵT. Define $\tau_\epsilon(i, \mathbf{b})$ to be the total time that vertex i and vertices that it is correlated with spend in configuration **b** during bad episodes. Then analogous to (13) we have

$$\sum_{\mathbf{b}} \sum_{i} \tau_\epsilon(i, \mathbf{b}) \leq O(k^m) N_\epsilon.$$

For a trajectory $\mathcal{T}$ where the good event G holds, the total expected regret in bad episodes can be bounded as

$$\mathbb{E}[\text{Reg}_\epsilon | \mathcal{T}] \leq \sum_{i} \sum_{\mathbf{b}} \sum_{h:h \text{ is bad}} 20B \sqrt{m \frac{\log(kT/\delta)}{\tau_\epsilon(i, \mathbf{b}, t_h)}} t(h) \quad (17)$$

$$\leq \sum_{i} \sum_{\mathbf{b}} 20B \sqrt{m \log(kT/\delta)} \sqrt{\tau_\epsilon(i, \mathbf{b})} \quad (18)$$

$$\leq O(Bk^{m+1}) \sqrt{m \log(kT/\delta)} \sqrt{N_\epsilon}. \quad (19)$$

Using the fact that $\mathbb{E}[\text{Reg}_\epsilon | \mathcal{T}] > \epsilon N_\epsilon$ we get that for a trajectory where the event G holds,

$$N_\epsilon \leq \frac{O(R^2 k^{2m+2} m \log(kT/\delta))}{\epsilon^2}.$$

Hence we get that conditioned on the good event G, the total expected regret accumulated in the bad episodes is at most

$$\mathbb{E}[\text{Reg}_\epsilon | G] \leq O\left(R^2 \, mk^{2m+2} \frac{\log(kT/\delta)}{\epsilon}\right).$$

Combining the above with the total expected regret accumulated in the good episodes, the loss of moving to different states, and the probability of the event G not holding we get

$$\text{Reg}(\mathcal{A}) \leq O\left(B^2 \, mk^{2m+2} \frac{\log(kT/\delta)}{\epsilon}\right) + \epsilon T + \frac{k(c+B)}{T^3} + O(k^m \log T).$$

Setting $\epsilon = \frac{1}{\sqrt{T}}$ and $\delta = \frac{1}{T^4}$, we have the final bound

$$\text{Reg}(\mathcal{A}) \leq O\left((c+B)^2 \, mk^{2m+2} \sqrt{T} \log(T)\right).$$

An important corollary of the above is the following

Corollary 1. *If $\mathcal{G}$ is a constant degree graph with correlation sets consisting of subsets of edges in $\mathcal{G}$, then there is a polynomial time algorithm that achieves a pseudo-regret bounded by $O(k^6(c+B)^2 \sqrt{T} \log T)$.*

D Adversarial Setting

Motivated by the discussion on adversarial manipulation in Sect. 3, we study a setting with no distributional assumptions about the arrival of complaints. We consider an adversarial model where, at each time step, multiple complaints arrive for the vertices in $\mathcal{G}$. Initially all the vertices in $\mathcal{G}$ are in an unfixed state and each vertex has a fixing cost of c_i. Each time, the algorithm can decide to fix a particular vertex, and as a result its neighbors get unfixed. At time step t, if criterion v_i is unfixed, then the algorithm incurs a loss of $\ell_{i(t)}$ (which depends on the current state of the system), otherwise the algorithm incurs no loss. For an algorithm $\mathcal{A}$, during T time steps, the total loss is

$$\text{Loss}(\mathcal{A}) = \sum_{i=1}^{k} \sum_{t=1}^{T} \ell_{i(t)} \cdot \mathbb{I}(s_t(i) = 0)$$
$$+ \sum_{i=1}^{k} \sum_{t=2}^{T} c_i \cdot \mathbb{I}(s_{t-1}(i) = 0, s_t(i) = 1). \quad (20)$$

Let OPT be the algorithm that, given the entire loss sequence in advance, makes the decisions to fix vertices. We define the *competitive ratio* [7] of $\mathcal{A}$ to be the maximum of Loss($\mathcal{A}$)/Loss(OPT) over all possible complaint sequences. Our main result is stated below.

Theorem 11. *Let $\mathcal{G}$ be a graph with fixing costs at least one. There is a polynomial-time algorithm with a competitive ratio of at most $2B+4$ on any sequence of complaints with loss values in $[0, B]$.*

Our algorithm for this setting is provided in Fig. 10. The intuition behind the algorithm is the following. For each criteria we consider it fixing cost, the accumulated loss since the last time it was fixed plus a measure of the cost of fixing its neighbors. We fix a criteria (and implicitly unfix its neighbors) once its accumulated loss is larger than both its fixing cost and a measure of the cost of fixing its neighbors (the decision in the algorithm is more refined). This allows us to "charge" the cost of fixing a criteria to its accumulated loss. In addition, we make sure that (future) fixing of the neighbors of the criteria can also be charged appropriately.

In particular, we study an adversarial model where at each time step multiple complaints arrive for the vertices in $\mathcal{G}$ via the choice made by an oblivious adversary. For a given vertex v_i and time step t, we denote by $\ell_{i(t)}$ the loss incurred if criterion v_i is unfixed at time t. Similar to the setting from the previous section, initially all the vertices in $\mathcal{G}$ are in unfixed state and each vertex has a fixing cost of c_i. At each time step the algorithm can decide to fix a particular vertex. As a result all its neighbors get unfixed. At time step t, if criterion v_i is unfixed then the the algorithm incurs a loss of $\ell_{i(t)}$. If v_i is fixed at time step t then algorithm incurs no loss. The overall loss incurred by the algorithm is the total fixing cost and the total loss incurred over the arrival complaints. As before, we will denote a configuration of the vertices in $\mathcal{G}$ using a vector $s \in \{0,1\}^k$ with $s(i) = 0$ representing an unfixed vertex. For an algorithm $\mathcal{A}$ processing the request sequence, During the course of T time steps, the total loss of processing the complaints is

$$\text{Loss}(\mathcal{A}) = \sum_{i=1}^{k} \sum_{t=1}^{T} \ell_{i(t)} \cdot \mathbb{I}(s_t(i) = 0) + \sum_{i=1}^{k} \sum_{t=2}^{T} c_i \cdot \mathbb{I}(s_{t-1}(i) = 0, s_t(i) = 1). \quad (21)$$

Define OPT to be the algorithm that given the entire loss sequence in advance, makes the optimal choice of decisions to fix vertices. Following standard terminology we define the *competitive ratio* of an algorithm $\mathcal{A}$ to be the maximum of Loss($\mathcal{A}$)/Loss(OPT) over all possible complaint sequences. We will design efficient online algorithms for processing the complaints that achieve a constant competitive ratio. Notice that in this setting, in order for the competitive ratio to be finite, we need to bound the range of the losses and the fixing costs of the vertices. We will assume that the cost of fixing each vertex is at least one and as before assume that the losses are bounded in the range $[0, B]$. For ease of exposition, in the rest of the discussion we will assume that at each time step complaints arrive for one of the vertices in $\mathcal{G}$. A simple reduction shows that an algorithm that is competitive with OPT in this setting remains so in the general setting with the same competitive ratio. We discuss this at the end of the section. Via this reduction we can consider the loss sequence to be of the form $((i_1, \ell_{i_1}), \ldots, (i_T, \ell_{i_T}))$ where i_t is the index of the criterion for which the tth complaint arrives and ℓ_{i_t} is the associated loss.

To get a better understanding of the above adversarial setting, consider the case when the graph $\mathcal{G}$ over the criteria has no edges, i.e., there are no conflicts. In this case, given a sequence of complaints, each with unit loss value, the optimal offline algorithm that has the entire loss sequence in advance can independently make a decision for each vertex. In particular, if the total loss of the complaints incurred at vertex v_i exceeds the fixing cost c_i then the optimal decision is to fix the vertex v_i, and otherwise simply incur the loss from the arriving complaints. In this case the online algorithm can also simply process each vertex independently. At each vertex the algorithm is faced with the classical *ski-rental* problem for which there exists a deterministic algorithm that is 2-competitive with optimal algorithm [28]. For each vertex i, the online algorithm simply waits till a total loss of c_i or more has been incurred on vertex i and then decides to fix it. It is easy to see that the total cost incurred by this strategy is at most twice the cost incurred by OPT.

However, the above algorithm will fail miserably in the presence of conflicts in the graph $\mathcal{G}$. As an example consider a graph with two vertices v_i and v_j that are connected by an edge. Let the fixing cost of v_i be 1 and the fixing cost of v_j be $C \gg 1$. Consider a sequence of complaints, each of unit loss, consisting of C complaints for v_j followed one complaint for v_i. If this sequence is repeated T times the optimal offline algorithm OPT incurs a loss of $C + T$ by fixing v_j and incurring losses due to v_i. However, the algorithm above will incur a cost of $(2C + 2)T$ thereby leading to an unbounded competitive ratio. Hence, in order to achieve a good competitive ratio one must make decisions not only based on the loss incurred at the given vertex v_i, but also the status of the vertices in the neighborhood of v_i. Our main result in this section is the algorithm in Fig. 10 that achieves a constant factor competitive ratio.

The algorithm described in Fig. 10 makes decisions based on local neighborhood information of a vertex. Intuitively, if a vertex is fixed only once or a few times in the optimal algorithm one would like to avoid fixing it too many times. In order to achieve this, each time a vertex v_i is fixed, it adds a barrier of $\kappa_i = c_i$ to the loss any of its neighbors need to incur before getting fixed. Hence, if a vertex is connected to a lot of fixed vertices then it has a high barrier to cross before getting fixed. During the course of the algorithm each unfixed vertex is in one of the two phases. In phase one, the vertex

> **Input:** The graph $\mathcal{G}$, fixing costs c_i, loss sequence $(i_1, \ell_{i_1}), \ldots, (i_T, \ell_{i_T})$.
>
> 1. For each $i \in [k]$, initialize τ_i, κ_i to 0.
> 2. Process the complaints in sequence and for each complaint (i, ℓ_i) such that v_i is unfixed do:
>   (a) $\tau_i = \tau_i + \ell_i$.
>   (b) While $\ell_i > 0$ and exists $j \in N(i)$ with $\kappa_j > 0$ do:
>   i. Set $\Delta = \min(\ell_i, \kappa_i)$ and reduce both κ_i and ℓ_i by Δ.
>   (c) If $\tau_i \geq \max\left(c_i, \sum_{j \in N(i)} \kappa_j\right)$ fix v_i. Set τ_i to 0 and κ_i to c_i. Set $\tau_j = 0$ for all $j \in N(i)$.

Fig. 10. Online algorithm for the adversarial setting.

is accumulating losses to pay for the barrier introduced by its neighbors (step 2(b) of the algorithm). In phase two, once the barrier has been crossed the vertex follows the standard ski-rental strategy independent of other vertices for making a decision as to fix or not. Notice that via step 2(b) of the algorithm, multiple neighbors of a vertex v_i can help bring down the barrier of c_i introduced by the action of fixing vertex v_i. This is necessary to ensure the online algorithm does not incur a large loss on a vertex by waiting too long to fix it.

As an example consider a graph $\mathcal{G}$ with k vertices and $k-1$ edges, where vertex v_0 is the central vertex connected to every other vertex. Let the fixing cost of vertex v_0 be a large value C, and the fixing cost of other vertices be one. We consider a sequence of C complaints, each with unit loss arriving for vertex v_0, followed by a sequence of C complaints for vertex v_1 and so on. In this case the optimal offline solution incurs a loss of $C + k$ by deciding to fix every vertex except v_0. After processing C complaints for v_0, the online algorithm will fix v_0 and incur a loss of $2C$. Next, during the course of processing C complaints for v_1, the algorithm fixes v_1 and incurs an additional loss of $C + 1$. More importantly, due to step 2(b), the barrier κ_0 introduced by vertex v_0 has been reduced to zero and hence the algorithm only incurs a loss of 2 per vertex for the remaining sequence for a total loss of $3C + 2k - 1$. Without the presence of step 2(b) each vertex will incur a loss of C leading to a large competitive ratio.

Notice that our algorithm in Fig. 10 is designed for a setting where in each time step complaints arrive for a single vertex in $\mathcal{G}$. If multiple vertices accumulate complaints in a time step, we can simply order them arbitrarily and run the algorithm on the new sequence. Let OPT be the optimal offline algorithm according to the chosen ordering of the complaints. Let OPT' be the optimal offline algorithm when processing multiple complaints per time step. Notice that for each time step, the loss of OPT cannot be larger than that of OPT' since any choice available to OPT' is available to OPT as well. Hence it is enough to design an algorithm that is competitive with OPT. In particular, we have the following theorem.

Theorem 12 (Theorem 11). *Let $\mathcal{G}$ be a graph with fixing costs at least one. Then, the algorithm of Fig. 10 achieves a competitive ratio of at most $2B + 4$ on any sequence of complaints with loss values in $[0, B]$.*

Proof. Recall that $\ell_{i(t)}$ denotes the loss incurred by vertex v_i at time t. We divide this loss into the amount that was used to reduce the κ_j value of one its neighbors and the rest. Formally, for every edge (i, j) we define $\delta^t_{i \to j}$ as follows. If in time step t, the complaint arrived for vertex i and step 2(b) was executed to reduce κ_j by Δ, then we define $\delta^t_{i \to j} = \Delta$. Otherwise we define $\delta^t_{i \to j}$ to be zero. We also define

$$\delta^t_{i \to i} = \ell_{i(t)} - \sum_{j \in N(i)} \delta^t_{i \to j}. \tag{22}$$

If vertex v_i is fixed f_i times during the course of the algorithm then we have that the total loss incurred by the algorithm can be written as

$$\text{Loss}(\mathcal{A}) = \sum_{i=1}^{k} f_i c_i + \sum_{i=1}^{k} \sum_{t=1}^{T} \left(\delta^t_{i \to i} + \sum_{j \in N(i)} \delta^t_{i \to j} \right). \tag{23}$$

Next we notice that each time a vertex v_i is fixed it accumulates a value of $\kappa_i = c_i$. Furthermore, the total loss incurred by vertices as a result of executing step 2(b) is upper bounded by the total κ value accumulated. Hence we have

$$\sum_{t=1}^{T} \sum_{i=1}^{k} \sum_{j \in N(i)} \delta^t_{i \to j} \leq \sum_{i=1}^{k} f_i c_i. \tag{24}$$

Substituting into (23) we have

$$\text{Loss}(\mathcal{A}) \leq \sum_{i=1}^{k} 2 f_i c_i + \sum_{i=1}^{k} \sum_{t=1}^{T} \delta^t_{i \to i}. \tag{25}$$

Next we bound the above loss for each vertex separately. For a given vertex v_i that is fixed f_i times by the algorithm, we can divide the time steps into $f_i + 1$ intervals consisting of an interval I_0 starting from $t = 0$ up to (and including) the first time v_i is fixed. The next f_i intervals correspond to the time spent by v_i between two successive fixes. Denoting these intervals as $I_0, I_1, \ldots$ we have that

$$2 f_i c_i + \sum_{i=1}^{k} \sum_{t=1}^{T} \delta^t_{i \to i} = \sum_{t \in I_0} \delta^t_{i \to i} + \sum_{t \in I_r} (2 c_i + \delta^t_{i \to i}). \tag{26}$$

Next we compare the above to the loss incurred by OPT for vertex v_i. Let $\ell^*_{i(t)}$ be the loss incurred by OPT for vertex v_i at time t. We will denote by s^*_t the state of the vertices at time t according to OPT.

We instead redefine the loss incurred by OPT for vertex v_i at time t to be

$$\tilde{\ell}_{i(t)} = \ell^*_{i(t)} + \sum_{j \in N(i)} \delta^t_{j \to i} \mathbb{I}(s^*_t(j) = 0). \tag{27}$$

Notice that

$$\sum_{i \in N(j)} \delta^t_{j \to i} \mathbb{I}(s^*_t(j) = 0) \leq \ell^*_{j(t)}.$$

Hence we get that

$$\sum_{i=1}^{k}\sum_{t=1}^{T} \tilde{\ell}_{i(t)} \leq \sum_{i=1}^{k}\left(\sum_{t=1}^{T} \ell_{i(t)}^* + \sum_{j \in N(i)} \ell_{j(t)}^*\right)$$

$$\leq 2 \cdot \text{Loss}(\text{OPT}).$$

Next we consider each interval in (25) separately. For any interval I_r we have that

$$\sum_{t \in I_r} \delta_{i \to i}^t \leq Bc_i. \tag{28}$$

This is because after incurring a loss of more than c_i, any additional loss incurred by v_i is due to step 2(b), since otherwise step 2(c) will be executed and v_i will be fixed.

Next consider interval I_0. The loss incurred by the algorithm on vertex v_i equals $\sum_{t \in I_0} \delta_{i \to i}^t \leq Bc_i$. Either OPT fixes v_i at least once during this interval or incurs the total loss. Either way we have that the loss incurred by OPT is at least

$$\min\left(c_i, \sum_{t \in I_0} \delta_{i \to i}^t\right) \geq \frac{\sum_{t \in I_0} \delta_{i \to i}^t}{B}. \tag{29}$$

Next consider an interval I_r between two successive fixes. The loss incurred by the algorithm for vertex v_i during this interval is at most

$$\sum_{t \in I_r} \delta_{i \to i}^t + 2c_i \leq (B+2)c_i.$$

If OPT fixes v_i at least once during this interval then it incurs a cost of c_i. If v_i remains unfixed in OPT during the course of the interval then OPT incurs a loss of at least c_i. This is because vertex v_i went from being unfixed to fixed during the second half of the interval and hence a total loss of at least c_i must have arrived for the vertex v_i during this interval.

Finally, suppose vertex v_i is fixed in OPT before the start of the interval and remains so throughout. Since v_i goes from being fixed to unfixed during the first half of the interval, we must have $\sum_{t \in I_r} \sum_{j \in N(i)} \delta_{j \to i}^t \geq c_i$. Furthermore, since v_i is fixed by OPT during this interval, OPT must incur a loss on all neighbors of j. In particular, from (27) we have

$$\sum_{t \in I_r} \tilde{\ell}_{i(t)} \geq \sum_{t \in I_r} \sum_{j \in N(i)} \delta_{j \to i}^t \mathbb{I}(s_t^*(j) = 0)$$

$$\geq c_i.$$

In either of the three cases we have that the loss $\sum_{t \in I_r} \tilde{\ell}_{i(t)}$ incurred by OPT is at least a $1/(B+2)$ fraction of the loss incurred by the algorithm. Summing over all the vertices and the corresponding intervals, we get that the total loss incurred by the algorithm can be bounded by

$$\text{Loss}(\mathcal{A}) \leq (B+2)\sum_{t=1}^{T}\sum_{i=1}^{k} \tilde{\ell}_{i(t)} \leq 2(B+2)\text{Loss}(\text{OPT}).$$

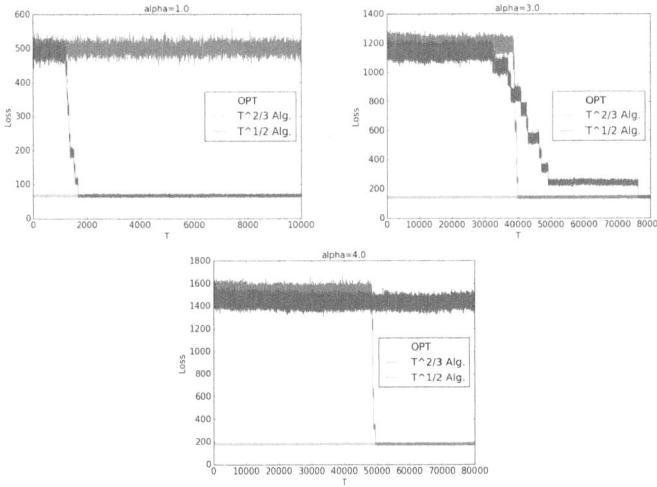

Fig. 11. Cumulative loss of the Algorithms of Fig. 3 and Fig. 9 on a graph with $k = 100$ criteria. $\alpha = 1, 3, 4$ determines the number of random pairs of vertices, αk, added into correlation sets.

E Experiments

In this appendix we provide more detail of the experiments discussed in Sect. 4 and also report additional results with the algorithm for the stochastic setting.

E.1 Experiments with Simulated Data

We first evaluated the performance of our stochastic setting algorithms (Sect. 3) on simulated data. We generated the graph $\mathcal{G}$ from the Erdős-Rényi model: $G(k, p)$ and set $p = 2\frac{\log k}{k}$, to ensure connectivity. We generated correlation sets ($\mathcal{C}$) of size two by picking αk pairs of vertices at random and adding them to $\mathcal{C}$, where α is a parameter. Hence, on average, each vertex is in α correlation sets. We also added to $\mathcal{C}$ singleton sets for each vertex in $\mathcal{G}$. The fixing cost of a vertex was sampled uniformly in $[1, 5]$. Parameters governing the loss distribution were drawn from a Beta distribution. We approximated the oracle for the optimization in (2) via a linear programming relaxation. Our algorithms of Fig. 3 and Fig. 9 admit complementary guarantees. The former admits a higher regret as a function of T, but only a polynomial dependence on α, i.e., the average number of correlation sets a vertex participates in. The latter incurs a smaller regret of $\tilde{O}(\sqrt{T})$ as a function of T at the expense of an exponential dependence on α. Figure 11 shows an empirical illustration of this: for smaller values of α, the $\tilde{O}(\sqrt{T})$ regret algorithm performs significantly better, while for larger values of α the $\tilde{O}(T^{2/3})$ regret algorithm is more desirable. We choose to compare the performance of our two proposed stochastic algorithms as we are not aware of any existing baselines for simultaneously optimizing multiple diverse metrics. Additionally, we focus on experiments with the stochastic model as it is hard to approximate the best offline algorithm in the adversarial setting of Appendix D.

We consider a simulated environment where the conflict graph $\mathcal{G}$ is generated from the Erdős-Renyi model: $G(k,p)$ where we set $p = 2\frac{\log k}{k}$. This ensures that with high probability $\mathcal{G}$ is connected. Next we generate correlation sets $\mathcal{C}$ consisting of pairs of vertices in $\mathcal{G}$ sampled uniformly at random. For a parameter $\alpha > 0$ that we vary, we choose αk pairs of vertices at random and add them as correlation sets in $\mathcal{C}$. Hence on average, each vertex participates in α correlation sets. We also add to $\mathcal{C}$ singleton sets for each vertex in $\mathcal{G}$. The fixing cost of each vertex is samples uniformly at random in the range $[1,5]$.

Next we describe the choice of parameters governing the loss distribution of the different states in the MDP. For a correlation set $\{i\}$ of size one corresponding to vertex v_i, we sample a parameter γ_i^1 from the beta distribution Beta$(0.5, 0.5)$. For a given state s with $s(i) = 1$, the loss generated due to $\{i\}$ is drawn from an exponential distribution with mean γ_i^1. For a given state s with $s(i) = 0$, the loss generated due to $\{i\}$ is drawn from an exponential distribution with mean $\lambda\gamma_i^1$, where $\lambda > 1$ is a parameter that we vary. For a correlation set $\{i,j\}$ of size two, we generate two parameters $\gamma_{i,j}^{1,1}$ and $\gamma_{i,j}^{1,0}$ from the beta distribution Beta$(0.5, 0.5)$ such that $\gamma_{i,j}^{1,0} > \gamma_{i,j}^{1,1}$. For a given state s with $s(i) = 1$ and $s(j) = 1$, the loss generated due to $\{i,j\}$ is drawn from an exponential distribution with mean $\gamma_{i,j}^{1,1}$. For states where $s(i) = 0$ and $s(j) = 1$ or vice-versa, the loss is generated from an exponential distribution with mean $\gamma_{i,j}^{1,0}$. Finally, for states where both $s(i) = 0$ and $s(j) = 0$, the loss is generated from an exponential distribution with mean $\lambda\gamma_{i,j}^{1,0}$.

In general, computation of the optimal state in (2) requires time exponential in k. In our experiments we approximate the optimal state by a linear programming relaxation of the optimization in (2) and use the appropriately rounded linear programming relaxation solution as a proxy for the optimal state.

For general m, our proposed algorithms in Fig. 3 and Fig. 9 have complementary strengths. While the algorithm in Fig. 3 incurs a higher regret as a function of the number of time steps T, its running time has a polynomial dependence on the parameter α, i.e., the number of correlation sets that a vertex participates in, on average. The algorithm in Fig. 9 incurs a smaller regret of $\tilde{O}(\sqrt{T})$ as a function of T at the expense of an exponential dependence on α. In Figs. 12 and 13 we empirically demonstrate this behavior where for small values of α, the $\tilde{O}(\sqrt{T})$-regret algorithm is much better, whereas for higher values of α the $\tilde{O}(T^{2/3})$-regret algorithm is more desirable.

For the case of $m = 1$ however, i.e., singleton correlation sets, the algorithm in Fig. 9 achieves a smaller regret and runs in polynomial time and hence is expected to outperform the explore-exploit based algorithm from Fig. 3. As can be seen from Fig. 14 this is indeed the case and the $\tilde{O}(\sqrt{T})$ regret algorithm significantly outperforms the $\tilde{O}(T^{2/3})$ regret algorithm.

E.2 Experiments with a Real-World Dataset

In this section we demonstrate via experiments how our framework and algorithms can be applied to real world data. In order to do this we study the UCI Adult dataset [31]. The dataset comprises of 48852 examples each represented using 124 features, after binarizing categorical features. Each data point corresponds to a person and the label is

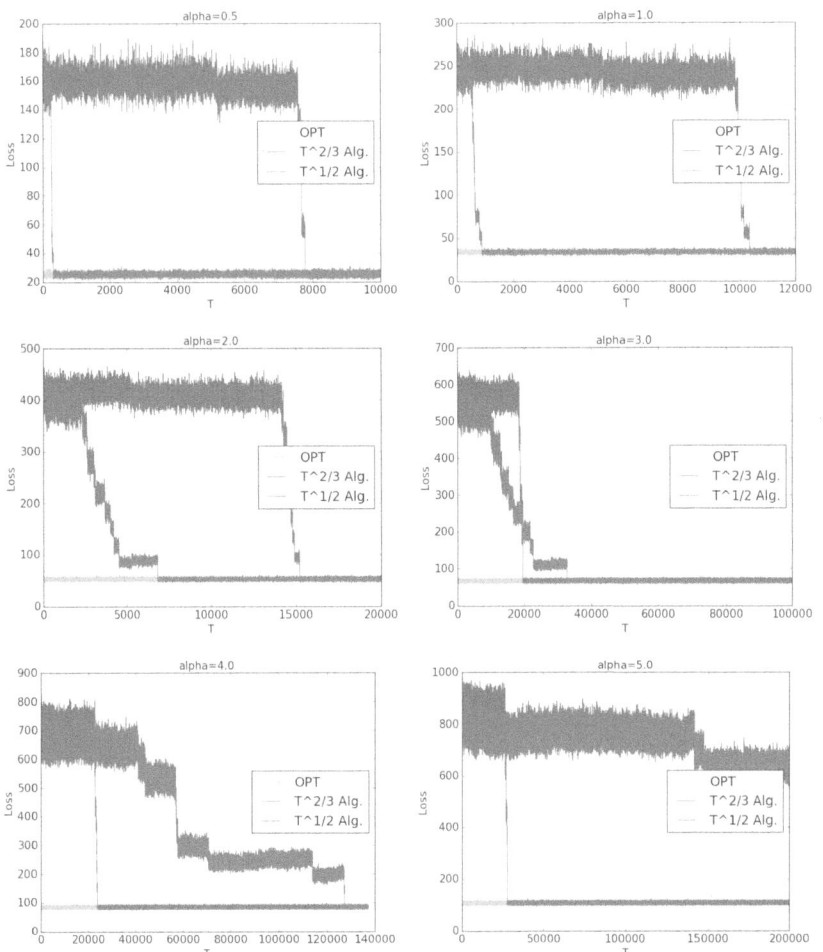

Fig. 12. The figure shows the total accumulated loss incurred by the Algorithms in Fig. 3 and Fig. 9 on a graph with $k = 50$ criteria. The parameter α controls the total number of correlation sets. For each value of α, we add αk random pairs of vertices into correlation sets.

a 0/1 value representing whether the income of the person is more or less than $50,000. The dataset contains information about sensitive attributes such as race and gender. We will simulate an online scenario where a classifier is making predictions on the income of individuals. At each time step a batch of complaints arrive, the system incurs a loss and responds by transitioning to a different state (and updating the classifier). We next describe how we instantiate various components of our stochastic model from Sect. 3.

Graph $\mathcal{G}$: We take race as a sensitive attribute that takes values in {black, white}, to obtain two sub-populations and consider two natural criteria namely the true positive rate and the AUC score. This leads to four vertices $tpr_w, tpr_b, auc_w, auc_b$. Furthermore,

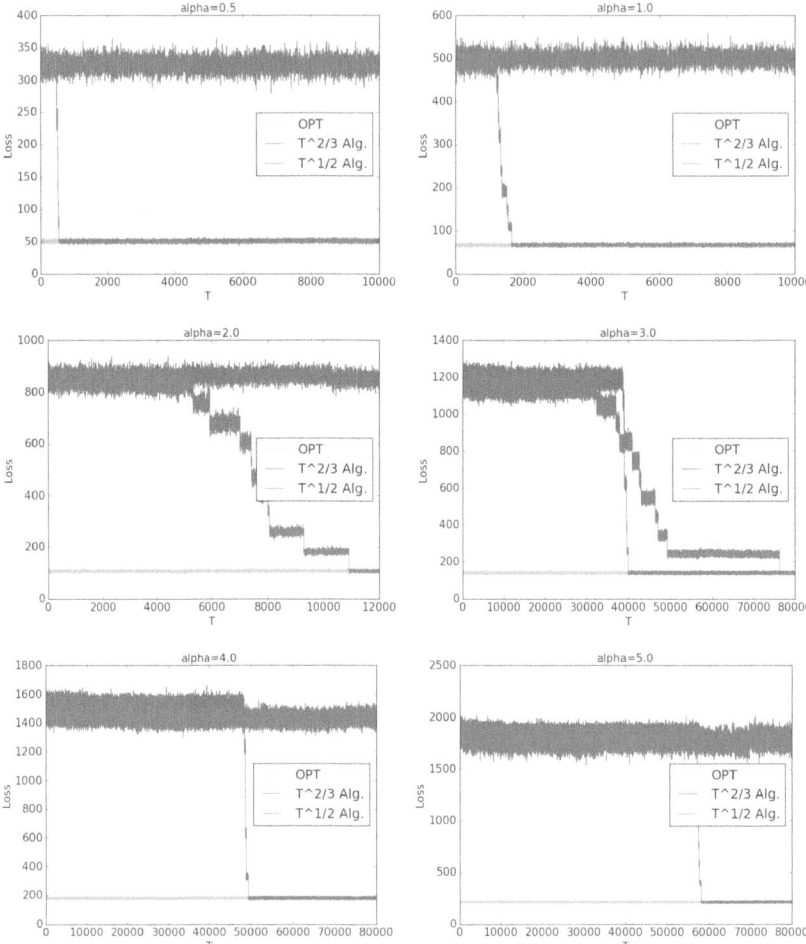

Fig. 13. The figure shows the total accumulated loss incurred by the Algorithms in Fig. 3 and Fig. 9 on a graph with $k = 100$ criteria. The parameter α controls the total number of correlation sets. For each value of α, we add αk random pairs of vertices into correlation sets.

we add the classifier accuracy as another criterion. This leads to total 5 vertices in the graph.

Losses and Correlation Sets: We consider correlation sets of size one, and hence the total loss incurred at any state is the sum of the losses incurred by each criterion. For the accuracy criterion we simply define the loss to be the error of the system (the classifier). We next describe how we define the loss for the tpr_w criterion. We first compute the overall true positive rate of the classifier and also the true positive rate on the white population. If the two deviate by more than a threshold τ, then we penalize the classifier linearly in the violation. Therefore the loss for tpr_w is defined as: $\max(0, |tpr_{overall} -$

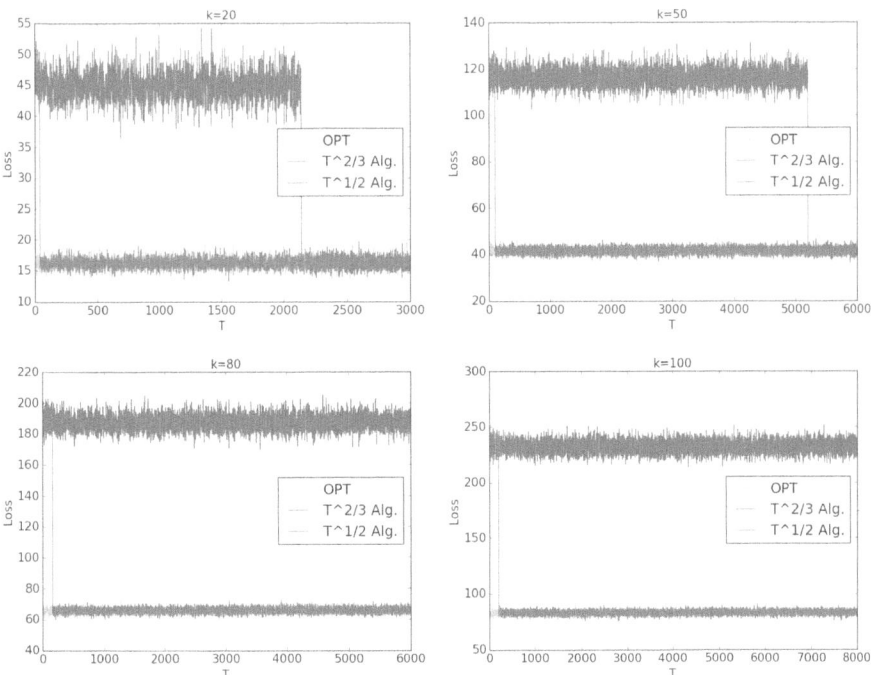

Fig. 14. The figure shows the total accumulated loss incurred by the Algorithms in Fig. 3 and Fig. 4 for the case of $m = 1$ and varying graph sizes.

$tpr_w| - \tau)$. The loss for all other criteria is defined the same way. In our experiments we choose $\tau = 0.005$.

Incompatibilities and State Transitions: To generate incompatibilities among criteria we compute a set of valid and invalid states as follows. For each state $s \in \{0, 1\}^5$, we solve a constrained optimization problem on a training set to compute a classifier. We then evaluate the classifier on the test set to compute the loss of each criterion. If the loss of any criterion is more than a specific threshold then we consider the state as an invalid state, otherwise the state is valid. In our experiments we set a threshold of 0.4 for the accuracy criterion. For the considered criteria we present results for two thresholds, 2τ and 6τ, the first one resulting in 4 valid states and other second one resulting in 7 valid states. To solve a constrained optimization problem we use the tensorflow constrained optimization toolkit [12, 13]. We use the default parameter settings provided by the toolkit. The toolkit is released under Apache license 2.0. If a state s has accuracy criterion set to 1, then we optimize for model accuracy subject to constraints for the other criteria that are set to 1 in s. If the accuracy criterion is set to 0 then we optimize for a constant loss function subject to constraints. Recall that our proposed algorithms function by fixing a criterion and as a result moving to another state. We obtain these state transitions as follows. If the algorithm asks to fix criterion v_i in state s, we set $s(i) = 1$

to go to the next state s'. While s' is invalid, we unfix the criterion (not including v_i) with the highest loss in the state s' to reach another state.

Fixing Cost: We simply take the fixing cost of each criterion to be 1.

Simulating Complaints: We divide the dataset into a set of 16000 examples that we use to update our classifier at each time step and the remaining *test* set to simulate the arrival of complaints. At each time step, we randomly select a batch of examples from the test set to generate complaints. This set of complaints is used to compute the loss of a given state at a given time step.

Benchmark and Results: We compare our Algorithm from Fig. 4 with an offline optimal solution that has been computed to find the state with the minimum average loss over the arrival sequence of complaints. The results are averaged over 10 independent runs.

The results are shown in Fig. 15 and Fig. 16. We show results for two values of the threshold parameters and in each case plot the loss of the algorithm as compared to the benchmark, as well as the states chosen by the algorithm, as a function of the number of time steps. As can be seen from Fig. 15 our algorithm quickly converges to the offline optimal solution after an initial exploration phase. To get a better understanding of the performance of the algorithm in the initial phases, in Fig. 16 we plot the same setting as in the case of Fig. 15, but with x-axis on a log-scale. For the case of threshold being 0.01, one can see that the state 0 results in much higher loss and, during exploration, the algorithm alternates in a periodic pattern between states 1 and 3 that have similar loss. A similar pattern holds for the case of the threshold being 0.03. It is important to note that the choice of the loss functions was important in this case and that we did not weight each criterion by the volume of the complaints. This demonstrates that our algorithms, when combined with a good choice of the loss function, can be useful in practice.

Compute Resources. All our experiments were performed on a machine containing a Tesla P100 GPU with 80 GB of RAM and four CPUs.

Hyperparameters. For the case of simulated data the hyperparameters have been mentioned in Sect. E.1. For the case of real data, apart from the hyperparameters mentioned in Sect. E.2, we used the default learning rates and optimizers provided by the tensorflow constrained optimization toolkit [12, 13]. We performed a random train/test split as detailed in Sect. E.2.

Assets. We used publicly available code from the tensorflow constrained optimization toolkit[2] and the publicly available UCI Adult Dataset[3].

---

[2] License at: https://github.com/google-research/tensorflow_constrained_optimization/blob/master/README.md.
[3] https://archive.ics.uci.edu/ml/datasets/adult.

F   Further Discussion on the COMPAS Example

Throughout the main sections, we have mentioned that the choice of the loss function is important in the effectiveness of our model. We briefly discussed this in Sect. 3. Below, we present a more detailed discussion of the effect of the loss function on our model, by using the COMPAS scenario from Sect. 1 as an example.

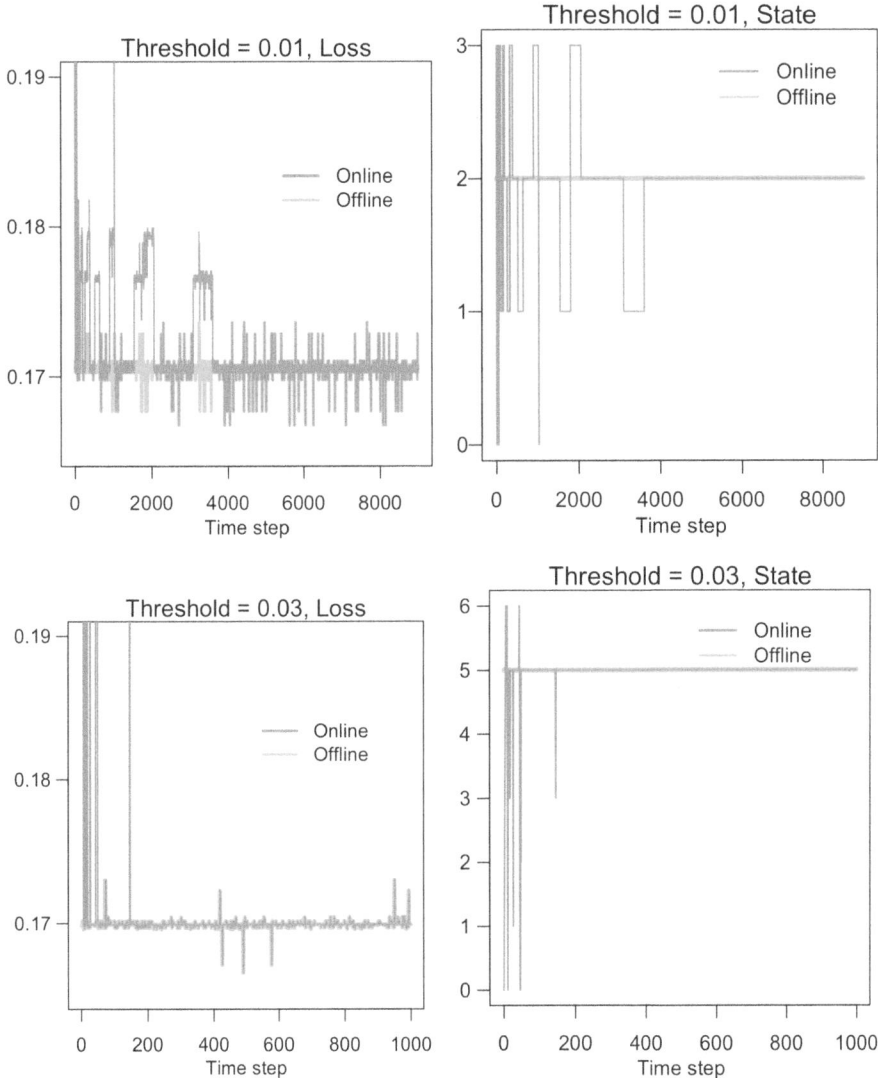

Fig. 15. The figure shows the performance of the Algorithm in Fig. 4 on the UCI Adult dataset. We present results for two threshold values, and in each case plot the loss of the offline solution and the online algorithm as well as the states chosen by the online algorithm, as a function of the time steps.

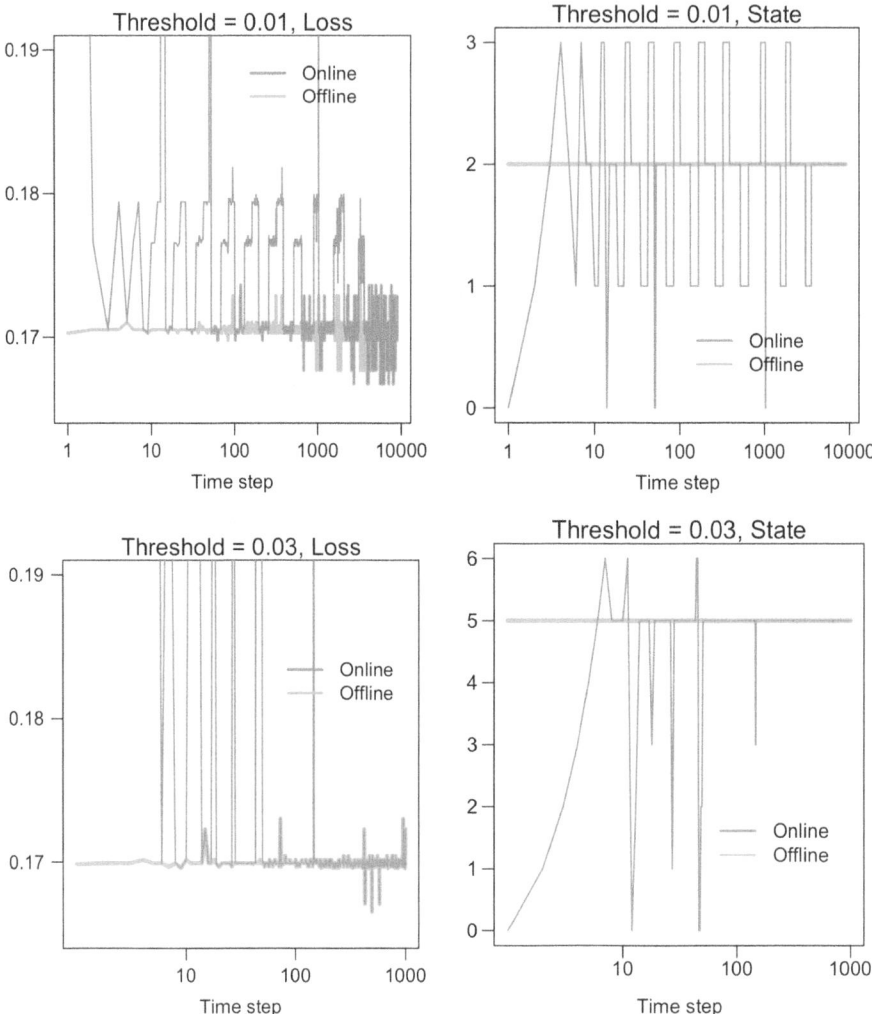

Fig. 16. The figure shows the performance (x-axis on a log scale) of the Algorithm in Fig. 4 on the UCI Adult dataset. We present results for two threshold values, and in each case plot the loss of the offline solution and the online algorithm as well as the states chosen by the online algorithm, as a function of the time steps.

Loss Function – COMPAS Illustration . Consider the COMPAS example with a graph $\mathcal{G}$ with four criteria namely, false positive rate on population A, false positive rate on population B, AUC score for population A and AUC score for population B. We want to understand what kinds of loss functions will result in an overall suboptimal system when our model and algorithms from Sect. 3. Suppose our algorithm take an action to fix a criterion and reach a state where the true positive rates and the AUC scores associated with the four criteria are: $[0.1, 0.8, 0.5, 0.5]$. Then a poor choice of the loss function would be $f_1 \cdot 0.1 + f_2 \cdot 0.8 + f_3 \cdot 0.5 + f_4 \cdot 0.5$, where f_i represents the fraction of complaints that trigger criterion i. Such a choice of the loss function will make our system vulnerable to the loudest voices in the system and as a result might not lead to a good solution at all. A more reasonable choice of the loss is $0.1 + 0.8 + 0.5 + 0.5$, that weighs each criteria equally and does not take into account the underlying size of the population. Another alternative is $\lambda_1 |0.1 - 0.8| + \lambda_2(|0.5 - 0.5|)$, that aims at keeping both the discrepancy in the false positive rate and the AUC scores small. Finally, the choice we make in our experiments of penalizing each criterion for the deviation from the value over the entire population, i.e., $\max(0, |tpr_{overall} - tpr_w| - \tau)$, also leads to good solutions empirically.

Another case where additive losses are a poor choice is if the criteria in $\mathcal{G}$ is not chosen carefully. For instance, consider a scenario in the COMPAS example where all except one of the criteria correspond to the performance of the system on population A. An additive loss would then naturally force the system to disproportionately favor population A over a period of time.

References

1. Agarwal, A., Beygelzimer, A., Dudík, M., Langford, J., Wallach, H.: A reductions approach to fair classification. arXiv preprint arXiv:1803.02453 (2018)
2. Angwin, J., Larson, J., Mattu, S., Kirchner, L.: Machine bias: there's software used across the country to predict future criminals. and it's biased against blacks 2016 (2019). https://www.propublica.org/article/machine-bias-risk-assessments-in-criminal-sentencing
3. Auer, P., Cesa-Bianchi, N., Fischer, P.: Finite-time analysis of the multiarmed bandit problem. Mach. Learn. **47**(2–3), 235–256 (2002)
4. Bastani, O., Zhang, X., Solar-Lezama, A.: Probabilistic verification of fairness properties via concentration. Proc. ACM Program. Lang. **3**(OOPSLA), 1–27 (2019)
5. Bechavod, Y., Jung, C., Wu, Z.S.: Metric-free individual fairness in online learning. arXiv preprint arXiv:2002.05474 (2020)
6. Bellamy, R.K., et al.: Ai fairness 360: an extensible toolkit for detecting, understanding, and mitigating unwanted algorithmic bias. arXiv preprint arXiv:1810.01943 (2018)
7. Borodin, A., El-Yaniv, R.: Online Computation and Competitive Analysis. Cambridge University Press, Cambridge (1998)
8. Cesa-Bianchi, N., Dekel, O., Shamir, O.: Online learning with switching costs and other adaptive adversaries. In: Advances in Neural Information Processing Systems, pp. 1160–1168 (2013)
9. Chen, L., Pu, P.: Critiquing-based recommenders: survey and emerging trends. User Model. User-Adap. Inter. **22**(1), 125–150 (2012)
10. Cortes, C., Mohri, M., Gonzalvo, J., Storcheus, D.: Agnostic learning with multiple objectives. Adv. Neural. Inf. Process. Syst. **33**, 20485–20495 (2020)
11. Coston, A., et al.: Fair transfer learning with missing protected attributes. In: Proceedings of the 2019 AAAI/ACM Conference on AI, Ethics, and Society, pp. 91–98 (2019)
12. Cotter, A., et al.: Training well-generalizing classifiers for fairness metrics and other data-dependent constraints. arXiv preprint arXiv:1807.00028 (2018)
13. Cotter, A., Jiang, H., Sridharan, K.: Two-player games for efficient non-convex constrained optimization. arXiv preprint arXiv:1804.06500 (2018)
14. Doroudi, S., Thomas, P.S., Brunskill, E.: Importance sampling for fair policy selection. Grantee Submission (2017)

15. Dwork, C., Immorlica, N., Kalai, A.T., Leiserson, M.: Decoupled classifiers for group-fair and efficient machine learning. In: Conference on Fairness, Accountability and Transparency, pp. 119–133 (2018)
16. Feller, A., Pierson, E., Corbett-Davies, S., Goel, S.: A computer program used for bail and sentencing decisions was labeled biased against blacks. it's actually not that clear. The Washington Post (2016)
17. Ghosh, B., Basu, D., Meel, K.S.: Justicia: a stochastic sat approach to formally verify fairness. arXiv preprint arXiv:2009.06516 (2020)
18. Gupta, M., Cotter, A., Fard, M.M., Wang, S.: Proxy fairness. arXiv preprint arXiv:1806.11212 (2018)
19. Gupta, V., Nokhiz, P., Roy, C.D., Venkatasubramanian, S.: Equalizing recourse across groups. arXiv preprint arXiv:1909.03166 (2019)
20. Hashimoto, T.B., Srivastava, M., Namkoong, H., Liang, P.: Fairness without demographics in repeated loss minimization. arXiv preprint arXiv:1806.08010 (2018)
21. Holstein, K., Wortman Vaughan, J., Daumé III, H., Dudik, M., Wallach, H.: Improving fairness in machine learning systems: what do industry practitioners need? In: Proceedings of the 2019 CHI Conference on Human Factors in Computing Systems, pp. 1–16 (2019)
22. Jabbari, S., Joseph, M., Kearns, M., Morgenstern, J., Roth, A.: Fairness in reinforcement learning. In: Proceedings of the 34th International Conference on Machine Learning, vol. 70, pp. 1617–1626. JMLR.org (2017)
23. Jaksch, T., Ortner, R., Auer, P.: Near-optimal regret bounds for reinforcement learning. J. Mach. Learn. Res. **11**(Apr), 1563–1600 (2010)
24. Jin, Y.: Multi-objective Machine Learning, vol. 16. Springer, Heidelberg (2006). https://doi.org/10.1007/3-540-33019-4
25. Jin, Y., Sendhoff, B.: Pareto-based multiobjective machine learning: an overview and case studies. IEEE Trans. Syst. Man Cybern. Part C (Appl. Rev.) **38**(3), 397–415 (2008)
26. Kaminskas, M., Bridge, D.: Diversity, serendipity, novelty, and coverage: a survey and empirical analysis of beyond-accuracy objectives in recommender systems. ACM Trans. Interact. Intell. Syst. (TiiS) **7**(1), 1–42 (2016)
27. Kannan, S., Roth, A., Ziani, J.: Downstream effects of affirmative action. In: Proceedings of the Conference on Fairness, Accountability, and Transparency, pp. 240–248 (2019)
28. Karlin, A.R., Manasse, M.S., Rudolph, L., Sleator, D.D.: Competitive snoopy caching. Algorithmica **3**(1–4), 79–119 (1988)
29. Kearns, M., Roth, A., Sharifi-Malvajerdi, S.: Average individual fairness: algorithms, generalization and experiments. arXiv preprint arXiv:1905.10607 (2019)
30. Kleinberg, J., Mullainathan, S., Raghavan, M.: Inherent trade-offs in the fair determination of risk scores. In: Innovations in Theoretical Computer Science Conference (ITCS) (2017)
31. Kohavi, R.: Scaling up the accuracy of naive-bayes classifiers: a decision-tree hybrid. In: KDD 1996, pp. 202–207 (1996)
32. Lamy, A., Zhong, Z., Menon, A.K., Verma, N.: Noise-tolerant fair classification. In: Advances in Neural Information Processing Systems, pp. 294–305 (2019)
33. Lin, X., et al.: A pareto-efficient algorithm for multiple objective optimization in e-commerce recommendation. In: Proceedings of the 13th ACM Conference on Recommender Systems, pp. 20–28 (2019)
34. Liu, L.T., Dean, S., Rolf, E., Simchowitz, M., Hardt, M.: Delayed impact of fair machine learning. arXiv preprint arXiv:1803.04383 (2018)
35. Marler, R.T., Arora, J.S.: Survey of multi-objective optimization methods for engineering. Struct. Multidiscip. Optim. **26**(6), 369–395 (2004)
36. Masthoff, J.: Group recommender systems: combining individual models. In: Ricci, F., Rokach, L., Shapira, B., Kantor, P.B. (eds.) Recommender Systems Handbook, pp. 677–702. Springer, Boston, MA (2011). https://doi.org/10.1007/978-0-387-85820-3_21

37. Menon, A.K., Williamson, R.C.: The cost of fairness in binary classification. In: Conference on Fairness, Accountability and Transparency, pp. 107–118 (2018)
38. Mohri, M., Sivek, G., Suresh, A.T.: Agnostic federated learning. In: International Conference on Machine Learning, pp. 4615–4625. PMLR (2019)
39. Mouzannar, H., Ohannessian, M.I., Srebro, N.: From fair decision making to social equality. In: Proceedings of the Conference on Fairness, Accountability, and Transparency, pp. 359–368 (2019)
40. Sener, O., Koltun, V.: Multi-task learning as multi-objective optimization. arXiv preprint arXiv:1810.04650 (2018)
41. Shah, A., Ghahramani, Z.: Pareto frontier learning with expensive correlated objectives. In: International Conference on Machine Learning, pp. 1919–1927. PMLR (2016)
42. Simchi-Levi, D., Xu, Y.: Phase transitions and cyclic phenomena in bandits with switching constraints. In: Advances in Neural Information Processing Systems, pp. 7521–7530 (2019)
43. Speicher, T., et al.: Potential for discrimination in online targeted advertising. In: Conference on Fairness, Accountability and Transparency, pp. 5–19. PMLR (2018)
44. Thomas, P.S., da Silva, B.C., Barto, A.G., Giguere, S., Brun, Y., Brunskill, E.: Preventing undesirable behavior of intelligent machines. Science **366**(6468), 999–1004 (2019)
45. Tsirtsis, S., Gomez-Rodriguez, M.: Decisions, counterfactual explanations and strategic behavior. arXiv preprint arXiv:2002.04333 (2020)
46. Wang, S., Guo, W., Narasimhan, H., Cotter, A., Gupta, M., Jordan, M.I.: Robust optimization for fairness with noisy protected groups. arXiv preprint arXiv:2002.09343 (2020)
47. Wen, M., Bastani, O., Topcu, U.: Fairness with dynamics. arXiv preprint arXiv:1901.08568 (2019)
48. Xiao, L., Min, Z., Yongfeng, Z., Zhaoquan, G., Yiqun, L., Shaoping, M.: Fairness-aware group recommendation with pareto-efficiency. In: Proceedings of the Eleventh ACM Conference on Recommender Systems, pp. 107–115 (2017)
49. Yu, B., Yuan, Y., Terveen, L., Wu, Z.S., Forlizzi, J., Zhu, H.: Keeping designers in the loop: communicating inherent algorithmic trade-offs across multiple objectives. In: Proceedings of the 2020 ACM Designing Interactive Systems Conference, pp. 1245–1257 (2020)

Trick Costs for $\alpha\mu$ and New Relatives

Samuel Bounan[1(✉)] and Stefan Edelkamp[2,3]

[1] École normale supérieure de Lyon, Lyon, Auvergne-Rhône-Alpes, France
samuel.bounan@ens-lyon.fr
[2] Department of Theoretical Computer Science and Mathematical Logic,
Faculty of Mathematics and Physics, Charles University, Prague, Czech Republic
edelkamp@ktiml.mff.cuni.cz
[3] Computer Science Department, Faculty of Electrical Engineering,
Czech Technical University in Prague, Prague, Czech Republic
edelkste@fel.cvut.cz

Abstract. In this paper we present a player for incomplete information card games with tricks scored by points or eyes. We factorize existing algorithms into a template that captures the main ingredients of such game trees. We then analyze three different algorithms, and the impact of the information given during the algorithm on the decisions it makes. We extend these algorithms to work with cost instead of winning vectors, and illustrate their effectiveness in finding good cards. We develop a new algorithm that tries to respect known information.

Keywords: Tree search · Game theory · Card games

1 Introduction

Many perfect-information board games such as Checkers [23], Oware [2], Connect4 [1], Awari [22], and Nine-Men-Morris [14] have been solved, or, as in Chess, Shogi or Go, computer AIs clearly outperform humans [26]. Therefore, research attention has shifted to incomplete information games. The partially observable board game Stratego has recently been analyzed with *DeepNash*, a model-free multiagent reinforcement learning algorithm [21]. It achieved an all-time top-3 rank on the Gravon games platform, competing with human expert players.

Card games remain an objective of research for decision making with imperfect information. After some variants of Poker have been solved or played to a satisfying degree [3,20], trick-taking games such as Skat [5,17], Hearts [28] and Bridge [6,13,18] have been identified as current AI challenges. One obstacle is that, given the large number of tricks and degree of uncertainty, a direct application of reinforcement learning as in Go [25] is less obvious.

Most recently, some card game AIs are beginning to challenge human supre- macy. Notably worldclass caliber play in Bridge [6,7], Spades [8], and Skat [10,11]. In these cutting-edge players, domain-dependent information is provided, such as winning probabilities extracted from human expert games [9]. We have implemented an efficient framework for general card games. General game playing has a long tradition: there have been several insightful international competitions using GDL or GDL-II (for incomplete information) [24] as the input language. While the players achieved a remarkable playing strength [12], the generation of moves is slow. Even faster frameworks like Google DeepMind's *OpenSpiel* are less efficient for card games than our framework with its concise card encodings.

Another problem is that for multi-player teams most general game playing algorithms are not yet competitive. We focus on trick-taking stage of card games. While we have implemented bidding and dog putting strategies for all the card games, they are not the subject of this work, as they are often based on domain-dependent conventions.

The paper is organized as follows. First we will define the building blocks for the formalization of the problem. Our notation is not taken from any specific work, but helps to design a naive algorithm. Section 4 refers to the construction of the driver template algorithm. Then, we discuss the efficiency of three algorithms. The first one, called Perfect Information Monte-Carlo, PIMC, is well-known [15]. The second one, $\alpha\mu$ is newer [7] and shown effective in Bridge. The third one is a new. In all three, we present the algorithm principles in a template algorithm thus adapt them to our model. Also, $\alpha\mu$ was only used with Boolean score vectors and with two players. We extend it to scores in $\mathbb{R}$ and to several players. Finally, we present some results we obtained in our implementation. Our code generalizes ideas from [10] to implement, while in a restricted setting with simplified bidding. The contribution of this paper are as follows. We provide an efficient framework. calculate the effect of truncating tree search with subsequent random playouts. quantify the information gain of $\alpha\mu$ wrt PIMC. successfully apply $\alpha\mu$ algorithm to support more players in a team and to general trick costs; propose a novel general incomplete-information algorithm as a compromise of different nodes in the backup.

2 Preliminaries

The *set of cards* $\mathcal{C}$ is dealt to p players at the beginning of the game. A *world* is any possible deal of the cards among the players, $\mathcal{W}$ the set of worlds, and for a world $w \in \mathcal{W}$, w_i denotes the hand of player i, $i \in \{1, \ldots, p\}$. Worlds change during a game. We call *state* the history of cards already played, and $\mathcal{S}$ the set of possible states.

In a state s, the *next player* to play a card is determined by a function $turn(s) \in \{1, \ldots p\}$ and the set of *legal cards* that can be played in s with a hand w_i by $legal(w_i, s) \subseteq \mathcal{C}$. In some states the game ends, this is determined by a Boolean function $over(s)$, and in these cases each player receives an amount of points given by a function $score(i, s) \in \mathbb{R}$. The goal of each player is to maximize its final score, and we assume rationality of everyone and common knowledge of rationality. In general in a state $s \in \mathcal{S}$ the world is neither perfectly known nor completely unknown and with a hand w_i player $i \in \{0, \ldots, p\}$ assumes a particular distribution on the set of worlds $D_{i, w_i, s}$.

To construct an artificial player, we (only) need to know in a state s, in a position i and with a hand h, what is the best card to play. In extension to articles that formalize tree searches, as in [4,19] we chose the following definition.

If we define $\sigma(i,h,s) \in \mathbb{R}$ as the best score that can be obtained, with $\arg\sigma$ to denote a probabilistic function that returns a card corresponding to the maximum score, uniformly chosen in $\mathcal{U}$ among cards that satisfy this property, we have

$$\sigma(i,h,s) = \max_{c \in legal(h,s)} \mathbb{E}_{w \hookrightarrow D(i,h,s)} \mathbb{E}_{f \hookrightarrow \mathcal{U}(\mathcal{F})} score(i,f)$$

where $\mathcal{F}$ is the set of possible optimal final states. With $s_1 = s$ and $s_{k+1} = s_k \cup \arg\sigma(turn(s_k), w_{turn(s_k)} \setminus s_k, s_k)$ for $k > 1$ we have

$$\mathcal{F} = \{f \in \mathcal{S} \mid \exists k \in \mathbb{N} \quad \mathbb{P}[s_k = f] \neq 0 \text{ and } over(f)\}.$$

Every object can be easily derived from the rules, except the knowledge distribution $D_{i,h,s}$. We will construct O the set of possible worlds with a non-zero probability for D.

Let w be a world in $\mathcal{W}$ with state s and hand h. We have the following constraints on w: a) $w_i = h$; and b) in every state before s, the players played optimally (because of rationality). Let n be chosen, so that s is an n-tuple. Let s_k be the card played in s at step $k < n$. From the second constraint, we know that in the state $s_{|k} = (s_1, \ldots s_k)$ player $turn(s_{|k})$ played a card s_{k+1} so that $\mathbb{P}[s_k = \arg\sigma(turn(s_{|k}), w_{turn(s_{|k})} \setminus s_{|k}, s_{|k})] \neq 0$. We have

$$O = \{w \in \mathcal{W}; w_i = h \text{ and } \forall k \in [n]$$
$$\mathbb{P}[s_{k+1} = \arg\sigma(turn(s_{|k}), w_{s_{|k}} \setminus s_{|k}, s_{|k}] \neq 0.\}$$

Because no additional knowledge can be used to infer a probability among this set of possible worlds O we have $D_{i,h,s} = \mathcal{U}(O)$. There are two problems with this definition of D. First, in practice players may differ from the perfect rationality, and restricting O to the worlds where everyone played optimally can quickly become a strong bad bias. Second, computing O is computationally costly (it needs to compute σ several times). To simplify $D_{i,h,s}$ we assume it follows an uniform distribution over the set of *legal* worlds P, that can be computed in constant time as $D_{i,h,s} = \mathcal{U}(P)$ with $w \in P$ iff $w_i = h$ and for all $k < n$ we have $s_{k+1} \in legal(w_{turn(s_{|k})}, s_{|k})$.

From a card played by a player i, as player j can often derive a set of cards that player i does not have (initially, player j's hand, maybe enlarged by certainties through the bidding process). We can represent the knowledge of a player j with sets $\neg K_i^j$ corresponding to the cards that player j knows that player i must not have. This ease the exploration and the search tree can be pruned based on the knowledge. When a card c has been played by player i, we can update $\neg K_i^j$ efficiently with Algorithm 1. With this reduction of D we study the general structure of the algorithm we want to design.

Algorithm 1. $update(\neg K_i^j, c, s)$

for $c' \in \mathcal{C} \setminus \neg K_i^j$ do
    if $c \notin legal(\{c, c'\}, s)$ then
        $\neg K_i^j \leftarrow \neg K_i^j \cup \{c'\}$

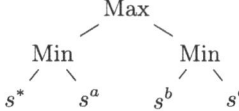

Fig. 1. Depth-1 minimax tree.

3 Minimax Truncated with Random Playouts

Most algorithms include a fast open-card solver to compute the game value. There are different approaches but they all use the minimax search tree as their basis. Note that due to the tricks taken and the teams the max or min modes may not alternate on each path. Finding the optimal result of a minimax tree can be computationally expensive, so an intuitive idea is to limit its depth by estimating the score of a node at the maximal depth by random playouts of the end of the game. In our case, the tree is not that deep, and we have more information on the knowledge of the players, starting with the knowledge that each leaf of the tree is at a constant depth. It is, therefore, possible to obtain a bound on the error made by approximating the optimal score with these playouts, in particular by taking the average of the score obtained by a number of random playouts (Fig. 1).

Suppose we have a perfect alternating minimax tree T. Given the outcomes at each leaf of T, we know that some of them cannot be the optimal minimax score. Let s^* denote the optimal score. In Fig. 3, we know that $s^a \geq s^*$ because Min would otherwise play for s^a. We also know that one of s^b and s^c is lower than s^* because Max chooses the left subtree in the optimal strategy. So we know that if we order the scores of the leaves $s_1 \leq s_2 \leq s_3 \leq s_4$, we have $s^* \in \{s_2, s_3\}$. More generally, we can compute the number of playouts g (resp. l) that have a score higher (resp. lower) than s^*. We will estimate the exact number of playouts greater and less than s^* that are needed to be certain that s^* is optimal for the minimax strategy. Thus, our bounds on the minimax score will be tight. Note that this is also the minimum number of leaves that an algorithm needs to explore to compute s^*. We study alternating minimax trees. For consistency, we denote by depth of a tree T the number (Max,Min) nodes to a leaf. To estimate l and g we additionally assume that the number of actions at a given depth d is constant, so that a_d denotes the number of possible actions at the root of a tree of depth d for Max, and b_d for Min.

First, we compute the number of playouts that have a score greater than s^*. We note $s(T)$, the optimal minimax score of a tree T. We have $s(T_{init}) = s^*$. To compute the number of playouts with a score greater than s^* we only need the assumption that $s(T_{init}) \geq s^*$. Indeed, if we know that $s(T) \geq s^*$, it means that Max chooses a card a such that for all possible choices b of Min, the score of the subtree of T after playing a and b has a score greater than s^*. Otherwise Min could choose an action that leads to a score that is strictly lower than s^*. It follows that $s(T) < s^*$, a contradion. Thus, we have $\exists a \ \forall b \ \ s(T_{a,b}) \geq s^*$.

We use the assumptions $s(T_{a,b}) \geq s^*$ to setup a recursion. The terminal case is when T has one node, in which case we have one playout that we know has a score greater than s^*. With the above assertion we can deduce that the number of playouts in a tree that have a score greater than s^* depends only on the number of actions Min that can made at each node of the tree. Here, we have made the assumption that in a tree with depth d, the number of possible actions for Min is a fixed number b_d. Thus the number of playouts that achieve a score greater than s^* only depends on the depth of the tree. We can, therefore, define it as a sequence g_d. The number g_d represents the number of playouts that have a score greater than the optimal minimax score for a tree of depth d. Following what we established earlier, we have $g_0 = 1$ and $g_d = b_d \ g_{d-1}$ for all $d > 0$. Finally, we get $g_d = \prod_{i=1}^{d} b_i$. Symmetrically, we can compute the number of playouts that achieve a score lower than s^* as $l_d = \prod_{i=1}^{d} a_i$. Assuming T is a perfect alternating minimax tree of depth d, the optimal score can be achieved by a playout that is neither among the $l_d - 1$ lowest playout scores nor in the $g_d - 1$ higher playout scores. So the total number of playouts is $l_d \ g_d$.

For a bound on the actual score we need to model the distribution of the playout scores. We assume that the number of points that can be won in T_d is S_d (we assume here that the minimum score that can be obtained is 0 but we can easily adjust the result with S_d in $s_{max}(T_d) - s_{min}(T_d)$). We will, therefore, model the distribution of score with a binomial law of parameters $n = S_d$, p unknown. We approximate this law by $\mathcal{N}(\mu, \sigma^2)$ with $\mu = S_d p$ and $\sigma = \sqrt{S_d p(1-p)}$. In this setting we can compute bounds on s^* directly. In fact, a lower bound on s^* is given by the value of the $l_d/(l_d \ g_d)$ quantile, as we showed above. We have $Q(l_d/(l_d \ g_d)) \leq s^* \leq Q((l_d \ g_d - g_d)/(l_d \ g_d))$ with Q being the quantile function of the normal distribution $Q(p) = \mu + \sigma\sqrt{2} \ f^{-1}(2p-1)$ and f being shorthand for erf. By estimating s^* as the midpoint between the two quantiles (μ in our model) we can estimate the error ϵ_d. For simplicity we assume that $g_d = l_d = n_d \geq 2$ (symm. Min / Max). Thus, we have $l_d/(l_d \ g_d) = 1/n_d$, so that the error is:

$$\epsilon_d = \frac{Q\left(\frac{n_d-1}{n_d}\right) - Q\left(\frac{1}{n_d}\right)}{2}$$

$$= \frac{\sigma\sqrt{2}\left(f^{-1}(1-\frac{2}{n_d}) - f^{-1}(\frac{2}{n_d}-1)\right)}{2}$$

$$= \frac{\sigma}{\sqrt{2}}\left(2 \cdot f^{-1}\left(1-\frac{2}{n_d}\right)\right) \text{ (symmetry of } f^{-1})$$

We set $\sigma = \sqrt{S_d p(p-1)} \leq \frac{\sqrt{S_d}}{2}$, so that $\epsilon_d \leq \sqrt{\frac{S_d}{2}} \ f^{-1}\left(1-\frac{2}{n_d}\right)$.

If we approximate s^* with the mean of the score obtained by a number m of random playouts $\hat{\mu}$, we have, using the Bienaymé-Tchebychev inequality for the

Algorithm 2. *naive*

>**function** $search(i, h, s)$
>$\quad res \leftarrow \{s\}$
>$\quad max \leftarrow \bot$
>$\quad$**for** $c \in legal(h,s)$ **do**
>$\quad\quad j \leftarrow turn((s,c))$
>$\quad\quad score' \leftarrow \bot$
>$\quad\quad$**for** $w \in P$ **do**
>$\quad\quad\quad F \leftarrow search(j, w_j, (s,c))$
>$\quad\quad\quad$**for** $x \in F$ **do**
>$\quad\quad\quad\quad score' + = score(i,x)/|F \times P|$
>$\quad\quad$**if** $score' > max$ **then**
>$\quad\quad\quad res \leftarrow \{c\}$
>$\quad\quad\quad max \leftarrow score'$
>$\quad\quad$**else if** $score' = max$ **then**
>$\quad\quad\quad res \leftarrow res \cup \{c\}$
>$\quad$**return** res
>
>**function** $chooseCard(h,s)$
>$\quad$**for** $c \in legal(h,s)$ **do**
>$\quad\quad F \leftarrow search(id, h, s)$
>$\quad\quad S[c] \leftarrow \mu_{f \in F}(score(id, f))$
>$\quad$**return** $\arg\max S$

normal approximation $P(|\hat{\mu} - \mu| \geq \epsilon_m) \leq \frac{\sigma^2}{m\epsilon_m^2} = \frac{S_d}{4m\epsilon_m^2}$. Adding the estimation error, we know that by choosing $m \geq \frac{S_d}{4\delta\epsilon_m^2}$ we have $P(|\hat{\mu} - s^*| \geq \epsilon_m + \epsilon_d) \leq \delta$.

For Belote we set $n_d = 2^d d!$ and $S_d = 10d$. if we want a 20%-approximation of the score, valid with a 95% chance, we can estimate the tree after the 5th trick, with $m = 8$. If we want to have a 15%-approximation valid with a 95% chance, we can cut after the 6th trick with $m = 3$.

4 Template Algorithm

Following $\sigma(i, a, s)$ we design a first naive Algorithm 2 to solve our problem, which branches on every maximum and expectation of σ. The problem with this algorithm is its time complexity. Let $T(d)$ denote the time complexity of the algorithm if d cards are still to be played before the game is over. We have $T(d) = |legal(h,s)| \cdot |P| \cdot T(d-1)$. If we assume (for the sake of clarity) the number of legal cards to be fixed during the game, and that the number of possible worlds to be constant, we have $T(d) = K^d$, where K is a constant. Because of this time complexity the naive algorithm is not practical.

The common methods to solve this problem use the same set of worlds for all nodes in the search tree. This was not the case in our naive algorithm, where each node branches to its possible set of worlds P. Let us assume a set of worlds W for whole tree. We will give each node a value representing the optimal scores

Algorithm 3. tree

function $tree(s, \alpha, W, parent)$
    $parent[myteam] \leftarrow myid$
    $r \leftarrow init_{myid}(s, W)$
    for $c \in \bigcup_{w \in W} legal(s, w_{myid})$ **do**
        $\alpha[myteam] \leftarrow \max_{myid}(r, \alpha[myteam])$
        if $\exists t \neq myteam : \alpha[t] <_{parent[t]} r$ **then**
            return $\bot_t$
        $W_c \leftarrow \{w \in W \mid c \in legal(s, w_{myid})\}$
        $v \leftarrow tree(\alpha, (s, c), W_c, parent)$
        if $v = \bot_t$ **then**
            if $t \neq team$ **then**
                return $\bot_t$
        else
            $r \leftarrow \max_{myid}(r, v)$
    return r

function $chooseCard(s, h)$
    $W \leftarrow genWorlds()$
    for every team t **do**
        $parent[t] \leftarrow \emptyset$
        $\alpha[t] \leftarrow \bot_t$
    return $\arg criterion_{c \in legal(s,h)}(tree((s, c), \alpha, W, parent))$

---

in state s for the worlds in W. The type of value depends on the specification of the algorithm, it will be some kind of array indexed by $w \in W$ with the optimal scores of each w. It should contain the scores of each player, as it will be the only value passed through the tree.

With this tree, it is often easy to cut some unused nodes using deep pruning, as in the $\alpha\beta$ pruning algorithm. This pruning is essential in the minimax tree. As shown in [16] we can hope to reach a complexity close to $K^{d/2}$. This is adapted to two player minimax games, with perfect information. It has been extended to other frameworks, for example with [18] to partially ordered values with a cache. We will propose here a new generalization, not very sophisticated, but flexible to different specifications.

5 Imperfect Information Tree Search

Algorithm 3 captures the structure of a MiniMax tree search algorithm with vector α and set of worlds W.

We have

- $<_i$ partial order on the pruning preferences of player i;
- $init_i(s, W)$, initialize the result for a player i, which has to be a lower bound on the actual value (depending on the worlds in W) of the node s for $<_i$

- max_i calculates the return value based on the values of the children. It is actually a fusion of the values of the children rather than a maximum over $<_i$.
- $genWorlds()$ generates the worlds over which to compute the values, uses the distribution D of the players.
- $criterion_i$ a total order representing the preferences of player i for the final choice of the card to play

The complexity of this algorithm is hard to study, the partial order causing pruning is difficult to approach and it makes the model deviating from the ones developed in [16]. In practice, however, we can see a significant improvement of time complexity with this pruning. Now that we have a generic algorithm for solving the problem, with an a priori reasonable time complexity, we will compare different specifications of this algorithm.

We will explore different specifications of the template we presented. Each one is specified for two teams. The score of a final state is the score made by the *Max* (declarer) team. We tried to adapt it to multiple teams following the work of [27], but it is a long time work. This could probably be done in a future work. The first one is a direct extension of the usual $\alpha\beta$ pruning algorithm for perfect information games. The second one is the one proposed in [7]. The last one is a novel contribution. In each case we will present the specifications of the functions we have previously invoked. We will not describe the function $init_i$ in detail, because we only found a trivial lower bound over the value of a node, based on the score already made by a team in our trick-taking card games.

Perfect-Information Monte Carlo. The idea of the Perfect-Information Monte Carlo (PIMC) algorithm is to generate random worlds according to D, and then to compute a score in each world *as if it were in perfect information* using the $\alpha\beta$ algorithm. The generation of the worlds follows directly from D. We generate a set of N worlds. Over the worlds in W, the *value* of a node is the score of each world in this setting of perfect information for everyone: $\mathbb{R}^N$. This hypothesis of perfect information in these worlds is not consistent with the settings. To illustrate it, this algorithm fits a game where everyone shows their hand, and plays openly after one card is played. Thus the generation of the worlds tries to capture the initial unknown, and then in each world, we play with perfect information. Let us talk about pruning. For the *Max* team a value a is greater than b if the score of each world $w \in W$ is greater in a than in b: $\forall k \in [N] a_k \geq b_k$ It is symmetric for *Min* nodes: $\forall k \in [N] a_k \leq b_k$ In this framework, we can derive the other function. The function $max_i(a, b)$ returns the maximum/minimum, according to the team of i, of each score of W. In fact with perfect information the players can choose the "objective" best card in each world. To make the final choice for the card to play, we set the criterion to prefer the best mean of the score, according to the team. The details are presented in Algorithm 4 (left).

$\alpha\mu$ Algorithm. The problem with PIMC is that it assumes perfect information for everyone. It leads to several imperfections and difficulties in decision-making as shown in [13]. An example of a situation where PIMC fails is shown in Fig. 2

Algorithm 4. PIMC (left) and $\alpha\mu$ (right)

function *genWorlds*
  $W \leftarrow \emptyset$
  for $k \in [N]$ **do**
    $w \hookrightarrow D$
    $W \leftarrow W \cup \{w\}$
  return W

function $max_i(a, b)$
  for $k \in [N]$ **do**
    if Max team **then**
      $r_k \leftarrow \max(a_i, b_i)$
    else
      $r_k \leftarrow \min(a_i, b_i)$
  return r

function $criterion_i(a, b)$
  $s_a \leftarrow \sum_{k=1}^{n} a_k$
  $s_b \leftarrow \sum_{k=1}^{n} b_k$
  if Max team **then**
    return $s_a > s_b$
  else
    return $s_a < s_b$

function $max_i(X, Y)$
  $R \leftarrow \emptyset$
  if root **then**
    $R \leftarrow X \cup Y$
  else
    for $(x, y) \in X \times Y$ **do**
      $r \in \mathbb{R}^N$
      for $k \in [N]$ **do**
        if Max team **then**
          $r_k \leftarrow \max(x_i, y_i)$
        else
          $r_k \leftarrow \min(x_i, y_i)$
      $R \leftarrow R \cup \{r\}$
  for $r_1 \neq r_2 \in R$ **do**
    if $\forall k \in [N] r_{1k} \leq r_{2k}$ **then**
      if root is Max **then**
        remove r_1 from R
      else
        remove r_2 from R
  return R

function $criterion_i(X, Y)$
  if Max team **then**
    $s_X \leftarrow \max_{x \in X} \sum_{k=1}^{n} x_k$
    $s_Y \leftarrow \max_{y \in Y} \sum_{k=1}^{n} y_k$
    return $s_X > s_Y$
  else
    $s_X \leftarrow \max_{x \in X} \sum_{k=1}^{n} x_k$
    $s_Y \leftarrow \max_{y \in Y} \sum_{k=1}^{n} y_k$
    return $s_X < s_Y$

(left). South's expected score over these two tricks is 0.5. In fact, South has a 50% chance of discarding the useless ace on the ♣A and keeping the useful ace. In this case South scores 1, otherwise 0. However PIMC would generate several worlds, some with North having the ♠2, some with North having the ♡2, and in each one South would score 1. In fact, if South knows the North's remaining card (this is the case in the PIMC algorithm, where perfect information is given), it is easy to keep the right ace, and to win the last trick. Thus PIMC would average the score over the world sample and predict score 1 for South, which is a mistake.

The idea with the $\alpha\mu$ algorithm presented in [7] is to include the imperfect information of the player who has to choose a card in the reasoning. The player at the root of the tree doesn't know which world is the good one, but the others do and play with perfect information. The algorithm fits a game where one player has to play a card, but everyone else sees his/her hand. Although it is not perfect in terms of the information given, it is an improvement over PIMC.

Fig. 2. In this game we follow the rules of Bridge except we only play with two enemy players. Trump is clubs. (left) North plays ♣A and South is to play. South does not know if the second card of North is ♠2 or ♡2, it is a 50% − 50% situation. (right) Trump is clubs. North does not know if the second card of South is ♠2 or ♡2. South has to play.

We study it using information theory. Suppose C cards are remaining and player i knows the location of $|C| - k$ of those C cards, k being unknown. We will model D as the uniform distribution on the unknown cards over all players. These unknown cards can be dealt p^k different times. Thus, for $w \in \mathcal{W}$ we have $D(w) = 1/p^k$ (with p the number of players) if w matches the known cards, and 0, otherwise.

Knowing that we are playing in a world w^* (which is coherent with the known cards of i) gives the information $-\log(D(w^*)) = k \log p$. Note that this is also the KL-divergence between the certain distribution where $w*$ has a probability of 1 and D. We next measure the total information $I(n)$ given to the players when n cards are to be played. Consider that n cards are unknown at this state, and all the other $C - n$ cards are known. For a solution given by PIMC we compute the information gain contained in this solution. In Formula (2) every call of σ adds some information to this gain: the perfect information given to the node of the state. We have

$$I_{pimc}(n) = n \log p + \sum_{k=0}^{n-1} I(k) = n \log p + \sum_{k=0}^{n-1} (n-k) k \log p$$
$$= \log p(n + n(n^2 - 1)/6) \geq \log p(n^3/6)$$

Now consider a solution of $\alpha\mu$. Perfect information is only given to the non-root player. So, assuming we start with the root, we have

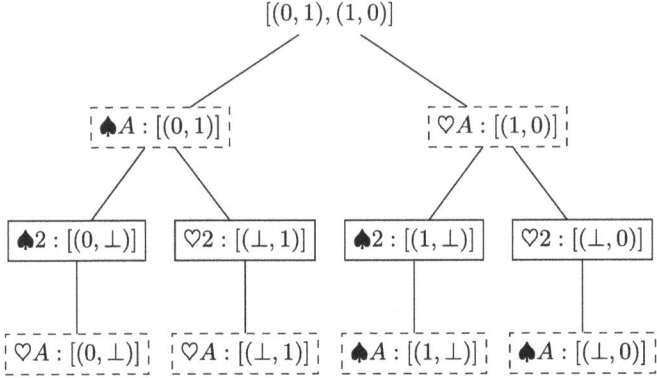

Fig. 3. Application of Fusion algorithm to the problem raised in Fig. 2 (left). The scores are the ones of South, viewed as a Max node. Max nodes are circled by continuous line, and Min nodes by dashed lines. Here world 1 is the world where North has ♠2 (the real world), and world 2 is the one where she has ♡2 (the fake world but South doesn't know it).

$$I_{\alpha\mu}(n) = 0 + \sum_{k=0}^{n-1}(n-k)\delta_{p_{depth=k},root}(k \log p)$$

$$= \log p \bigg(\sum_{k=0}^{n}(n-k)k - \sum_{i=0}^{n/p}(n-ip)ip \bigg)$$

$$= \log p \bigg(\frac{n(n^2-1)}{6} - p^2 \frac{\frac{n}{p}((\frac{n}{p})^2-1)}{6} \bigg)$$

$$= \log p \bigg(\frac{n^3}{6} \bigg)\bigg(1 - \frac{1}{p} + \frac{p-1}{n^2}\bigg).$$

Since the expression $\frac{p-1}{n^2}$ is small when we are not at the end of the game (in Bridge in middle game with 7 tricks left, we have $p = 4$, $n = 4 \cdot 7$), the information saved with $\alpha\mu$ is about a factor $(1 - 1/p)$. We noticed that with $\alpha\mu$ we achieve a reduction of the information gain by a factor $(1 - 1/p)$. This is not surprising, since we have removed the information gift of one player.

Let us introduce the functions used in $\alpha\mu$. This algorithm solves the problem of Fig. 2 (left). In Fig. 5 we see that the uncertainty about the outcome of the game is preserved. The value computed, $[(0, 1), (1, 0)]$ represents the fact that one strategy leads to a nonzero score in one world, and another strategy leads to a nonzero score in the other world.

In $\alpha\mu$ *generateWorlds* is the same as in PIMC; max_{root} computes the union of the values of the children, and removes the elements that are dominated (i.e., if the scores associated with one strategy are all better than another, it removes the latter); max_i is the union of the product of the strategies of the children. One element is computed by taking an element from all the children (we construct one possible strategy for root), and by taking in each world the best score for i, as in PIMC (if root follows the constructed strategy, then the score would be the one computed, with perfect information of i). Again, the set is simplified by removing the dominated elements for $root$; $criterion$ selects the node containing the strategy with the best expected score over the different worlds in W.

5.1 Fusion Algorithm

Even with $\alpha\mu$ we are giving away too much information to the players other than root. This can lead to bugs like the one shown in Fig. 2 (right). If South plays ♠2, s/he will score 1. If s/he plays ♣A s/he has 50% chance of scoring 2 and 50% chance of scoring 1. So ♣A is the better choice. $\alpha\mu$ supposes perfect information of North. Thus if South plays ♣A she will score 1, and if s/he plays ♠2 s/he will also score 1. Both cards are equal, and $\alpha\mu$ could make the mistake of choosing ♠2.

To solve this problem, we need to incorporate the imperfect information of all the players into our algorithm. This is the goal of the fusion algorithm. We choose to fix the value of a node to be the expected score of each world if everyone plays optimally according to their knowledge. At a node with two children, player i will choose the value of the child that has the best expectation, according to his/her team, over the worlds that are possible according to his/her knowledge. If the two children have the same expected value, player i has 50% chance of playing each one or them. So the value of the node should be a mixture of both (the middle of the score segment).

For example, if a Max node has two children $[2, 5, 3]$ and $[3, 7, 0]$, its value would be $[2.5, 6, 1.5]$ The problem with this simple idea is that different players have different knowledge, and different possible worlds. So each set of possible worlds P is different for each player. Moreover for a player, his/her set of possible worlds depends on its hand, that is unknown. We can't take one set W for the whole tree and assume it represents all the players, as we did before.

We compute the values over subsets of W that correspond to the different P possible for a player. For example, if both Max and Min can have two hands, there are four possible worlds $\{Max_1 Min_1, Max_1 Min_2, Max_2 Min_1, Max_2 Min_2\}$. Suppose Max is the root, and you want to merge two values at a Min node: $[0, 2, 3, 1]$ and $[2, 0, 2, 3]$. If Min has hand 1, the worlds she will consider are the first and third (you do not know the hand of Max. So the first value is better because it has an expected score of 1.5 instead of 2. But if Min has hand 2 the two values have the same expected score over worlds 2 and 4, which is 1.5. Thus Min has a 50% chance of playing each value and we mix the scores of these worlds. We come up with the value $[0, 1, 3, 2]$. A more detailed example is shown in Fig. 4. This algorithm solves the issue raised in Fig. 2 as shown in Fig. 5. At the root, South has the choice of two values, over the actual world 1.5 and 1. Because $1.5 > 1$ s/he will then play ♣A.

In practice, taking any set W won't work to have this system of subsets of possible worlds. To have a correct number of worlds for each hand of each

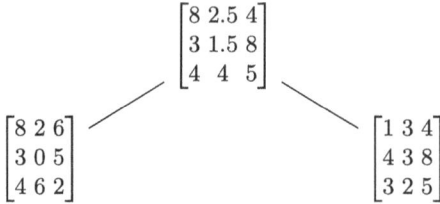

Fig. 4. Fusion of two values by a Max node. The row correspond to the different hands possible for Min, and the columns the one for Max (m_{ij} represents the score of the world where Min has hand i and Max hand j).

player, we need to have similarities between the worlds of W. To do this, we use a small set of initial worlds and we create the set W by taking all the possible permutations of the hands with this initial set. So in Algorithm 5 *generateWorlds* samples some worlds and then returns all the permutations of those worlds. max_i follows the scheme explained before. $criterion_i$ selects the value with the best average score over the possible worlds of i. We prune if all the scores of one value are worse than another one.

Algorithm 5. Fusion
---
   function $max_i(a, b)$
      for h possible hand of i **do**
         $score_a \leftarrow \sum_{w \in P_h} a_w$
         $score_b \leftarrow \sum_{w \in P_h} b_w$
         if $score_a < score_b$ **then**
            for $w \in P_h$ **do**
               if Max team **then** $r_w \leftarrow b_w$ else $r_w \leftarrow a_w$
         else
            for $w \in P_h$ **do**
               if Max team **then** $r_k \leftarrow a_w$ else $r_k \leftarrow b_w$
      return r

---

6 Implementation

For the implementation, we have adapted the above formalization to the code. This allows us to keep some flexibility in the rules used later. We put the organization of the card game into one file. This file is for the general game, and when rule-dependent functions are needed, we call them in the specific rule file associated with them.

A simple modification in the Makefile selects which game is played and which rule file is read when these rule-dependent functions are used. It uses a game class that stores the information about the game being played: the contract,

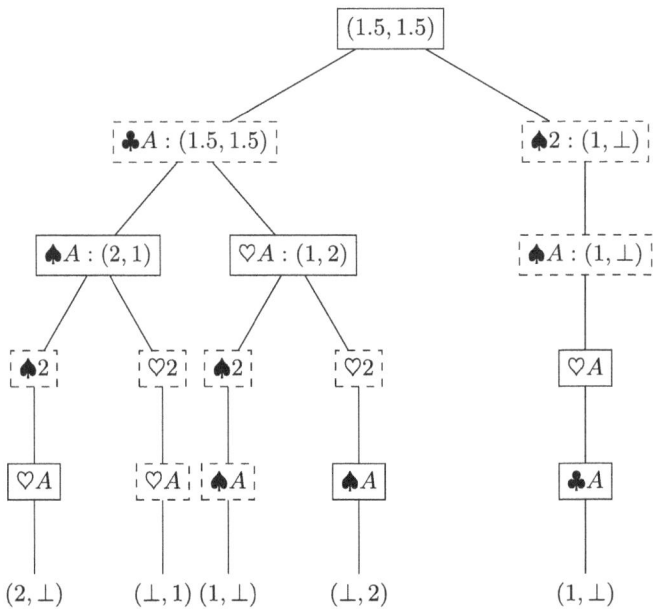

Fig. 5. Application of the Fusion algorithm to the problem raised in Fig. 2 (right). The scores are the ones of South, viewed as a Max node. Here world 1 is the world where South has ♠2 (the real world), and world 2 is the one where she has ♡2 (the fake world but North doesn't know it).

the cards already played, who has to play, the scores of the teams, what the current trick consists of, and who is the leader of that trick. We have a *player* class, which contains the information that a player has (his hand, what he knows about the other hands, etc.). The communication with the player is realized in the methods of this class. The selection of the card to play is implemented by a function pointer, which is an attribute of the player class. Several functions have been created to select a card. We define this attribute of the player class to define its strategy.

We have collected all these different strategies in a directory. These are mainly the strategies described above. Since they often follow the same pattern, as we have explained, we used the algorithm of Fig. 3. We then used an abstract class for each tree strategy. This abstract class is then specified into precise classes corresponding to each strategy that follows the template. In each class, the specification-dependent functions, such as $\max_i$, are implemented as attributes. To invoke the function that selects a card, we call the template with the class of the selected specification.

To encode the knowledge, the player class has an attribute *have_not*, which is an array containing the cards that this player knows that the other players do not have. This knowledge is used to generate the worlds used in the strategy functions, so that the generated worlds respect what the player knows about the others.

As for the basic types, we used bit vectors (mainly a set of unsigned integers) to represent the set of cards. For example, in Belote with its 32 cards, we used one unsigned integer. If a bit in a representing vector is set to true, then the corresponding card in the deck is contained in the set.

You can start the games by choosing which hands are played, which strategies are used by each player, and how many games are played. We ran several experiments with this structure.

7 Experiments

We tested the approach in two different games: Belote and Bridge. While Bridge is well-studied, our core interest is in (La) Belote, a French national card game for four players. For the player's strategy we implemented the three algorithms we discussed and a random card strategy, to have some kind of naive player. We then compared pairs of algorithms. In both cases there are two teams, so we assigned one strategy to one team and the other to the second team. We played matches of 100 different deal (same across the different experiments). With each deal, we play once as it is, and then swap the team's positions, so that it is fair. We measure the score made by the team of the first player to play. In each experiment we finally collect the 200 scores across these games. We add the 100 scores made by one strategy, and the 100 scores made by the other, and deduce the percentage of the total amount of points per game that on strategy wins over the other.

In each algorithm there are some hyperparameters to fix: n_{sample} is the number of worlds used in the algorithm. *depth-leaf* in $\alpha\mu$ and the fusion algorithm is the depth at which we stopped the specification to end the tree with PIMC (this is done for complexity issues, it is also done in [7]. *depth-rd* is the depth at which pimc is stopped to end the tree with the an average of the random playouts.

For Belote, we derived a 10% approximation of the score, that was valid with 95% probability with these random playouts. For Bridge, we imposed a low value of *depth-rd* for matters of complexity, and the bounds on the approximation made were not consistent. We tried to keep the same hyperparameters for all the algorithms, but Fusion needed more worlds than the others (Figs. 6 and 7).

All algorithms are better than random. We can also see that the $\alpha\mu$ algorithm is slightly better than the others. Let us note that all these algorithms are designed to play against smart players so the differences of the scores against random are not significant (PIMC being better than $\alpha\mu$ in beating random in Bridge is not relevant). Fusion algorithm is slightly worse than the others. However, pruning in the Fusion algorithm needs to be improved.

In Belote vs. another AI PIMC ($n_{sample} = 10$, *depth-rd* $= 25$) we found a 9.7% gain in the max. score on average.

	rd	pimc	$\alpha\mu$	fusion
rd	-1.8	-19.3	-22.1	-18.5
pimc	19.3	-3.3	-0.2	1.5
$\alpha\mu$	22.1	0.2	2.4	0.8
fusion	18.5	-1.5	-0.8	3.5

Fig. 6. Belote; team 0 in rows, team 1 in columns. Pimc: $n_{sample} = 10$. $\alpha\mu$: $n_{sample} = 10$, $depth\text{-}leaf = 5$, $depth\text{-}rd = 20$. Fusion: $n_{sample} = 24$, $depth\text{-}leaf = 5$, $depth\text{-}rd = 20$.

	rd	pimc	$\alpha\mu$	fusion
rd	1.5	-25.2	-21.8	-21.6
pimc	25.2	-0.7	-1.8	5.0
$\alpha\mu$	21.8	1.8	-2.1	2.7
fusion	21.6	-5.0	-2.7	0.5

Fig. 7. Bridge; team 0 in rows, team 1 in columns. Pimc: $n_{sample} = 10$. $\alpha\mu$: n_{sample}, $depth\text{-}leaf = 5$, $depth\text{-}rd = 9$, Fusion: $n_{sample} = 24$, $depth\text{-}leaf = 5$, $depth\text{-}rd = 20$.

8 Conclusion

We presented an efficient framework for imperfect information card games and found a factorization of the search in a template that captures the main ingredients of the search. We then analyzed three different algorithms and derived the impact of the information given during the algorithm on the decisions it makes. Finally, we designed a new algorithm that tries to respect known information more. We also estimated the effects of truncating the search tree with random rollouts, and the information gain obtained by $\alpha\mu$. To continue this work, we need to explore more deeply the possibilities for a better pruning of the tree. Last, but not least we will aim at card games with *talon* that do not yet fit the setting.

Acknowledgments. This research was partly funded by AFOSR project Flexible and Resilient Auton. Systems (FRAS) and by the Czech Science Foundation, grant number 22-30043S.

References

1. Allis, L.V.: A knowledge-based approach to connect-four. The game is solved: White wins. Master's thesis, Vrije Univeriteit, The Netherlands (1998)
2. Blanvillain, X.: Oware is strongly solved. In: Browne, C., Kishimoto, A., Schaeffer, J. (eds.) Computers and Games. LNCS, vol. 13865, pp. 87–96. Springer, Cham (2022). https://doi.org/10.1007/978-3-031-34017-8_8
3. Bowling, M., Burch, N., Johanson, M., Tammelin, O.: Heads-up limit hold'em poker is solved. Commun. ACM **60**(11), 81–88 (2017)
4. Brennan, A., Kharroubi, S., O'Hagan, A., Chilcott, J.: Calculating partial expected value of perfect information via Monte Carlo sampling algorithms. Med. Decis. Making **27**(4), 448–470 (2007)
5. Buro, M., Long, J.R., Furtak, T., Sturtevant, N.R.: Improving state evaluation, inference, and search in trick-based card games. In: IJCAI, pp. 1407–1413 (2009)
6. Cazenave, T., Ventos, V.: The $\alpha\mu$ Search Algorithm for the Game of Bridge (2019)
7. Cazenave, T., Ventos, V.: The $\alpha\mu$ search algorithm for the game of bridge. In: Cazenave, T., Teytaud, O., Winands, M.H.M. (eds.) MCS 2020. CCIS, vol. 1379, pp. 1–16. Springer, Cham (2021). https://doi.org/10.1007/978-3-030-89453-5_1
8. Cohensius, G., Meir, R., Oved, N., Stern, R.: Bidding in spades. In: ECAI, pp. 387–394 (2020)
9. Edelkamp, S.: Challenging human supremacy in skat. In: Proceedings of the Twelfth International Symposium on Combinatorial Search (SOCS), pp. 52–60 (2019)
10. Edelkamp, S.: Knowledge-Based Paranoia Search in Trick-Taking. CoRR, abs/2104.05423 (2021)
11. Edelkamp, S.: Improving computer play in Skat with hope cards (2023)
12. Finnsson, H.: Cadia-player: a general game playing agent. Ph.D. thesis (2007)
13. Frank, I., Basin, D.: Search in games with incomplete information: a case study using bridge card play. Artif. Intell. **100**(1–2), 87–123 (1998)
14. Gasser, R.: Harnessing computational resources for efficient exhaustive search. Ph.D. thesis, ETH Zürich (1995)
15. Ginsberg, M.: Step toward an expert-level bridge-playing program. In: IJCAI, pp. 584–589 (1999)
16. Knuth, D.E., Moore, R.W.: An analysis of alpha-beta pruning. Artif. Intell. **6**(4), 293–326 (1975)
17. Kupferschmid, S.: Entwicklung eines Double-Dummy Skat Solvers mit einer Anwendung für verdeckte Skatspiele. Master's thesis, University of Freiburg (2006)
18. Li, J., Zanuttini, B., Cazenave, T., Ventos, V.: Generalisation of alpha-beta search for AND-OR graphs with partially ordered values. Research Report, GREYC CNRS UMR 6072, Universite de Caen (2022)
19. Long, J.R., Sturtevant, N.R., Buro, M., Furtak, T.: Understanding the success of perfect information Monte Carlo sampling in game tree search. In: Twenty-Fourth AAAI Conference on Artificial Intelligence (2010)
20. Moravčík, M., et al.: DeepStack: Expert-Level Artificial Intelligence in No-Limit Poker. CoRR, abs/1701.01724 (2017)
21. Perolat, J., et al.: Mastering the game of stratego with model free multiagent reinforcement learning. Science **378**(6623), 990–996 (2022)
22. Romein, J.W., Bal, H.E.: Solving awari with parallel retrograde analysis. Computer **36**(10), 26–33 (2003)
23. Schaeffer, J., et al.: Solving checkers. In: IJCAI, pp. 292–297 (2005)

24. Schiffel S., Thielscher, M.: Reasoning about general games described in GDL-II. In: Proceedings of the AAAI Conference on Artificial Intelligence, pp. 846–851. AAAI Press (2011)
25. Silver, D., et al.: Mastering the game of go with deep neural networks and tree search. Nature **529**, 484–489 (2016)
26. Silver, D. et al.: Mastering Chess and Shogi by Self-Play with a General Reinforcement Learning Algorithm. Tech Report, arxiv, 1712018 (2017)
27. Sturtevant, N.R., Korf, R.E.: On pruning techniques for multi-player games. AAAI/IAAI **49**, 201–207 (2000)
28. Sturtevant, N.R., White, A.M.: Feature construction for reinforcement learning in hearts. In: van den Herik, H.J., Ciancarini, P., Donkers, H.H.L.M.J. (eds.) CG 2006. LNCS, vol. 4630, pp. 122–134. Springer, Heidelberg (2007). https://doi.org/10.1007/978-3-540-75538-8_11

A Differential Approach for Several NP-hard Optimization Problems

Sangram K. Jena[1], K. Subramani[1]([✉]), and Alvaro Velasquez[2]

[1] LDCSEE, West Virginia University, Morgantown, WV, USA
{sangramkishor.jena,k.subramani}@mail.wvu.edu
[2] Department of Computer Science, University of Colorado Boulder, Boulder, CO, USA
alvaro.velasquez@colorado.edu

Abstract. In this paper, we study a variety of **NP-hard** optimization problems, such as MAXCUT, MAXkSAT, and MAXNAE2SAT from the perspective of obtaining exact solutions. We derive differentiable functions for each of these problems using the dataless neural networks framework. Recently, it was shown that a single differentiable function on a dataless neural network can capture the Maximum Independent Set problem. Inspired by this design, we design dataless neural networks for a host of combinatorial optimization problems. We also establish the correctness of our derivations in a rigorous fashion.

Keywords: MAXCUT · MAXkSAT · MAXNAE2SAT · dNN

1 Introduction

NP-hard optimization problems have applications in almost every domain, such as scheduling, routing, telecommunications, planning, transportation, and decision-making processes [3,7]. These problems do not admit a polynomial-time efficient solution unless some well-established complexity-theoretic conjectures fail. While it is challenging to solve such problems optimally, several techniques can approximate [18] optimal solutions or even solve the problems efficiently in non-trivial exponential time [9]. One method to handle these problems is by using neural network (NN) frameworks. However, this method requires extensive training of neural networks to generate the required solution. To overcome these problems, recently, an efficient method was developed using dataless neural network (dNN) frameworks [2]. Unlike NN frameworks, dNN frameworks do not require any data other than input instances.

Let f be a conventional neural network parameterized by trainable parameters θ. Furthermore, assume that θ can be trained on some dataset $\{(x_i, y_i)\}$. Let x_i be an instance of a differentiable **NP-hard** optimization problem χ, and y_i be the values of the optimal solution of χ. To make the output $f(x_i; \theta)$ of f as close to y_i as possible, the parameters θ are typically updated using backpropagation by minimizing a differentiable loss function $L(x_i, f(x_i; \theta))$. The parameters are updated by the backpropagation in the direction of $\theta := \theta - \alpha \cdot \partial L(x_i, f(x_i; \theta))/\partial \theta$. Here, α controls the learning rate. The dataless neural networks are based on the concept that there is no data for training the neural network. So, what we have as an output of the neural network is simply $f(e_n; \theta) = f(\theta)$, where e_n is the all-ones vector representing a trivial input to the neural network. Thus, instead of finding patterns in some data set, dataless neural networks attempt to find the optimal solution to a given discrete optimization problem by enforcing a certain structure on f and θ.

The rest of this paper is organized as follows: In Sect. 2, we formally define the problems and the notations used in the paper. Section 3 describes related work in the literature. In Sect. 4, we design a dNN for the MAXCUT problem. Section 5 discusses a dNN for the MAXkSAT problem. We design a dNN in Sect. 6 for the MAXNAE2SAT problem. Finally, we conclude in Sect. 7 by summarizing our results and discussing avenues for future work.

2 Statement of Problems

In this section, we define the problems and some of the notations used in this paper.

Definition 1. *MAXCUT: Given a graph $G = (V, E)$, find a set $C \subseteq E$ that partitions the set V into two subsets S and T such that the number of edges between S and T is maximized.*

Let $X = \{x_1, x_2, \ldots, x_n\}$ denote a collection of n Boolean variables. A literal is either a variable or the negation of a variable. A clause C is a disjunction of literals over the variable set X, for example, $C = (x_1 \lor x_2 \lor x_3 \lor x_4)$. A formula is said to be in Conjunctive Normal Form (CNF) if it is the conjunction of clauses. For instance, $(x_1 \lor x_2) \land (\bar{x}_1 \lor x_2 \lor \bar{x}_3) \land (x_2 \lor \bar{x}_3 \lor \bar{x}_4)$ is a CNF formula with three clauses over the variable set $X = \{x_1, x_2, x_3, x_4\}$. A CNF formula is said to be in kCNF form if each clause consists of exactly k literals.

Definition 2. *SAT: Given a CNF formula Φ, does there exist an assignment that satisfies each clause, i.e., sets at least one literal to* **true** *in each clause?*

Definition 3. *MAXSAT: Let Φ denote a formula in CNF with m clauses over n variables, find an assignment for the variables, which maximizes the number of satisfied clauses in Φ.*

Definition 4. *MAXkSAT: Let Φ denote a formula in kCNF with m clauses over n variables, find an assignment for the variables, which maximizes the number of satisfied clauses in Φ.*

Definition 5. *NAESAT: Given a CNF formula Φ, does there exist an assignment that nae-satisfies each clause, i.e., each clause is satisfied and at least one literal from each clause is set to* **false**?

Note that unit clauses cannot be nae-satisfied. Only clauses of size at least two can be nae-satisfied.

Definition 6. *MAXNAEkSAT: Given a kCNF formula Φ with m clauses over n variables, find an assignment for the variables of Φ that maximizes the number of nae-satisfied clauses in Φ.*

In our design of dataless neural networks, we use a rectified linear (ReLU) activation function. It is a piecewise linear function that outputs the input directly if it is positive; otherwise, it outputs zero, i.e., $\sigma(x) = max(0, x)$. For any positive integer n, $[n] := \{1, 2, \ldots, n\}$. Unless mentioned otherwise, $|\cdot|$ represents the absolute value or modulus.

The principal contributions of this paper are as follows:

1. A differential approach for the MAXCUT problem (see Sect. 4).
2. A differential approach for the MAXkSAT problem (see Sect. 5).
3. A differential approach for the MAXNAE2SAT problem (see Sect. 6).

3 Related Work

This section discusses some of the related work available in the literature for the neural network (NN) and dataless neural network (dNN). We discuss the available results for NNs and dNNs with respect to many combinatorial optimization problems (COPs). The most interesting COPs are **NP-hard**. It is well-known that such problems do not have polynomial time-efficient algorithms unless some established complexity-theoretic conjectures fail. Despite being inefficient, these problems have real-time applications in almost every domain, such as routing, scheduling, planning, telecommunications, transportation, and decision-making processes [3,7]. Due to the importance of these problems, a lot of research is going on to address them with different efficient, approximate solvers [12]. Broadly, these solvers are categorized into heuristic algorithms [1], approximation algorithms [4], and conventional branch-and-bound methods [16]. Such approaches may produce suboptimal solutions. Some of the other well-studied approaches to dealing with **NP-hard** problems use parameterized [5,8,15] and exact exponential algorithmic techniques [9,10].

However, another approach to address the COPs is to use the concept of machine learning [3]. The use of reinforcement learning to automate the search of the heuristics for COPs is discussed in [6,14]. These models require training based on the problems. More specifically, they rely on supervised learning using datasets of the combinatorial structures of interest drawn from some distribution of problem instances. In [2], the authors introduced dNNs for which no data is required for training. By designing a single differentiable function, they captured the well-known combinatorial optimization problem, the maximum independent set (MIS) problem. They also designed a similar dNN structure for the maximum clique (MC) and minimum vertex cover (MVC) problems related to the MIS problem. In [11], we developed dNNs tailored for solving the maximum dissociation set, k-coloring, and maximum cardinality distance matching problems.

The literature discusses several powerful heuristic solvers for the MIS problem. One of the heuristic solvers is ReduMIS [12]. It consists of two components. The first component is an iterative implementation of a series of graph reduction techniques. The second component is the use of an evolutionary algorithm. These methods usually involve extensive training of neural networks (NNs) using large graph datasets for which solutions are known. Another method for the MIS problem, similar to the method of dNN for the MIS problem discussed in [2], was developed in [17]. The method discussed in [17] uses a graph neural network and does not require training data. More specifically, its output is represented by the probability of each node being in the solution. In contrast to the method discussed in [2], it uses a loss function to adjust its parameter that encodes the graph of interest. Furthermore, the approach discussed by Alkhouri et al. [2] uses n trainable parameters where n is the number of vertices in the input graph. However, the number of tunable parameters used by the approach discussed in [17] is large in size. It uses n parameters in its last layer only. In [2], the authors also showed some experimental results by comparing them with the best heuristics available in the literature. They evaluated success by taking the solution size obtained by ReduMIS as a benchmark. They also showed that their experimental results perform as well or outperform the state-of-the-art learning-based methods discussed in [13].

4 MAXCUT

In this section, we design a dataless neural network (dNN) for the MAXCUT problem. Note that the MAXCUT problem is a special case of the MAXNAE2SAT problem, where all the literals in the CNF formula are positive. Thus, the following dNN for the MAXCUT problem works for the MAXNAE2SAT problem with positive literals.

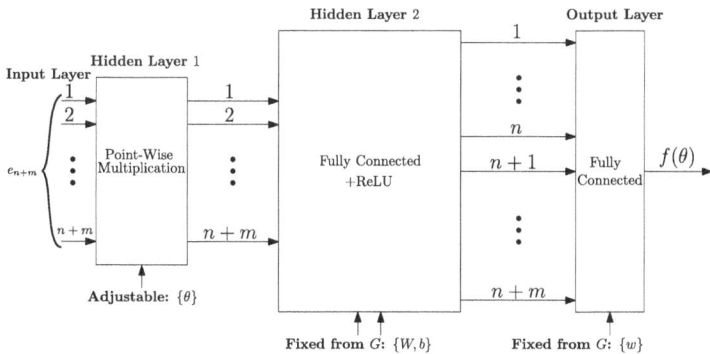

Fig. 1. Block diagram of dNN.

Let $G = (V, E)$ be a graph with n vertices over m edges. We construct a dNN f with trainable parameters $\theta \in [0, 1]^{n+m}$ with respect to G. That means for each vertex $v \in V$, there is a corresponding trainable parameter θ_v in f. For each edge $e_i \in E$, there is a corresponding trainable parameter θ_{e_i} in f. The input to the dNN is an all-ones vector e_{n+m} which does not depend upon any data. The output of the dNN is $f(e_{n+m}; \theta) = f(\theta) \in \mathbb{R}$. There are four layers in the dNN for the MAXCUT problem. The four layers are categorized as one input layer, two hidden layers, and one output layer (see the block diagram in Fig. 1 for the proposed network). The input layer e_{n+m} is connected with the first hidden layer through an elementwise product of the trainable parameters θ. The first hidden layer is connected to the second hidden layer by the binary matrix $W \in \{0,1\}^{n \times (n+m)}$. The binary matrix is only dependent on G. At the second hidden layer, there exists a bias vector $b \in \{0, -\frac{1}{2}\}^{2 \cdot m}$. There is a fully connected weight matrix $w \in \{-1, -n\}^{2 \cdot m}$ in the second hidden layer to the output layer. Note that all the parameters are defined as a function of G. The output of f is given as follows:

$$f(e_{n+m}; \theta) = f(\theta) = w^T \cdot \sigma((W^T \cdot (e_{n+m} \odot \theta)) + b). \tag{1}$$

Here $\odot$ is the element-wise Hadamard product that represents the operation of the first hidden layer of the constructed network. The fully-connected second hidden layer consists of the fixed matrix W and a bias vector b with a ReLU activation function $\sigma(x) = max(0, x)$. The last layer is another fully-connected layer and is expressed in vector w.

On the other hand, we prove that when a max-cut $C \subseteq E$ in G is found, $f(\theta)$ attains its minimum value. Therefore, $f(\theta)$ is an equivalent differentiable function of the max-cut generated in G. Moreover, C can be constructed from θ as follows. Let $\theta^* = argmin_{\theta \in [0,1]^{n+m}} f(\theta)$ be an optimal solution to f. Let $I : [0,1]^m \to 2^E$ be a max-cut corresponding to θ such that $I(\theta) = \{e \in E \mid \theta_e^* \geq \alpha\}$, for $\alpha > 0$. We show that $|I(\theta^*)| = |C|$. We choose the edges in the max-cut C in G corresponding to the indices of θ whose value exceeds a threshold (say α). From an input graph $G = (V, E)$, the fixed parameters of f can be constructed as follows: In the binary matrix W, the first $n \times n$ submatrix represents the vertices V of G. Its weights are set equal to the identity matrix I_n (see the 5×5 submatrix in Fig. 2 (b) corresponding to the 5 vertices of G in Fig. 2 (a)). Furthermore, the remaining m columns of W represent the edges of G and for each edge $e_i = uv \in E$, the value of $u = v = 1$ in the column (see the 6 columns e_1 to e_6 in Fig. 2 (b) corresponding to the 6 edges of G in Fig. 2 (a)).

The bias vector consists of two parts, each with m entries resulting in $2 \cdot m$ entries. For each edge of G, the corresponding value of the first m entries is $-\frac{1}{2}$ in the biased vector b. For each edge, the corresponding value of another m entries in the bias vector is set to 0. Finally, the value of -1 is assigned in the entries corresponding to the edges of G in vector w. For another m entries, the value is set to $-n$ in w. Hence, the parameters W, b, and w are defined as follows:

$$\begin{aligned} W(i,i) &= 1, v_i \in V, i \in [n], \\ W(i, n+k) &= W(j, n+k) = 1, \forall e_k = v_i v_j \in E, k \in [m], \end{aligned} \tag{2}$$

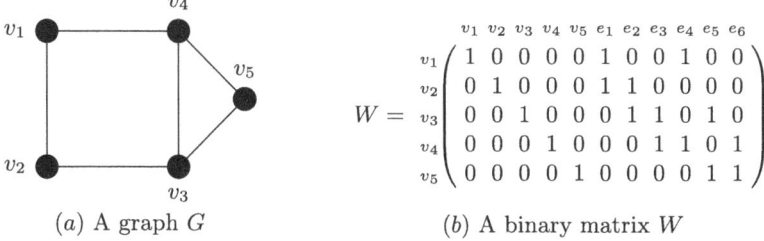

(a) A graph G (b) A binary matrix W

Fig. 2. Representation of a binary matrix W corresponding to G.

$$b(i) = -\frac{1}{2}, w(i) = -1, e_i \in E, i \in [m],$$
$$b(m+k) = 0, w(m+k) = -n, k \in [m]. \qquad (3)$$

The function in (1) can be rewritten as follows:

$$f(\theta) = -\sum_{e_i=uv \in E} (\sigma(\theta_{e_i} - \frac{1}{2}) - n \cdot \sigma(\theta_{e_i} - |\theta_u - \theta_v|)). \qquad (4)$$

An example of the above discussed dNN construction is presented in Fig. 3.

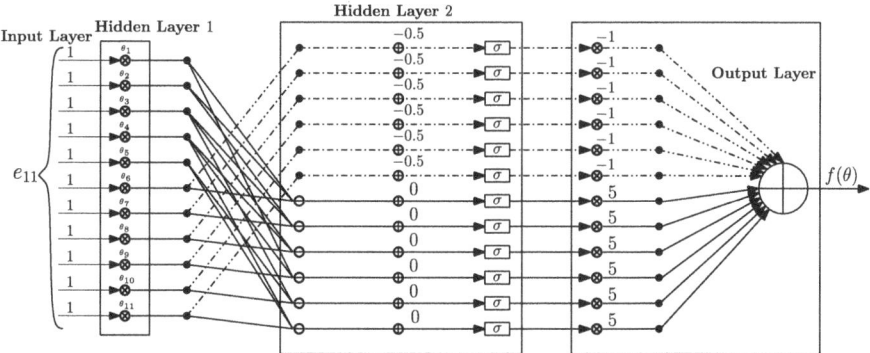

Fig. 3. Construction of dNN f corresponding to the graph in Fig. 2 (a) for the MAX-CUT problem.

With the above dNN construction, we prove the following theorem to establish a relation between the solution of the MAXCUT problem and the minimum value of f.

Theorem 1. *Let $G = (V, E)$ be a graph with m edges over n vertices. Let f be the corresponding dNN of G. G has a max-cut of size k, if and only if the minimum value of f is $-\frac{k}{2}$.*

Proof. Let $C \subseteq E$ be a max-cut of size k in G that partitions V into two sets S and T. For each $v \in V$, if $v \in S$, then set $\theta_v = 1$. Otherwise, set $\theta_v = 0$. For each edge $e_i \in E$, set $\theta_{e_i} = 1$, if $e_i \in C$; otherwise set $\theta_{e_i} = 0$. Consider the output f represented in Fig. 4 for an arbitrary edge $e_i = uv \in E$.

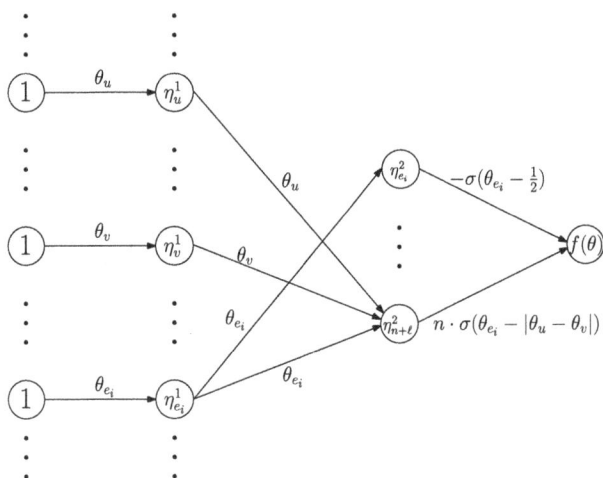

Fig. 4. Output with respect to an arbitrary edge.

As per the construction of the dNN, the edge values denote the outputs of the preceding nodes in the network. Furthermore, the ith neurons in the first hidden layer are denoted by η_i^1 and the second hidden layer is denoted by η_i^2. If $e_i \in C$, then the values of both θ_u and θ_v are not equal. Moreover, $\theta_{e_i} = 1$. The above θ values of the corresponding edge e_i will contribute $-\frac{1}{2}$ to f. Furthermore, if $e_i = uv \notin C$, then the values of both θ_u and θ_v are equal and $\theta_{e_i} = 0$. The above θ values of the corresponding edge e_i will contribute 0 to f. There are k edges present in C. Therefore, it will contribute $-\frac{k}{2}$ to f.

Conversely, assume that the minimum value of the output function f is $f(\theta) = -\frac{k}{2}$. We construct a max-cut C for G of size k from f as follows: From the construction of the dNN, it is clear that, for each edge $e_i = uv \in E$, if $\theta_{e_i} = 1$, then the values of θ_u and θ_v are not equal. Otherwise, f does not achieve its minimum value. To prove this, assume that the values of θ_u and θ_v are equal and $\theta_{e_i} = 1$. It follows that the neuron $\eta_{n+\ell}^2$ contributes $n \cdot (\theta_{e_i} - |\theta_u - \theta_v|) > 0$ to the output $f(\theta)$. This is a contradiction to the fact that f achieves its minimum value. We can simply assign the value of $\theta_{e_i} = 0$ and reduce the value of f further. So, it is clear that for any edge $e_i = uv \in E$, if $\theta_{e_i} = 1$, then the values of θ_u and θ_v are not equal. Each such edge $e_i \in E$, which θ values of both the end vertices are not equal and $\theta_{e_i} = 1$ contributes $-\frac{1}{2}$ to $f(\theta)$ through η_i^2. Furthermore, it contributes a value of 0 to $f(\theta)$ through $\eta_{n+\ell}^2$. That means there are k entries of value 1 in θ corresponding to edges. For each $e_i = uv \in E$, consider the θ_u and θ_v values. If $\theta_u = 0$ ($\theta_v = 0$), then assign the vertex u (v)

in set S. Otherwise, assign the vertex u (v) in set T. For each edge $e_i \in E$, if $\theta_{e_i} = 1$, then assign e_i in the set C. Observe that C is a max-cut in G of size k that divides V into two sets S and T.

5 MAXkSAT

In this section, we design a dNN for the MAXkSAT problem. First, we discuss a dNN for the MAX2SAT problem and then we generalize it for the MAXkSAT problem. Let Φ be an instance of the MAX2SAT problem. Furthermore, assume that Φ has m clauses over n variables. We construct a dNN f with trainable parameters $\theta \in [0,1]^{n+m}$ with respect to Φ as follows: For each variable x_i, create a corresponding trainable parameters θ_{x_i}. For each clause c_i, there is a corresponding trainable parameter θ_{c_i}. The input to the dNN is an all-ones vector e_{n+m} which does not depend upon any data. The output of the dNN is $f(e_{n+m}; \theta) = f(\theta) \in \mathbb{R}$. Similar to the dNN structure of MAXCUT, the dNN for the MAX2SAT consists of four layers (see Fig. 1 for the block diagram of the dNN structure). The dNN consists of one input layer, two hidden layers, and one output layer. The input layer e_{n+m} is connected with the first hidden layer through an elementwise product of the trainable parameters θ. The first hidden layer is connected to the second hidden layer by the binary matrix $W \in \{0,1\}^{2 \cdot n \times (2 \cdot n + m)}$. Here, $2 \cdot n$ is the total number of literals size for n variables. That means for a variable x_i, the binary matrix has two entries: one for literal x_i and another for literal $\bar{x}_i$. Note that the binary matrix is only dependent on Φ. At the second hidden layer, there exists a bias vector $b \in \{0, -\frac{1}{2}\}^{2 \cdot m}$. There is a fully connected weight matrix $w \in \{-1, n\}^{2 \cdot m}$ in the second hidden layer to the output layer. Note that all the parameters are defined as a function of Φ as follows:

$$f(e_{n+m}; \theta) = f(\theta) = w^T \sigma(W^T(e_{n+m} \odot \theta) + b). \quad (5)$$

Note that we represent the value of the variable $\bar{x}_i$ as the complement of the neuron θ_{x_i}. That means, if θ_{x_i} is assigned as one then $\bar{\theta}_{x_i}$ represents zero. So, the function in (5) can be rewritten as follows:

$$f(\theta) = -\sum_{c_i \in \Phi}(\sigma(\theta_{c_i} - \frac{1}{2})) + n \cdot \sum_{c_i = (\bar{x}_i \vee x_j) \in \Phi}(\sigma(\theta_{c_i} - (\bar{\theta}_{x_i} + \theta_{x_j}))). \quad (6)$$

Here the clause $c_i = (\bar{x}_i \vee x_j)$ has taken arbitrarily to highlight the utilization of the negative literals. We prove the following theorem using the above dNN:

Theorem 2. *Let Φ be an instance of MAX2SAT with m clauses over n variables. Let f be the corresponding dNN of Φ. There exists an assignment for Φ that satisfies k clauses, if and only if the minimum value of f is $-\frac{k}{2}$.*

Proof. Let $\mathcal{A}$ be an assignment for Φ that satisfies k clauses. For each variable x_i that is assigned to **true**, set the corresponding $\theta_{x_i} = 1$. If it is assigned to **false**, then set $\theta_{x_i} = 0$. For each clause c_i that is satisfied, set the corresponding

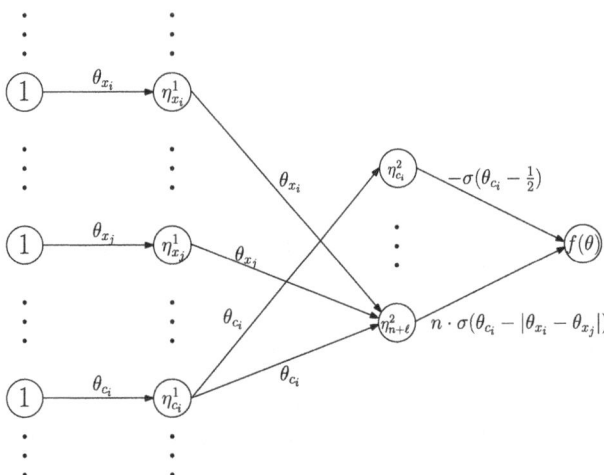

Fig. 5. Output with respect to an arbitrary clause.

$\theta_{c_i} = 1$. If it is not satisfied, then set $\theta_{c_i} = 0$. Consider the output f represented in Fig. 5 for an arbitrary clause $c_i = (x_i \vee x_j) \in \Phi$.

As per the construction of the dNN, the clause values denote the outputs of the preceding nodes in the network. Furthermore, the ith neurons in the first hidden layer are denoted by η_i^1 and the second hidden layer is denoted by η_i^2. If clause c_i is satisfied, then the value of both θ_{x_i} and θ_{x_j} cannot be 0. Moreover, $\theta_{c_i} = 1$. The above θ values of the corresponding clause c_i will contribute $-\frac{1}{2}$ to f. Furthermore, if c_i is not satisfied, then the values of θ_{x_i}, θ_{x_j}, and θ_{c_i} are 0. The above θ values of the corresponding clause c_i will contribute 0 to f. There are k clauses that are satisfied by assignment $\mathcal{A}$. Therefore, it will contribute $-\frac{k}{2}$ to f.

Conversely, assume that the minimum value of the output function f is $f(\theta) = -\frac{k}{2}$. We construct an assignment $\mathcal{A}$ for Φ that satisfies k clauses from f as follows: From the construction of the dNN, it is clear that, for each clause $c_i = (x_i \vee x_j)$, if $\theta_{c_i} = 1$, then $\theta_{c_i} - (\theta_{x_i} + \theta_{x_j}) \leq 0$. Otherwise, f does not achieve its minimum value. To prove this, assume that $\theta_{c_i} - (\theta_{x_i} + \theta_{x_j}) > 0$. It follows that the neuron $\eta_{n+\ell}^2$ contributes $n \cdot \sigma(\theta_{c_i} - (\theta_{x_i} + \theta_{x_j})) > 0$ to the output $f(\theta)$. This is a contradiction to the fact that f achieves its minimum value. We can simply assign the value of $\theta_{c_i} = 0$ and reduce the value of f further. So, it is clear that for any clause $c_i = (x_i \vee x_j)$, if $\theta_{c_i} = 1$, then $\theta_{c_i} - (\theta_{x_i} + \theta_{x_j}) \leq 0$. Each such clause, which θ value is one, contributes $-\frac{1}{2}$ to $f(\theta)$ through η_i^2. Furthermore, it contributes a value of 0 to $f(\theta)$ through $\eta_{n+\ell}^2$. That means there are k entries of value 1 in θ corresponding to clauses. For each $\theta_{c_i} = 1$, consider the θ_{x_i} and θ_{x_j} values. If $\theta_{x_i} = 0$ ($\theta_{x_j} = 0$), then assign the truth value of variable x_i (x_j) as **false**. Otherwise, assign the truth value of variable x_i (x_j) as **true**. Similarly, assign the truth values of the variables when $\theta_{c_i} = 0$. Observe that it will lead to an assignment $\mathcal{A}$ for Φ that satisfies k clauses.

Corollary 1. *The above-discussed dNN for the MAX2SAT problem can be generalized for the MAXkSAT problem.*

Proof. Observe that the above-discussed dNN for MAX2SAT can capture MAXkSAT by changing the entries of the binary matrix W. For each clause in Φ, the corresponding column in W matrix has k entries as one instead of two in case of MAX2SAT. The values of the parameters θ, b, and w are the same in the dNN. In this case, $f(\theta)$ can be represented as follows:

$$f(\theta) = -\sum_{c_i \in \Phi} (\sigma(\theta_{c_i} - \frac{1}{2})) + n \cdot \sum_{c_i = (x_1 \vee \cdots \vee x_k) \in \Phi} (\sigma(\theta_{c_i} - (\theta_{x_1} + \cdots + \theta_{x_k}))). \quad (7)$$

The construction and the proof for the MAX2SAT problem will follow for the MAXkSAT problem.

6 MAXNAE2SAT

In this section, we design a dNN for the MAXNAE2SAT problem. Observe that the dNN for the MAXCUT problem captures the MAXNAE2SAT problem when all the literals of the formula Φ are positive. We design a dNN for the MAXNAE2SAT problem, where the problem instance Φ consists of both positive and negative liters.

Let Φ be an instance of the MAXNAE2SAT problem with m clauses over n variables. We construct a dNN f with trainable parameters $\theta \in [0,1]^{n+m}$ with respect to Φ as follows: For each variable x_i, there is a corresponding trainable parameter θ_{x_i}. For each clause c_i, there is a corresponding trainable parameter θ_{c_i}. The input to the dNN is an all-ones vector e_{n+m} which does not depend upon any data. The output of the dNN is $f(e_{n+m}; \theta) = f(\theta) \in \mathbb{R}$. Similar to the above-discussed dNN structures, the dNN for the MAXNAE2SAT consists of one input layer, two hidden layers, and one output layer. The input layer e_{n+m} is connected with the first hidden layer through an elementwise product of the trainable parameters θ. The first hidden layer is connected to the second hidden layer by the binary matrix $W \in \{0,1\}^{2 \cdot n \times (2 \cdot n + m)}$. At the second hidden layer, there exists a bias vector $b \in \{0, -\frac{1}{2}\}^{2 \cdot m}$. There is a fully connected weight matrix $w \in \{-1, n\}^{2 \cdot m}$ in the second hidden layer to the output layer. Observe that all the parameters are defined as a function of Φ. The output of the dNN f corresponding to the MAXNAE2SAT problem is given by (5). The second hidden layer consists of the fixed matrix W and a bias vector b with a ReLU activation function $\sigma(x) = max(0, x)$. The last layer consists of the vector w.

Therefore, we can rewrite the function in (5) as follows:

$$f(\theta) = -\sum_{c_i \in \Phi} (\sigma(\theta_{c_i} - \frac{1}{2})) + n \cdot \sum_{c_i = (\bar{x}_i \vee x_j) \in \Phi} (\sigma(\theta_{c_i} - |\bar{\theta}_{x_i} - \theta_{x_j}|)). \quad (8)$$

We have used the clause $c_i = (\bar{x}_i \vee x_j)$ for showing the utilization of negative literals by the function and it is taken arbitrarily. With the above dNN construction, we prove the following theorem.

Theorem 3. *Let Φ be an instance of MAXNAE2SAT with m clauses over n variables. Let f be the corresponding dNN of Φ. There exists an assignment for Φ that nae-satisfies k clauses, if and only if the minimum value of f is $-\frac{k}{2}$.*

Proof. Let $\mathcal{A}$ be an assignment for Φ that nae-satisfies k clauses. For each variable x_i that is assigned to **true**, set the corresponding $\theta_{x_i} = 1$. If it is assigned to **false**, then set $\theta_{x_i} = 0$. For each clause c_i that is nae-satisfied, set the corresponding $\theta_{c_i} = 1$. If it is not nae-satisfied, then set $\theta_{c_i} = 0$. Consider the output f represented in Fig. 6 for an arbitrary clause $c_i = (x_i \vee x_j) \in \Phi$.

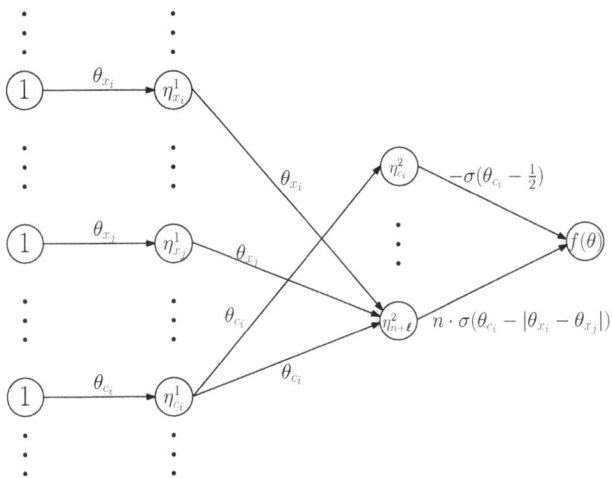

Fig. 6. Output with respect to an arbitrary clause.

As per the construction of the dNN, the clause values denote the outputs of the preceding nodes in the network. Furthermore, the ith neurons in the first hidden layer are denoted by η_i^1 and the second hidden layer is denoted by η_i^2. If clause c_i is nae-satisfied, then the values of both θ_{x_i} and θ_{x_j} cannot be same. Moreover, $\theta_{c_i} = 1$. The above θ values of the corresponding clause c_i will contribute $-\frac{1}{2}$ to f. Furthermore, if c_i is not nae-satisfied, then the values of both θ_{x_i}, θ_{x_j} is either 0 or 1. However, the value of θ_{c_i} is 0. The above θ values of the corresponding clause c_i will contribute 0 to f. There are k clauses that are nae-satisfied by assignment $\mathcal{A}$. Therefore, it will contribute $-\frac{k}{2}$ to f.

Conversely, assume that the minimum value of the output function f is $f(\theta) = -\frac{k}{2}$. We construct an assignment $\mathcal{A}$ for Φ that nae-satisfies k clauses from f as follows: From the construction of the dNN, it is clear that, for each clause $c_i = (x_i \vee x_j)$, if $\theta_{c_i} = 1$, then $\theta_{c_i} - |\theta_{x_i} - \theta_{x_j}| \leq 0$. Otherwise, f does not achieve its minimum value. To prove this, assume that $\theta_{c_i} - |\theta_{x_i} - \theta_{x_j}| > 0$. It follows that the neuron $\eta_{n+\ell}^2$ contributes $n \cdot \sigma(\theta_{c_i} - |\theta_{x_i} - \theta_{x_j}|) > 0$ to the output $f(\theta)$. This is a contradiction to the fact that f achieves its minimum value. We can simply assign the value of $\theta_{c_i} = 0$ and reduce the value of f further. So, it is clear that for any clause $c_i = (x_i \lor x_j)$, if $\theta_{c_i} = 1$, then $\theta_{c_i} - |\theta_{x_i} - \theta_{x_j}| \leq 0$. Each such clause, which θ value is one, contributes $-\frac{1}{2}$ to $f(\theta)$ through η_i^2. Furthermore, it contributes a value of 0 to $f(\theta)$ through $\eta_{n+\ell}^2$. That means there are k entries of value 1 in θ corresponding to clauses. For each $\theta_{c_i} = 1$, consider the θ_{x_i} and θ_{x_j} values. If $\theta_{x_i} = 0$ ($\theta_{x_j} = 0$), then assign the truth value of variable x_i (x_j) as **false**. Otherwise, assign the truth value of variable x_i (x_j) as **true**. Similarly, assign the truth values of the variables when $\theta_{c_i} = 0$. Observe that it will lead to an assignment $\mathcal{A}$ for Φ that nae-satisfies k clauses.

7 Conclusion

In this paper, we considered several **NP-hard** optimization problems, such as MAXCUT, MAXkSAT, and MAXNAE2SAT problems. We formulated differentiable functions with respect to these problems. We also proved the correctness of our formulations by showing the functions achieve their minimum value when the solution of a problem instance is found. It is worth noticing that the derivations of our designed differentiable functions use the framework of dataless neural networks. In future work, we intend to perform an implementation-based experiment of our designed method using a variety of dataless neural networks. An important avenue of research is to design a transformation procedure to convert arbitrary integer programs into dNNs. Another avenue of research is to implement the dNNs studied in this paper and compare them with state of the art solvers for the same.

Disclosure of Interests. This research was supported in part by the Defense Advanced Research Projects Agency through grant HR001123S0001-FP-004.

References

1. Akiba, T., Iwata, Y.: Branch-and-reduce exponential/fpt algorithms in practice: a case study of vertex cover. Theoret. Comput. Sci. **609**, 211–225 (2016)
2. Alkhouri, I.R., Atia, G.K., Velasquez, A.: A differentiable approach to the maximum independent set problem using dataless neural networks. Neural Netw. **155**, 168–176 (2022)
3. Bengio, Y., Lodi, A., Prouvost, A.: Machine learning for combinatorial optimization: a methodological tour d'horizon. Eur. J. Oper. Res. **290**(2), 405–421 (2021)
4. Boppana, R., Halldórsson, M.M.: Approximating maximum independent sets by excluding subgraphs. BIT Numer. Math. **32**(2), 180–196 (1992)
5. Cygan, M., et al.: Parameterized Algorithms. Springer, Heidelberg (2015). https://doi.org/10.1007/978-3-319-21275-3
6. Drori, I., et al.: Learning to solve combinatorial optimization problems on real-world graphs in linear time. In: IEEE International Conference on Machine Learning and Applications, pp. 19–24 (2020)
7. Festa, P.: A brief introduction to exact, approximation, and heuristic algorithms for solving hard combinatorial optimization problems. In: International Conference on Transparent Optical Networks, pp. 1–20 (2014)

8. Flum, J., Grohe, M.: Parameterized complexity theory. In: Texts in Theoretical Computer Science. Springer, Heidelberg (2006). https://doi.org/10.1007/3-540-29953-X
9. Fomin, F.V., Kratsch, D.: Exact exponential algorithms. In: Texts in Theoretical Computer Science. Springer, Heidelberg (2010). https://doi.org/10.1007/978-3-642-16533-7
10. Gaspers, S.: Exponential Time Algorithms - Structures, Measures, and Bounds. VDM (2010)
11. Jena, S.K., Subramani, K., Velasquez, A.: Differentiable discrete optimization using dataless neural networks. In: International Conference on Combinatorial Optimization and Applications, pp. 3–15 (2023)
12. Lamm, S., Sanders, P., Schulz, C., Strash, D., Werneck, R.F.: Finding near-optimal independent sets at scale. In: Proceedings of the Eighteenth Workshop on Algorithm Engineering and Experiments, pp. 138–150 (2016)
13. Li, Z., Chen, Q., Koltun, V.: Combinatorial optimization with graph convolutional networks and guided tree search. Adv. Neural Inf. Process. Syst. **31** (2018)
14. Mazyavkina, N., Sviridov, S., Ivanov, S., Burnaev, E.: Reinforcement learning for combinatorial optimization: a survey. Comput. Oper. Res. **134**, 105400 (2021)
15. Niedermeier, R.: Invitation to Fixed-Parameter Algorithms. Oxford University Press, Oxford (2006)
16. San Segundo, P., Rodríguez-Losada, D., Jiménez, A.: An exact bit-parallel algorithm for the maximum clique problem. Comput. Oper. Res. **38**(2), 571–581 (2011)
17. Schuetz, M., Brubaker, J., Katzgraber, H.G.: Combinatorial optimization with physics-inspired graph neural networks. Nat. Mach. Intell. **4**(4), 367–377 (2022)
18. Vazirani, V.: Approximation Algorithms. Springer, Heidelberg (2002). https://doi.org/10.1007/978-3-662-04565-7

On the Computational Complexities of Finding Selected Refutations of Linear Programs

K. Subramani[✉] and Piotr Wojciechowski

LDCSEE, West Virginia University, Morgantown, WV, USA
{k.subramani,pwojciec}@mail.wvu.edu

Abstract. In this paper, we establish the computational complexities of selected forms of refutations of linear programs. Linear programming is in the complexity class **P** and hence, it must have short affirmative and disqualifying certificates. One of the more celebrated lemmata in linear programming is Farkas' lemma, which establishes that both "yes" and "no" certificates can be thought of as solutions to complementary linear programs. Since then, it has been established that if a linear program is feasible, then it must have a solution which is bounded by a polynomial function of the input size. The latter observation, coupled with Farkas' lemma, immediately establishes that linear programming is in **NP** ∩ **coNP**. Our goal is to study the computational complexities of determining various constrained refutations for a given linear programming problem. This paper focuses on three distinct refutation forms, viz., **read-once**, **tree-like** and **dag-like**. We establish that checking if a linear program has a read-once refutation is **NP-complete**, even when it is defined by Binary Two Variable Per Inequality (BTVPI) constraints. Furthermore, the problems of finding the **shortest** tree-like and dag-like refutations are **NPO-complete** and **NPO PB-complete** respectively.

Keywords: Linear programming · Read-once refutation · Tree-like refutation · Dag-like refutation

1 Introduction

In this paper, we focus on the problems of checking if an unsatisfiable linear program has specific types of refutations under the ADD refutation system (see Sect. 2). A refutation is a "no"-certificate that attests to the infeasibility of the linear programming instance. Certificates are important from the perspective of enhancing trust in the underlying software system. Arbitrary certificates are difficult to verify, and hence there is significant interest in providing certificates of certain constrained forms. This paper focuses on three types of refutations, viz., **read-once**, **tree-like** and **dag-like**. Recall that in a read-once refutation, constraints (input or derived) cannot be reused. However, if a constraint can be rederived without reusing constraints from the original system, then it can be used as many times as it can be derived. In a tree-like refutation, derived constraints cannot be reused. However, constraints can be rederived. In a dag-like refutation, any constraint can be reused without needing to be rederived.

The notion of refutations is associated with a set of inference rules that can be used to produce contradictions. As mentioned before, the only inference rule in our refutation system is the ADD rule, discussed in Sect. 2. As we will establish later, the problem of checking if a linear program has a read-once refutation is **NP-complete**, even for very restricted forms. Furthermore, the problem of finding the length of a shortest tree-like refutation is **NPO-complete**, and the problem of finding the length of a shortest dag-like refutation is **NPO PB-complete**.

The rest of the paper is organized as follows: In Section 2, we formally define the problems under consideration in this paper. The notions of refutations and Farkas variables are also discussed in this section. Section 3 provides a brief discussion of the motivation for our work and related approaches in the literature. Read-once refutations are analyzed in Sect. 4. In Sect. 5, we examine tree-like refutations. Section 6 contains our analysis of dag-like refutations. We conclude in Sect. 7 by summarizing our contributions and identifying avenues for future research.

2 Statement of Problems

In this section, we introduce the concepts examined in this paper and define the problems under consideration.

In this paper, we examine proofs of infeasibility for linear programs.

Definition 1. *A* **linear program** *(LP) is a conjunction of constraints in which each constraint is an inequality of the form* $\mathbf{a}_j \cdot \mathbf{x} \leq b_j$ *where* $\mathbf{a}_j \in \mathbb{Z}^n$, $b_j \in \mathbb{Z}$, *and each variable* x_i *can take any real value.*

Example 1. System (1) is a linear program.

$$3 \cdot x_1 + 5 \cdot x_2 - 4 \cdot x_3 \leq -2 \quad -2 \cdot x_2 + 7 \cdot x_3 \leq 4 \qquad (1)$$

In this paper, we also look at LPs where each constraint has at most two non-zero coefficients.

Definition 2. *A* **Two Variable per Inequality (TVPI)** *constraint is a constraint with at most two non-zero coefficients.*

We can further restrict constraints by limiting the values which the non-zero coefficients can take.

Definition 3. *A* **Unit Two Variable per Inequality (UTVPI)** *constraint is a TVPI constraint such that each non-zero variable has a coefficient belonging to the set* $\{\pm 1\}$.

Definition 4. *A* **Binary Two Variable per Inequality (BTVPI)** *constraint is a TVPI constraint such that each non-zero variable has a coefficient belonging to the set* $\{\pm 1, \pm 2\}$.

Note that every UTVPI constraint is a BTVPI constraint and that every BTVPI constraint is a TVPI constraint. An LP in which every constraint is a BTVPI constraint is known as a BTVPI Constraint System (BCS).

Refutations are defined by the inference rules that can be used. Refutations of linear programs can use one inference rule. This rule corresponds to the summation of two constraints and is defined as follows:

$$\textbf{ADD}: \frac{\sum_{i=1}^{n} a_i \cdot x_i \leq b_1 \qquad \sum_{i=1}^{n} a_i' \cdot x_i \leq b_2}{\sum_{i=1}^{n} (a_i + a_i') \cdot x_i \leq b_1 + b_2} \quad (2)$$

We refer to Rule (2) as the ADD rule.

Example 2. Consider the constraints $3 \cdot x_1 + 5 \cdot x_2 - 4 \cdot x_3 \leq -2$ and $-2 \cdot x_2 + 7 \cdot x_3 \leq 4$. Applying the ADD rule to these constraints results in the constraint $3 \cdot x_1 + 3 \cdot x_2 + 3 \cdot x_3 \leq 2$.

It is easy to see that Rule (2) is sound in that any assignment satisfying the hypotheses **must** satisfy the consequent. Furthermore, the rule is **complete** in that if the original system is linear infeasible, then repeated application of Rule (2) will result in a contradiction of the form: $0 \leq b$, $b < 0$. The completeness of the ADD rule was established by Farkas [9], in a lemma that is famously known as Farkas' Lemma for systems of linear inequalities [18].

Definition 5. *A* **linear refutation** *is a sequence of applications of the ADD rule that results in a contradiction of the form* $0 \leq b$, $b < 0$.

Example 3. Consider the LP represented by System (3).

$$3 \cdot x_1 + 5 \cdot x_2 - 4 \cdot x_3 \leq -2 \quad -2 \cdot x_2 + 7 \cdot x_3 \leq 4 \quad -x_1 - x_2 - x_3 \leq -1 \; (3)$$

This system has the following linear refutation:

1. Apply the ADD rule to $3 \cdot x_1 + 5 \cdot x_2 - 4 \cdot x_3 \leq -2$ and $-2 \cdot x_2 + 7 \cdot x_3 \leq 4$ to get $3 \cdot x_1 + 3 \cdot x_2 + 3 \cdot x_3 \leq 2$.
2. Apply the ADD rule to $3 \cdot x_1 + 3 \cdot x_2 + 3 \cdot x_3 \leq 2$ and $-x_1 - x_2 - x_3 \leq -1$ to get $2 \cdot x_1 + 2 \cdot x_2 + 2 \cdot x_3 \leq 1$.
3. Apply the ADD rule to $2 \cdot x_1 + 2 \cdot x_2 + 2 \cdot x_3 \leq 1$ and $-x_1 - x_2 - x_3 \leq -1$ to get $x_1 + x_2 + x_3 \leq 0$.
4. Apply the ADD rule to $x_1 + x_2 + x_3 \leq 0$ and $-x_1 - x_2 - x_3 \leq -1$ to get $0 \leq -1$.

In this paper, we study read-once refutations, tree-like refutations, and dag-like refutations.

Definition 6. A **read-once** *refutation is a refutation in which each constraint, C, can be used in only one inference.*

This applies to constraints present in the original formula and those derived as a result of previous inferences. Note that in a read-once refutation, a constraint can be reused if it can be rederived. However, it must be rederived from a different set of input constraints.

Definition 7. *A **tree-like** refutation is a refutation in which each derived constraint can be used at most once.*

Note that in tree-like refutations, the input constraints can be used multiple times. Thus, any derived constraint can be derived multiple times as long as it is rederived each time it is used. Tree-like refutation is a **complete** refutation procedure [2].

Definition 8. *A **dag-like** refutation is a refutation in which each constraint can be used multiple times.*

It follows that dag-like refutation procedures are **complete** as well.
For any refutation, we can define the length of that refutation.

Definition 9. *The **length** of a refutation R of a constraint system is the number of inferences made in R.*

Note that the number of inferences in a refutation depends on the type of refutation. Recall that to reuse a constraint, a tree-like refutation has to rederive that constraint, however a dag-like refutation does not. This rederivation increases the number of inferences in the refutation. Thus, while every dag-like refutation can be converted into a tree-like refutation, the resultant refutation will need to include these rederivations. This can result in a much longer refutation.

In this paper, we study the following problems:

1. The **Read-once Refutation (ROR)** problem: Given an infeasible linear program **L**, does **L** have a read-once linear refutation?
2. The **Optimal Length Tree-like Refutation (OLTR)** problem: Given an infeasible linear program **L**, what is the length of the shortest tree-like refutation of **L**?
3. The **Optimal Length Dag-like Refutation (OLDR)** problem: Given an infeasible linear program **L**, what is the length of the shortest dag-like refutation of **L**?

Lemma 1. *Let $\mathbf{L}$ be the infeasible linear program $\mathbf{A} \cdot \mathbf{x} \leq \mathbf{b}$ with m constraints over n variables and let $\mathbf{y} \in \mathbb{Z}^m$ be a vector such that $\mathbf{y} \geq \mathbf{0}$, $\mathbf{y} \cdot \mathbf{A} = \mathbf{0}$, and $\mathbf{y} \cdot \mathbf{b} < 0$. $\mathbf{L}$ has a tree-like refutation of length $(\sum_{j=1}^{m} y_j - 1)$ and a dag-like refutation of length at most $(\sum_{j=1}^{m} 2 \cdot \log(y_j + 1) + m - 1)$.*

Proof. Since **L** is an infeasible LP, the vector **y** is guaranteed to exist by Farkas' Lemma [9]. We will now use **y** to generate both a tree-like and a dag-like refutation of **L**.

For $j = 1 \ldots m$, let l_j be the constraint $\mathbf{a_j} \cdot \mathbf{x} \le b_j$. Note that this is the constraint associated with the Farkas variable y_j.

For any constant $c \in \mathbb{Z}^+$, the constraint $c \cdot \mathbf{a_j} \cdot \mathbf{x} \le c \cdot b_j$ can be derived from $\mathbf{a_j} \cdot \mathbf{x} \le b_j$ by the following tree-like derivation:

1. Apply the ADD rule to $\mathbf{a_j} \cdot \mathbf{x} \le b_j$ and $\mathbf{a_j} \cdot \mathbf{x} \le b_j$ to get $2 \cdot \mathbf{a_j} \cdot \mathbf{x} \le 2 \cdot b_j$.
2. Apply the ADD rule to $2 \cdot \mathbf{a_j} \cdot \mathbf{x} \le 2 \cdot b_j$ and $\mathbf{a_j} \cdot \mathbf{x} \le b_j$ to get $3 \cdot \mathbf{a_j} \cdot \mathbf{x} \le 3 \cdot b_j$.
3. $\vdots$
4. Apply the ADD rule to $(c-1) \cdot \mathbf{a_j} \cdot \mathbf{x} \le (c-1) \cdot b_j$ and $\mathbf{a_j} \cdot \mathbf{x} \le b_j$ to get $c \cdot \mathbf{a_j} \cdot \mathbf{x} \le c \cdot b_j$.

Note that this derivation uses the ADD rule $(c-1)$ times.

Thus, **L** has the following tree-like refutation:

1. For each constraint l_j, the constraint $y_j \cdot \mathbf{a_j} \cdot \mathbf{x} \le y_j \cdot b_j$ can be generated by applying the ADD rule $(y_j - 1)$ times.
2. For each $i = 2 \ldots m$, apply the ADD rule to $(\sum_{i=1}^{j-1} y_i \cdot \mathbf{a_i}) \cdot \mathbf{x} \le \sum_{i=1}^{j-1} y_i \cdot b_i$ and $y_j \cdot \mathbf{a_j} \cdot \mathbf{x} \le y_j \cdot b_j$ to get $(\sum_{i=1}^{j} y_i \cdot \mathbf{a_i}) \cdot \mathbf{x} \le \sum_{i=1}^{j} y_i \cdot b_i$.
3. After $(m-1)$ applications of the ADD rule this derives the constraint $(\mathbf{y} \cdot \mathbf{A}) \cdot \mathbf{x} \le (\mathbf{y} \cdot \mathbf{b})$. Recall that $\mathbf{y} \cdot \mathbf{A} = \mathbf{0}$ and $\mathbf{y} \cdot \mathbf{b} < 0$. Thus, this is a contradiction.

Note that this refutation uses a total of $(\sum_{j=1}^{m} y_j - 1)$ inferences. Thus, **L** has a tree-like refutation of length $(\sum_{j=1}^{m} y_j - 1)$ as desired.

For any constant $c \in \mathbb{Z}^+$, the constraint $c \cdot \mathbf{a_j} \cdot \mathbf{x} \le c \cdot b_j$ can be derived from $\mathbf{a_j} \cdot \mathbf{x} \le b_j$ by the following dag-like derivation:

1. Apply the ADD rule to $\mathbf{a_j} \cdot \mathbf{x} \le b_j$ and $\mathbf{a_j} \cdot \mathbf{x} \le b_j$ to get $2 \cdot \mathbf{a_j} \cdot \mathbf{x} \le 2 \cdot b_j$.
2. Apply the ADD rule to $2 \cdot \mathbf{a_j} \cdot \mathbf{x} \le 2 \cdot b_j$ and $2 \cdot \mathbf{a_j} \cdot \mathbf{x} \le 2 \cdot b_j$ to get $4 \cdot \mathbf{a_j} \cdot \mathbf{x} \le 4 \cdot b_j$.
3. $\vdots$
4. Apply the ADD rule to $2^{\lfloor \log c \rfloor - 1} \cdot \mathbf{a_j} \cdot \mathbf{x} \le 2^{\lfloor \log c \rfloor - 1} \cdot b_j$ and $2^{\lfloor \log c \rfloor - 1} \cdot \mathbf{a_j} \cdot \mathbf{x} \le 2^{\lfloor \log c \rfloor - 1} \cdot b_j$ to get $2^{\lfloor \log c \rfloor} \cdot \mathbf{a_j} \cdot \mathbf{x} \le 2^{\lfloor \log c \rfloor} \cdot b_j$.
5. Let $S \subseteq \{0, \ldots, \lfloor \log c \rfloor\}$ be such that $\sum_{i \in S} 2^i = c$. Note that we have already derived the constraint $2^i \cdot \mathbf{a_j} \cdot \mathbf{x} \le 2^i \cdot b_j$ for each $i \in S$. Thus, applying the ADD rule $(|S| - 1)$ times lets us derive the constraint $(\sum_{i \in S} 2^i) \cdot \mathbf{a_j} \cdot \mathbf{x} \le (\sum_{i \in S} 2^i) \cdot b_j$. This is precisely the constraint $c \cdot \mathbf{a_j} \cdot \mathbf{x} \le c \cdot b_j$.

Note that this derivation uses the ADD rule at most $(2 \cdot \log(c+1))$ times. The remainder of the refutation is the same as the tree-like refutation of **L**. This refutation uses at most $(\sum_{j=1}^{m} 2 \cdot \log(y_j + 1) + m - 1)$ inferences. Thus, **L** has a dag-like refutation of length at most $(\sum_{j=1}^{m} 2 \cdot \log(y_j + 1) + m - 1)$ as desired.

Note that the size of **y** is polynomial in the size of **L**. Thus, the length of this dag-like refutation is also polynomial in the size of **L**.

Observe that if $\mathbf{y} \in \{0,1\}^m$, then both of these refutations are read-once.

2.1 Complexity Classes

We now define the complexity classes **NPO** and **NPO PB** used in this paper. We begin by defining the complexity class **NPO** [17].

Definition 10. *The complexity class* **NPO** *is the set of optimization problems such that:*

1. *The set of instances can be recognized in polynomial time.*
2. *Solutions are polynomially sized and can be verified in polynomial time.*
3. *The objective function can be computed in polynomial time.*

We next define the complexity class **NPO PB** [13].

Definition 11. **NPO PB** *is the set of* **NPO** *problems for which the value of the objective function is polynomial in the size of the input.*

Finally, we introduce the notion of **PTAS** reductions [17].

Definition 12. *A* **PTAS** *reduction from problem A to problem B, is a trio of functions f, g, and α computable in polynomial time, such that:*

1. *f maps instances of problem A to instances of problem B.*
2. *g takes an instance x of problem A, an approximate solution to the corresponding problem $f(x)$ in B, and an error parameter ϵ and produces an approximate solution to x.*
3. *α maps error parameters for solutions to instances of problem A to error parameters for solutions to problem B.*
4. *If the solution y to $f(x)$ (an instance of problem B) is at most $(1+\alpha(\epsilon))$ times worse than the optimal solution, then the corresponding solution $g(x,y,\epsilon)$ to x (an instance of problem A) is at most $(1+\epsilon)$ times worse than the optimal solution.*

Definition 13. *A problem P is* **NPO PB-hard** *under* **PTAS** *reductions, if every problem in* **NPO PB** *can be reduced to P by a* **PTAS** *reduction.*

Unless otherwise stated, we assume that **NPO PB-hardness** is specified with respect to **PTAS** reductions.

The set of problems which are in the class **NPO PB** and are **NPO PB-hard** are called **NPO PB-complete**. Additionally, for every **NPO PB-complete** problem P there exists an $\epsilon > 0$ such that P cannot be approximated to within a factor of $O(n^\epsilon)$ unless $\mathbf{P} = \mathbf{NP}$ [3]. Thus, if any **NPO PB-complete** problem can be approximated to within a polylogarithmic factor, then $\mathbf{P} = \mathbf{NP}$.

An example of an **NPO PB-complete** problem is Bounded Minimum 0-1 Programming problem. This problem is formulated as follows:

Given an integer program $\mathbf{A} \cdot \mathbf{x} \geq \mathbf{b}$, $\mathbf{x} \in \{0,1\}^n$, find the minimum value of $\mathbf{1} \cdot \mathbf{x}$. This specific form of Minimum 0-1 Programming is known to be **NPO PB-complete** [13].

2.2 Minimum 0–1 Integer Programming

To establish some of the completeness results in this paper, we utilize reductions from the Minimum 0–1 Integer Programming Problem.

In the minimum 0–1 integer programming problem, you are given an integer program of the form:

$$\min \sum_{i=1}^{n} c_i \cdot x_i \quad \mathbf{A} \cdot \mathbf{x} = \mathbf{b} \quad \mathbf{x} \in \{0,1\}^n.$$

where the variables are restricted to be in the set $\{0,1\}$, this problem is known to be **NPO-complete** [17].

We also utilize a restricted version of the minimum 0–1 integer programming problem where the coefficients are restricted to have values polynomial in the size of the input instance. Since the coefficients are bounded, the problem now lies in the complexity class **NPO PB**. Note that this problem is **NPO PB-complete** even in the case where $\mathbf{c} = \mathbf{1}$ [13].

The principal contributions of this paper are as follows:

1. Establishing that the ROR problem is **NP-complete** for BCSs (see Sect. 4).
2. Establishing that the OLTR problem for LPs is **NPO-complete** (see Sect. 5).
3. Establishing that the OLDR problem for LPs is **NPO PB-complete** (see Sect. 6).

3 Motivation and Related Work

In this section, we motivate our work and describe existing work for related problems.

Constraint systems (both linear programs and integer programs) are heavily used in the field of software verification [5–7]. A constraint system corresponding to a piece of software can be derived and then combined with constraints corresponding to the negation of the specifications. If the resultant system of constraints is infeasible, then the software is consistent with its specifications. Although this approach is intuitive and straightforward, it may become impractical because of the high number of constraints that are generated. A constraint based approach to program verification has also been attempted for rule-based programming [4]. Rule-based programming has gained interest in the software industry over the past years, because of the growing use of Business Rules Management Systems. Hence, a demand for verification of rule programs has emerged. Also, in [11] it is shown how the constraint-based approach can be used to model a wide spectrum of program analysis using disjunctions and conjunctions of linear inequalities. Linear programs have also been used as a finer grained abstraction for sequential programs offering an effective model checking procedure [1].

In linear programs, refutations are closely related to theorems of the alternative. Typically, theorems of the alternative connect pairs of linear constraint systems and have the following form: Given two linear systems **A** and **B**, exactly one of them is feasible. System **A** is called the primal system and System **B** is called the dual system. It is not hard to see that theorems of the alternative provide certificates of infeasibility. One famous theorem of the alternative is Farkas' Lemma [18]. Farkas' lemma is one of several lemmata that consider pairs of linear systems in which exactly one element of the pair is feasible. These lemmas are collectively referred to as "Theorems of the Alternative" [16]. The coefficients generated by Farkas' Lemma are called the Farkas variables corresponding to the system $\mathbf{A} \cdot \mathbf{x} \le \mathbf{b}$ and they serve as a witness that certifies the linear infeasibility of this system. In general, the Farkas variables can assume any rational value for a given constraint system. Read-once refutations have also been studied for clausal formulas [12,15].

Observe that theorems of the alternative provide natural certificates of infeasibility. For instance, if we are required to prove that a difference constraint system is infeasible, then we can produce a negative cost cycle in the corresponding constraint network [8]. In previous work, we showed that difference constraints always have a read-once refutation and this refutation can be determined in polynomial time [19]. Likewise, in [20], we showed that a system of Unit Two Variable Per Inequality constraints need not have a read-once refutation, but that the presence or absence of one can be detected in polynomial time. This paper completes the **trichotomy** by showing that checking if a BCSs has a read-once refutation is **NP-complete**.

4 Read-Once Refutations

In this section, we examine the ROR problem for linear programs.

First, we consider the problem of checking whether a linear program has a read-once refutation. It was already shown in [14] that for systems of Horn constraints, the problem of finding read-once linear refutations is **NP-complete**. In this paper, we show that the problem of finding read-once refutations for linear programs with at most two variables per constraint is still **NP-complete** for BCSs. This is in sharp contrast to UCSs, for which the problem of finding read-once linear refutations is solvable in polynomial time [20].

Theorem 1. *The ROR problem for BCSs is* **NP-complete**.

Proof. The ROR problem is clearly in **NP**, since we can guess a vector $\mathbf{y} \in \{0,1\}^m$ and check that $\mathbf{y} \cdot \mathbf{A} = \mathbf{0}$ and $\mathbf{y} \cdot \mathbf{b} < 0$. Note that this vector corresponds to a set of constraints that, when summed together, produces a contradiction. This summation can be represented as a sequence of applications of the ADD rule where each constraint is used at most once. This is precisely a read-once refutation.

We establish **NP-hardness** through a reduction from the Exact Cover by 3-Sets (X3C) problem. We recall the definition of the X3C problem.

X3C is one of the core six problems proved to be **NP-complete** in [10]. An instance of the X3C problem consists of a set X with $3 \cdot n$ elements and a collection **C** of 3-element subsets of X.

The query is: Can we find a $\mathbf{C}' \subseteq \mathbf{C}$, such that every element of X occurs in exactly one member of $\mathbf{C}'$? Note that the query is asking whether there is a subset of $\mathbf{C}'$ that *exactly* covers all the elements of X.

Assume that we are given the following instance of X3C:

1. Set $X = \{x_1, x_2, \ldots, x_{3 \cdot n}\}$.
2. Sets $C_1, C_2, \ldots C_m \subseteq S$, with $|C_i| = 3$, for $i = 1, 2, \ldots, m$.

From **I**, we construct the BCS **B**.

1. Corresponding to each set $C_j = (x_{j_1}, x_{j_2}, x_{j_3})$, create the constraints $l_{j,1}$: $x_{j_1} - 2 \cdot w_j \leq 0$, $l_{j,2}$: $w_j + x_{j_2} \leq 0$, and $l_{j,3}$: $w_j + x_{j_3} \leq 0$. Note that these constraints are equivalent to the constraint $x_{j_1} + x_{j_2} + x_{j_3} \leq 0$.
2. Create the constraints l_1 : $-x_1 + 2 \cdot y_1 \leq -1$, l_2 : $-y_1 - x_2 \leq 0$, l_3 : $2 \cdot y_2 - y_1 \leq 0$, l_4 : $-y_2 - x_3 \leq 0$, l_5 : $2 \cdot y_3 - y_2 \leq 0$, ..., $l_{6 \cdot n - 5}$: $2 \cdot y_{3 \cdot n - 2} - y_{3 \cdot n - 3} \leq 0$, $l_{6 \cdot n - 4}$: $-y_{3 \cdot n - 2} - x_{3 \cdot n - 1} \leq 0$, and $l_{6 \cdot n - 3}$: $-y_{3 \cdot n - 2} - x_{3 \cdot n} \leq 0$. Note that these constraints are equivalent to the constraint $-x_1 - \ldots - x_{3 \cdot n} \leq -1$.

The resulting BCS is:

$$\mathbf{B}: \bigwedge_{j=1}^{m} (l_{j,1} \wedge l_{j,2} \wedge l_{j,3}) \wedge \bigwedge_{i=1}^{6 \cdot n - 3} l_i \tag{4}$$

We now argue that there is a one-to-one correspondence between Exact Covers of X by 3-Sets and RORs of the constructed linear program.

Consider any exact cover $\mathbf{C}'$ of X by 3-Sets from the given subsets $C_1, \ldots, C_m$. It is clear that $|\mathbf{C}'| = n$. Without loss of generality, we can assume that $C_1, \ldots C_n$ are the subsets picked.

First, let us consider the constraints l_1 through $l_{6 \cdot n - 3}$. By construction, each x_i occurs in exactly one constraint with coefficient -1. Additionally, each y_i occurs in one constraint with coefficient 2 and in two constraints with coefficient -1. Thus, summing these constraints results in the constraint d_0 : $-x_1 - x_2 - \ldots - x_{3 \cdot n} \leq -1$.

Now focus on the corresponding linear constraints. Corresponding to each C_j, we have the constraints $l_{j,1}$, $l_{j,2}$, and $l_{j,3}$. Summing these constraints results in the constraint d_j $x_{j_1} + x_{j_2} + x_{j_3} \leq 0$. Since the C_js form an exact cover, each variable x_i will occur precisely once across all the d_js, $j = 1, 2, \ldots, n$. When we sum the constraints d_j, $j = 1, 2, \ldots, n$ with the constraint d_0. Every variable is canceled and we get the contradiction $0 \leq -1$. Since each constraint was used at most once, this is a read-once refutation.

Now assume that the linear program represented by System (4) has a read-once refutation. This means that some subset of the constraints in $\mathbf{L}$, when added together, produces a contradiction.

By construction, l_1 is the only constraint with a negative defining constant. Thus, l_1 must be part of this refutation; indeed, it is part of *every* refutation (read-once or otherwise).

Any read-once refutation of **L** must eliminate the $2 \cdot y_1$ term from l_1. By construction, the only constraints with $-y_1$ are l_2 and l_3. Thus, both of these constraints must be used in the refutation. Summing all three constraints together results in the constraint $-x_1 - x_2 + 2 \cdot y_2 \geq -1$.

Any read-once refutation of **L** must eliminate the $2 \cdot y_2$ term from this constraint. By construction, the only constraints with $-y_2$ are l_4 and l_5. Thus, both of these constraints must be used in the refutation. Summing all three constraints together results in the constraint $-x_1 - x_2 - x_3 + 2 \cdot y_3 \geq -1$. This continues until all y_is are eliminated in this fashion. This results in the constraint $d_0 : -x_1 - x_2 - \ldots - x_{3 \cdot n} \leq -1$.

Let us consider the set of constraints corresponding to C_j. By construction, these are the only constraints with the variable w_j. Observe that in $l_{j,1}$, w_j has coefficient -2, while in $l_{j,2}$ and $l_{j,3}$ it has coefficient 1. Thus, any read-once refutation of L must use either all three of these constraints, or none of these constraints. Otherwise, a w_j term will be left in the final summation. This means that, from the perspective of read-once refutation, the constraints $l_{j,1}$, $l_{j,2}$ and $l_{j,3}$ are equivalent to the constraint $d_j : x_{j_1} + x_{j_2} + x_{j_3} \leq 0$. Let $\mathbf{D} = \bigwedge_{j=1}^{m} d_j$. We can

Since all the variables in d_0 must be canceled by the refutation, the remaining constraints in the refutation correspond to a subset of $\mathbf{D}$ such that each x_i occurs in exactly one constraint. Let $\mathbf{D'}$ denote this set of constraints. When we look at the subsets C_j corresponding to the constraints $d_j \in \mathbf{D'}$, it is clear that they form an exact cover by 3-Sets of the set S.

5 Tree-Like Refutations

In this section, we determine the complexity of finding short tree-like refutations in general linear programs.

Theorem 2. *The OLTR problem is* **NPO-complete** *under* **PTAS** *reductions.*

Proof. Note that a tree-like refutation of a linear program can be represented by the coefficients generated from Farkas' Lemma [9]. Thus, a tree-like refutation R of a linear program **L** is polynomially sized in terms of the size of **L**. Additionally, from Lemma 1, the length of a tree-like refutation can be computed in polynomial time. Thus, the OLTR problem is in **NPO**. Now we need to show **NPO-hardness**.

This will be accomplished by a reduction from the Minimum 0-1 Programming problem.

Consider the following instance of the Minimum 0-1 programming problem:

$$\min \sum_{i=1}^{n} c_i \cdot x_i \quad \mathbf{A} \cdot \mathbf{x} = \mathbf{b} \quad \mathbf{x} \in \{0,1\}^n.$$

Assume without loss of generality that $\mathbf{c} \geq 1$. Even in this form the Minimum 0-1 programming problem is **NPO-complete** [17].

Let $\mathbf{D}$ be the $n \times n$ matrix such that $d_{i,i} = c_i - 1$ and $d_{i,j} = 0$ for $i \neq j$. Corresponding to the Minimum 0-1 programming instance, we can construct the following linear program:

$$\mathbf{y} \cdot \mathbf{A} - \mathbf{z} \cdot \mathbf{D} \leq \mathbf{0} \qquad -\mathbf{z} \leq \mathbf{0} \qquad -\mathbf{y} \cdot \mathbf{b} \leq -1.$$

We make the following observations about any tree-like linear refutation of the LP:

1. The constraint $-\mathbf{y} \cdot \mathbf{b} \leq -1$, is the only constraint in the system with a negative defining constant. Thus, it must be part of the refutation.
2. Let $\mathbf{y} \cdot \mathbf{a}_i - (c_i - 1) \cdot z_i \leq 0$ be the ith constraint of $\mathbf{y} \cdot \mathbf{A} - \mathbf{z} \cdot \mathbf{D} \leq \mathbf{0}$, and let x_i be 1 if this constraint is used in the refutation and 0 otherwise.
3. Summing all constraints in the refutation results in the constraint $0 \leq -1$. Thus, after canceling the $\mathbf{z}$ variables, we must have that each term of $-\mathbf{y} \cdot \mathbf{b}$ is canceled by a term of $(\mathbf{y} \cdot \mathbf{A}) \cdot \mathbf{x}$. Thus, we must have that $\mathbf{A} \cdot \mathbf{x} = \mathbf{b}$. Thus, $\mathbf{x}$ is a valid solution to the original 0-1 program.
4. For each constraint $\mathbf{y} \cdot \mathbf{a}_i - (c_i - 1) \cdot z_i \leq 0$, an additional $(c_i - 1)$ copies of the constraint $-z_i \leq 0$ must be used to cancel the z_i term introduced in the refutation. Thus, the length of the tree-like refutation is $\sum_{i=1}^{n} c_i \cdot x_i + 1$.

All that remains is to establish that a **PTAS** reduction exists from the Minimum 0-1 Programming problem to the OLTR problem. This will be done by establishing the existence of the functions f, g, and α.

1. The function f: We provided a method for constructing a linear program $\mathbf{L}$ from an instance of the Minimum 0–1 Programming problem $\mathbf{M}$. This forms the function f required for the **PTAS** reduction.
2. The function g: We provided a method to take a tree-like refutation of $\mathbf{L}$ and construct a feasible solution to $\mathbf{M}$. This forms the function g required for the **PTAS** reduction.
3. The function α: Let k^* be the optimal solution to $\mathbf{M}$. $\mathbf{L}$ has a tree-like refutation of length $(k^* + 1)$. Additionally, if $\mathbf{L}$ had a shorter tree-like refutation, then $\mathbf{M}$ would have a solution for which the optimization function has a lower value. Thus, the OLTR of $\mathbf{L}$ has length $(k^* + 1)$. Let $\alpha(\epsilon) = \frac{\epsilon}{2}$.

   Let R be a tree-like refutation of $\mathbf{L}$ of length $(k+1)$. The function g produces a solution to $\mathbf{M}$ with value k. Since the feasibility of the zero vector can be tested in polynomial time, we can assume without loss of generality that $k^* \geq 1$. Additionally, note that $k \geq k^*$. If $\frac{k+1}{k^*+1} \leq 1 + \alpha(\epsilon) = 1 + \frac{\epsilon}{2}$, then

$$\frac{k}{k^*} = \frac{k \cdot (k^*+1)}{k^* \cdot (k^*+1)} \leq \frac{k \cdot (k^*+1) + (k - k^*) \cdot (k^*-1)}{k^* \cdot (k^*+1)} = \frac{2 \cdot k \cdot k^* - k^{*2} + k^*}{k^* \cdot (k^*+1)}$$

$$= \frac{2 \cdot k - k^* + 1}{k^* + 1} = \frac{2 \cdot (k+1) - (k^*+1)}{k^*+1} = 2 \cdot \frac{k+1}{k^*+1} - 1 \leq 2 \cdot (1 + \frac{\epsilon}{2}) - 1 = 1 + \epsilon.$$

Thus, the OLTR problem for linear programs is **NPO-complete**.

6 Dag-Like Refutations

In this section, we determine the complexity of finding short dag-like refutations in general linear programs.

Theorem 3. *The OLDR problem is* **NPO PB-complete** *under* **PTAS** *reductions.*

The proof of Theorem 3 is similar to the proof of Theorem 2 and will be provided in the journal version of the paper.

7 Conclusion

This paper was concerned with establishing the computational complexities of three constrained refutation schemes for linear programming. As has been discussed extensively in the literature, general refutations may be difficult to verify and hence the search for constrained certificates is of significant interest. We studied three types of refutations under the ADD refutation rule. Our investigations established that the problem of checking if a linear program has a read-once refutation is **NP-complete**. Furthermore, the problems of finding the shortest tree-like and dag-like refutations are **NPO-complete** and **NPO PB-complete**. Our results essentially rule out the existence of efficient approximation schemes with "good" error bounds, unless **P=NP**.

Disclosure of Interests. This research was supported in part by the Defense Advanced Research Projects Agency through grant HR001123S0001-FP-004.

References

1. Armando, A., Castellini, C., Mantovani, J.: Software model checking using linear constraints. In: Davies, J., Schulte, W., Barnett, M. (eds.) ICFEM 2004. LNCS, vol. 3308, pp. 209–223. Springer, Heidelberg (2004). https://doi.org/10.1007/978-3-540-30482-1_22
2. Beame, P., Pitassi, T.: Simplified and improved resolution lower bounds. In: 37th Annual Symposium on Foundations of Computer Science, pp. 274–282. IEEE, Burlington (1996)
3. Berman, P., Schnitger, G.: On the complexity of approximating the independent set problem. Inf. Comput. **96**(1), 77–94 (1992)
4. Berstel, B., Leconte, M.: Using constraints to verify properties of rule programs. In: Proceedings of the 2010 International Conference on Software Testing, Verification, and Validation Workshops, pp. 349–354 (2008)
5. Ceberio, M., Acosta, C., Servin, C.: A constraint-based approach to verification of programs with floating-point numbers. In: Proceedings of the 2008 International Conference of Software Engineering Research and Practice, pp. 225–230 (2008)
6. Collavizza, H., Reuher, M.: Exploration of the capabilities of constraint programming for software verification. In: Proceedings of the 2006 International Conference on Tools and Algorithms for the Construction and Analysis of Systems (2006)

7. Collavizza, H., Rueher, M., Van Hentenryck, P.: CPBPV: a constraint-programming framework for bounded program verification. In: Stuckey, P.J. (ed.) CP 2008. LNCS, vol. 5202, pp. 327–341. Springer, Heidelberg (2008). https://doi.org/10.1007/978-3-540-85958-1_22
8. Cormen, T.H., Leiserson, C.E., Rivest, R.L., Stein, C.: Introduction to Algorithms. MIT Press, Cambridge (2001)
9. Farkas, G.: Über die Theorie der Einfachen Ungleichungen. Journal für die Reine und Angewandte Mathematik **124**(124), 1–27 (1902)
10. Garey, M.R., Johnson, D.S.: Computers and Intractability: A Guide to the Theory of NP-Completeness. W. H. Freeman Company, San Francisco (1979)
11. Gulwani, S., Srivastava, S., Venkatesan, R.: Program analysis as constraint solving. In: Proceedings of the 2008 ACM SIGPLAN Conference on Programming Language Design and Implementation. ACM, New York (2008)
12. Iwama, K., Miyano, E.: Intractability of read-once resolution. In: Proceedings of the 10th Annual Conference on Structure in Complexity Theory (SCTC 1995), pp. 29–36. IEEE Computer Society Press, Los Alamitos (1995)
13. Kann, V.: Polynomially bounded minimization problems that are hard to approximate. Nordic J. Comput. **1**(3), 317–331 (1994)
14. Kleine Büning, H., Wojciechowski, P., Chandrasekaran, R., Subramani, K.: Restricted cutting plane proofs in horn constraint systems. In: Herzig, A., Popescu, A. (eds.) FroCoS 2019. LNCS (LNAI), vol. 11715, pp. 149–164. Springer, Cham (2019). https://doi.org/10.1007/978-3-030-29007-8_9
15. Kleine Büning, H., Wojciechowski, P.J., Subramani, K.: Finding read-once resolution refutations in systems of 2CNF clauses. Theor. Comput. Sci. **729**, 42–56 (2018)
16. Nemhauser, G.L., Wolsey, L.A.: Integer and Combinatorial Optimization. John Wiley & Sons, New York (1999)
17. Orponen, P., Mannila, H.: On approximation preserving reductions: complete problems and robust measures. Department of Computer Science, University of Helsinki, Technical report (1987)
18. Schrijver, A.: Theory of Linear and Integer Programming. John Wiley and Sons, New York (1987)
19. Subramani, K.: Optimal length resolution refutations of difference constraint systems. J. Autom. Reason. (JAR) **43**(2), 121–137 (2009)
20. Subramani, K., Wojciechowki, P.: A polynomial time algorithm for read-once certification of linear infeasibility in UTVPI constraints. Algorithmica **81**(7), 2765–2794 (2019)

Extending the Tractability of the Clique Problem via Graph Classes Generalizing Treewidth

Philippe Jégou[✉]

Aix Marseille Univ, CNRS, LIS, Marseille, France
philippe.jegou@univ-amu.fr

Abstract. The study of the Clique problem in algorithmic graph theory is important both because it is a central problem in complexity theory, almost at the same level as SAT [8], but also because its practical resolution has many applications, notably in Artificial Intelligence (e.g. checking consistency of a binary CSP is equivalent to check the size of maximum clique in its microstructure [7]). A great deal of work has therefore been carried out, and this paper attempts to extend the results obtained in this field. It starts from the observation that this problem is tractable in polynomial time on graphs whose treewidth [10] are bounded by a constant (see [5]). Although this type of result is very interesting from a theoretical point of view, it often remains limited in terms of application. So, we propose here an extension of such these approaches based on the definition of graph classes wider than those of bounded treewidth for which we would be able to propose polynomial time algorithms. These graph classes denoted $\mathcal{C}_W^k$ include graphs of treewidth $W + k$, but also contain, even if k and W are constants, graphs known to be of unbounded treewidth. These $\mathcal{C}_W^k$ classes are introduced, their fundamental properties are given and the associated algorithms are presented and analysed.

Keywords: Clique Problem · Tractable Classes · Treewidth

1 Introduction

In Artificial Intelligence algorithms, the Clique Problem in a graph plays a significant role, as it can be found in a large number of application domains. For example, we know that a binary CSP has a solution if and only if its microstructure expression has a clique whose size corresponds to the number of variables in the instance. Beyond this domain, this problem, which is NP-complete, has a wide range of applications in optimization, and is, along with SAT, one of the central problems of Complexity Theory. So this is a problem that can legitimately be the subject of a study in its own right. Its statement is very simple: given a (undirected) graph G and an integer k, the question is whether G has a complete subgraph of k (or more) vertices. The associated optimization problem, called Maximum Clique Problem, consists of determining the largest clique in a graph.

If there are tractable classes (i.e. tractable in polynomial time), as with many other difficult problems, one way of tackling it is to use FPT-type approaches [6], even if Clique is not FPT. But this is made possible by using the notion of tree-decomposition of graphs and the associated notion of treewidth [10]. This notion now plays a fundamental role in many fields, notably in the efficient processing of graphical models (CSP, WCSP, cost function networks, Bayesian networks, etc.). Fundamental results with first Arnborg [2], then Courcelle's meta-theorem [5], have shown that for the case where a graph has a treewidth bounded by a constant, many problems, including the Clique problem, can then be handled in polynomial time.

In this paper, we address these issues by defining a theoretical tool related to the notion of tree-decomposition, for constructing an infinite number of recursively defined graphs. Under certain assumptions, these classes allow us to propose polynomial time algorithms for solving the Clique problem. These classes of graphs, called $\mathcal{C}_W^k$, are defined by considering two parameters, k and W. We show that any graph G in $\mathcal{C}_W^k$ is recognizable in $O(n^{2k} \times T(G,W))$ where T depends on G and W, W being the treewidth of a graph accessible after k operations on G. We show that if this parameter W is a constant, then $T(G,W)$ is a polynomial factor, which means that instance belonging to $\mathcal{C}_W^k$ can be processed in polynomial time. In the approaches based on tree-decomposition, the difficulty often arises from the fact that classes of graphs have treewidths that are potentially unbounded by constants, as it is the case for planar graphs and complete bipartite graphs. This problem seems to disappear for some of these classes, such as the two mentioned above, with the use of $\mathcal{C}_W^k$ classes. For example, we show that all planar graphs belong to class $\mathcal{C}_3^1$, even though the treewidth of these graphs is in $\Theta(\sqrt{n})$ for graphs of n vertices. We also show that any graph of treewidth $W + k$ belongs to class $\mathcal{C}_W^k$, which leads us to hope that this double parameterization offers a more accurate view of instance difficulty than that indicated by treewidth alone.

This paper is organized as follows. In the next section, we introduce notations and basic notions. In the third section, we define $\mathcal{C}_W^k$ classes and present two fundamental properties, one of which is used in the next section where we exhibit an algorithm for recognizing these classes. In the fifth section, we present an algorithm for solving the clique problem for the case of graphs belonging to a $\mathcal{C}_W^k$ class, and the analysis of its complexity allows us to highlight tractable cases while the sixth section show that some classes of graphs of unbounded treewidth appear in classes $\mathcal{C}_W^k$ such that k and W are constants. Finally, to conclude this paper, we discuss the relevance of these new classes as a general tool complementary to tree-decomposition.

2 Preliminaries

First of all, we remind a few notations and definitions. Let $G = (V, E)$ be a finite undirected graph with V the set of vertices and E the set of edges. An edge of E is denoted $\{x,y\}$ with $x,y \in V$. We use n to denote the number of vertices (so $|V| = n$) and e to denote the number of edges (so $|E| = e$). Given a subset of vertices $X \subseteq V$, the graph $G[X]$ denotes the subgraph of G induced by the subset of vertices X. A clique of a graph is a subset K of the vertices such that every two vertices $x,y \in K$ is an edge, i.e. $\{x,y\} \in E$. A complete graph is a graph $G = (V,E)$ such that V is a clique of G (It is usually referred to as K_n). Given a graph $G = (V,E)$, for each vertex $x \in V$, the neighborhood of x is the set $N(x) = \{y \in V : \{x,y\} \in E\}$. If $\sigma = [x_1, x_2, \ldots, x_n]$ is an ordering of V, the successors of a vertex x_i are the elements of the set $N^+(x_i) = \{x_j \in N(x) : i < j\}$. Now, we recall the definition of the *tree-decomposition* of a graph [10] and the its associated parameter called *treewidth*:

Definition 1. *A* tree-decomposition *of a graph* $G = (V,E)$ *is a pair* (B,T) *where* $T = (I,F)$ *is a tree (I is a set of nodes and F a set of edges) and* $B = \{B_i : i \in I\}$ *a family of subsets of V such as every* $B_i \in B$ *(called* bag*) corresponds to a node i of T and satisfies:*

1. $\cup_{i \in I} B_i = V$,
2. $\forall \{x,y\} \in E, \exists i \in I$ *such that* $\{x,y\} \subseteq B_i$, *and*
3. $\forall i,j,k \in I$, *if k is on a path between i and j in T, then* $B_i \cap B_j \subseteq B_k$

The width *of a tree-decomposition is equal to* $max_{i \in I} |B_i| - 1$ *and the* treewidth *of G denoted* $tw(G)$ *is equal to the minimum width among all the tree-decompositions of G.*

3 A Generalization of Treewidth by Defining New Graph Classes

Here we define an infinite number of recursively defined graph classes that generalize graphs of a given treewidth, and therefore also graphs of bounded treewidth. We will see that such a definition can be seen as a generalization of graph treewidth.

Definition 2. *Let* $W \in \mathbb{N}$ *be a constant. The class of graphs* $\mathcal{C}_W^0$ *is the set of graphs G whose treewidth is at most W, that is* $tw(G) \leq W$. *Given a natural number* $k > 0$, *the class of graphs* $\mathcal{C}_W^k$ *is the set of graphs* $G = (V,E)$ *such that there is an ordering* $\sigma = [x_1, x_2, \ldots, x_n]$ *of V, such that, for $i = 1,2,\ldots,n$, the subgraph* $G[N_\sigma^+(x_i)]$ *belongs to* $\mathcal{C}_W^{k-1}$. *Such an ordering σ is called* $\mathcal{C}_W^k$ scheme.

We illustrate this definition below, showing in particular that the same graph can belong to different classes of type $\mathcal{C}_W^k$. This is simply made possible by the use of a double parameterization with k and W.

First, the treewidth of the graph in Fig. 1 is 5. Indeed, we can easily construct a triangulation ordering such that any vertex will have at most five subsequent neighbors in this ordering and thus, this triangulation will construct a chordal graph whose largest cliques will be of size 6 at most. Such a triangulation ordering is given by the labeling of the vertices in Fig. 1. Since the treewidth of this graph is 5, this graph trivially belongs to $\mathcal{C}_5^0$. Moreover, for this graph, whatever the ordering σ considered on its vertices, the neighborhood of each vertex has a treewidth at most equal to two. Moreover, there will always be at least one vertex whose neighborhood has a treewidth equal to two. So, for each vertex x, the subgraph $G[N_\sigma^+(x)] \in \mathcal{C}_2^0$. Therefore, this graph belongs to $\mathcal{C}_2^1$ since it admits a $\mathcal{C}_2^1$ scheme. Although this is more difficult to observe directly, we can also show that this graph belongs to the class $\mathcal{C}_1^2$. We can see that this graph also belongs to many other classes, by simply changing the value of parameter W. For example, it belongs to any $\mathcal{C}_W^0$ class, taking $W \geq 5$.

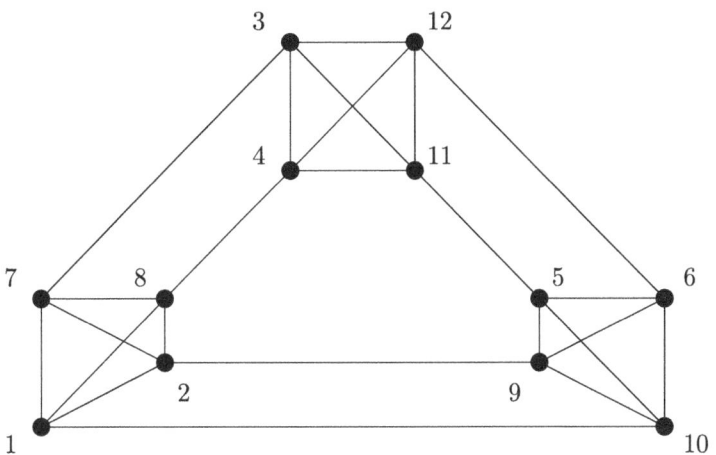

Fig. 1. A graph of the class $\mathcal{C}_2^1$ and whose treewidth is 5.

We give now a fundamental property about these classes.

Theorem 1. $\forall G$, if $tw(G) \leq W + k$, then $G \in \mathcal{C}_W^k$.

Proof. We prove this property by induction on k.

First, for the base case, consider $G \in \mathcal{C}_W^0$, i.e. $k = 0$. By definition, $tw(G) \leq W$. Note that this result holds also for $k = 1$. Indeed, for $G = (V, E)$ such that $tw(G) \leq W + 1$, there is a tree-decomposition of G such as, for all bags B_j, $|B_j| \leq W + 2$. So, there is an ordering $\sigma = [x_1, x_2, \ldots, x_n]$ on V, such that, $\forall i, 1 \leq i \leq n$, $tw(G[N_\sigma^+(x_i)]) \leq W$. To ensure this, it is sufficient to see that if one consider a leaf B_j of the tree-decomposition, its size is at most $W + 2$. In this bag B_j, consider a vertex x which is not included in a neighboring bag. So, its neighborhood $N(x)$ is exactly $B_j \backslash \{x\}$ whose the size is at most $W + 1$, and thus, the treewidth of $G[N(x)]$ is at most W. A such vertex x can be considered as the first one in the ordering σ. A similar reasoning can be extended to all the bags of such a tree-decomposition, and also, to all the vertices of G to complete the ordering σ. Therefore, the ordering σ is then a $\mathcal{C}_W^1$ scheme of G.

Now, assume that the property is true for all k', such as $0 \leq k' < k$, that is, if $tw(G) \leq W + k'$, then $G \in \mathcal{C}_W^{k'}$. We prove that this property also holds for k. Let $G = (V, E)$ be a graph such as $tw(G) \leq W + k$. So, there is a tree-decomposition of G such as, for all bags B_j, $|B_j| \leq W + k + 1$. By the same reasoning as for $k = 1$, we can define an ordering on $\sigma = [x_1, x_2, \ldots, x_n]$ on V, such that, $|N_\sigma^+(x_i)| \leq W + k$, and then, $\forall i, 1 \leq i \leq n$, $tw(G[N_\sigma^+(x_i)]) \leq W + k - 1$. So, using the induction hypothesis, $G \in \mathcal{C}_W^{k-1}$ and by definition, a such ordering σ is a $\mathcal{C}_W^k$ scheme of G, and thus $G \in \mathcal{C}_W^k$. QED.

This basic property on $\mathcal{C}_W^k$ classes thus allows us to consider that this type of class makes it possible to generalize the notion of treewidth of graph using the additional parameter k. Indeed, we can reduce any graph of treewidth at most w to graphs of parameters W and k, with $W + k = w$, knowing that the graphs of treewidth at most w already belong to the base class $\mathcal{C}_w^0$. And therefore, we can even estimate that the notion of tree decomposition of a given treewidth is generalized by classes with double parameters W and k.

The following property shows the hereditary structure of all these classes, that is, for a graph of a given class $\mathcal{C}_W^k$, all its subgraphs belong to this class. We will see later that this property is of interest from an algorithmic point of view.

Theorem 2. $\forall W \geq 0$, $\forall k \geq 0$, $\forall G = (V, E) \in \mathcal{C}_W^k$, then $\forall X \subseteq V, G[X] \in \mathcal{C}_W^k$.

Proof. We prove this property by induction on k. For $k = 0$, the property holds since graphs that belong to $\mathcal{C}_W^0$ are graphs whose the treewidth is bounded by W and since it is well known that every subgraph of a graph of a given treewidth has at most the same treewidth. So every subgraph $G[X]$ of a graph $G \in \mathcal{C}_W^0$ belongs too to $\mathcal{C}_W^0$.

Suppose now that this property holds $\forall k', 0 \leq k' < k$. Let $G = (V, E)$ belonging to $\mathcal{C}_W^k$ and $x \in V$. We prove that $G[V \setminus \{x\}] = G'$ belongs to $\mathcal{C}_W^k$. Let the ordering $\sigma = [x_1, x_2, \ldots, x_n]$ be the associated $\mathcal{C}_W^k$ scheme of G. Consider now σ' the ordering $[x_1, \ldots x_{i-1}, x_{i+1}, \ldots x_n]$ of $V \setminus \{x\}$ where $x = x_i$. We show that the ordering σ' is a $\mathcal{C}_W^k$ scheme of G':

- $\forall j, i + 1 \leq j \leq n$, it is clear that with respect to the orderings σ and σ', $G[N^+(x_j)] = G'[N^+(x_j)]$ and since G belongs to $\mathcal{C}_W^k$, then $G'[N^+(x_j)] = G[N^+(x_j)]$ belongs to $\mathcal{C}_W^{k-1}$.
- $\forall j, 1 \leq j \leq i - 1$, it is clear that with respect to the orderings σ and σ', $G'[N^+(x_j)]$ is a (non-necessarily strict) subgraph of $G[N^+(x_j)]$ which belongs to $\mathcal{C}_W^{k-1}$. So, by induction hypothesis, every subgraph of a graph belonging to $\mathcal{C}_W^{k-1}$ belongs to $\mathcal{C}_W^{k-1}$. So, $G'[N^+(x_j)]$ belongs to $\mathcal{C}_W^{k-1}$.

So, $\forall j \neq i, 1 \leq j \leq n$ with respect to the ordering σ', the induced subgraph $G'[N^+(x_j)]$ belongs to $\mathcal{C}_W^{k-1}$, and then, the ordering σ' is a $\mathcal{C}_W^k$ scheme of G' and thus, $G' \in \mathcal{C}_W^k$.QED.

This second theorem is useful for testing membership of a $\mathcal{C}_W^k$ class, which is the question addressed in the next section.

4 Recognition of Graphs of a Given Class $\mathcal{C}_W^k$

The question of recognizing graphs of a given class $\mathcal{C}_W^k$ is now asked. We show first that this is feasible, and we will then see that the complexity is of course linked to the assumptions we can make about the parameter W. Indeed, determining the treewidth of a graph is well known to be an NP-hard problem [3]. So, to know whether a graph is in a $\mathcal{C}_W^0$ class, we first have to determine whether its treewidth is at most W. On the other hand, if W is a constant, we can rely on efficient algorithms, since the problem is then known to be tractable (i.e. its time complexity is polynomial [4]). In the sequel, let $T(G, W)$ denotes the time complexity to check if the treewidth of a graph G is bounded by W. The value of $T(G, W)$ can be different, depending on whether the value of W is bounded by a constant or not.

Consider an integer W and a graph G. Recognizing whether G belongs to the class $\mathcal{C}_W^0$ can be done directly by calculating the treewidth of G and checking whether it is less than or equal to W. Now consider an integer k. If $k = 1$, a necessary condition for G to belong to $\mathcal{C}_W^1$ is that there exists a first vertex x_1 of G such that its neighborhood has a treewidth less than or equal to W. If no vertex satisfies this condition, then G does not belong to $\mathcal{C}_W^1$. On the other hand, if such a vertex does exist, then it can be the first vertex in a $\mathcal{C}_W^1$ scheme σ. Furthermore, if G belongs to $\mathcal{C}_W^1$, by the heredity property presented in the previous section, the subgraph of $G[V \setminus \{x_1\}]$ belongs to $\mathcal{C}_W^1$ and there exists a vertex x_2 such that its neighborhood in $G[V \setminus \{x_1\}]$ induces a graph whose the treewidth is less than or equal to W. This process can be repeated until all the vertices in G have been eliminated. If all the vertices of G are indeed eliminated, then G belongs to $\mathcal{C}_W^1$. If all the vertices of G are not eliminated, then G does not belong to $\mathcal{C}_W^1$. This approach defines an algorithm whose complexity is bounded by $O(n^2 \times T(G, W))$ since the $\mathcal{C}_W^1$ scheme σ which is computed needs to order the n vertices of G, and for each one, we have to check for the treewidth of at most n sugraphs.

Such an approach can be generalized to other values of k to define the algorithm called *RecoG-CkW* for the recognition of graphs of the class $\mathcal{C}_W^k$. If an input graph belongs to $\mathcal{C}_W^k$, this algorithm returns *True* and an associated scheme σ, otherwise, it returns *False*. In this algorithm, to simplify notations, we consider $N_u(x)$ to denote the subset of unnumbered vertices appearing in the neighbourhood of a vertex x in G. Note that at line 6, the recursive call of *RecoG-CkW* in the condition matches to the test $G[N_u(x)] \in \mathcal{C}_W^{k-1}$.

Theorem 3. *For $k \geq 0$, if $G \in \mathcal{C}_W^k$, the algorithm RecoG-CkW returns True and a $\mathcal{C}_W^k$ scheme (if $k > 0$), otherwise, it returns False.*

Proof. We prove this result by induction on k. This property trivially holds for $k = 0$ since the algorithm is limited to testing the treewidth of G (line 2).

Now, consider a graph $G = (V, E)$ and $k > 0$, and assume that *RecoG-CkW* is correct when calling with a parameter k' such that $k' < k$. If $G \in \mathcal{C}_W^k$, necessarily, G possesses a vertex x such that $G[N_u(x)]$ belongs to $\mathcal{C}_W^{k-1}$. This

Algorithm 1. RecoG-CkW

Input: $G = (V, E)$: **graph**; $k \geq 0$, W : **integer**;
Output: σ : **ordering**; CkW : **boolean**;
// if $G \in \mathcal{C}_W^k$, returns True and σ is the scheme,
// else, returns False
1: **if** $k = 0$ **then**
2:    $CkW \leftarrow (tw(G) \leq W)$;
3: **else**
4:    $\sigma \leftarrow [\ ]$; $CkW \leftarrow True$; $i \leftarrow 1$;
5:    **while** CkW **and** $i \leq n$ **do**
6:       **if** $\exists x \in V$: RecoG-CkW$(G[N_u(x)], k-1, W)$ **then**
7:          $\sigma[i] \leftarrow x$;
8:          $i \leftarrow i + 1$;
9:          $V \leftarrow V \setminus \{x\}$;
10:       **else**
11:          $CkW \leftarrow False$;
12:       **end if**
13:    **end while**
14: **end if**
15: **return** CkW;

vertex will be the first one in a $\mathcal{C}_W^k$ scheme and by induction hypothesis, the recursive call RecoG-CkW$(G[N_u(x)], k-1, W)$ returns True. Using the heredity of graphs belonging to $\mathcal{C}_W^k$, the same reasoning can be applied to the subgraph induced by $G[V \setminus \{x\}]$, And so on, until all the vertices of G are deleted. This will lead to the generation of an ordering σ on V, and ends the loop with CkW being true, which will then be the value returned by *RecoG-CkW*.

On the other hand, if $G \notin \mathcal{C}_W^k$, no $\mathcal{C}_W^{k-1}$ scheme can be find. So, during the **while** loop, necessarily, after some deletions of vertices x, none of the vertices belonging to the resulting set V such that $G[N_u(x)] \in \mathcal{C}_W^{k-1}$ will be found, and then, the **while** loop stops after the assignment of CkW to False, which will be the value returned by *RecoG-CkW*.

For the termination of the algorithm, simply note that with each recursive call, the value of k decreases strictly, so if the algorithm does not stop before then, the case $k = 0$ will be reached and the execution will stop. Thus, in all cases, this algorithm terminates and the returned result is correct. QED.

The following property indicates the time complexity of this algorithm.

Theorem 4. *The time complexity of the algorithm RecoG-CkW is $O(n^{2k} \times T(G, W))$.*

Proof. The worst case time complexity occurs for graphs that belongs to $\mathcal{C}_W^k$. In this case, the **while** loop will be executed n times. In this case, the cost of the algorithm is given by the cost of the condition in line 6. The condition contained in the *if* conditional statement potentially imposes the execution of an additional loop to find a vertex verifying the condition, and this loop will run from the current value i to n, which is bounded by n. This leads to a multiplicative factor n which is cumulated to that of the **while** loop, giving a multiplicative factor equal to n^2. So, find a suitable vertex x induces a factor n^2 that must be considered with the cost of the recursive call to $RecoG\text{-}CkW$ associated to the condition $G[V\setminus x] \in \mathcal{C}_W^{k-1}$. So, the time complexity $C(k)$ is given by a recurrence relation based on k:

- For $k=0$, the cost $C(k) = C(0)$ is $O(T(G,W))$ because the time complexity in line 2 is $O(T(G,W))$
- For $k \geq 1$, the cost $C(k)$ is $O(n^2 \times C(k-1))$

Thus, from this recurrence relation, we can easily prove that the time complexity of the algorithm $RecoG\text{-}CkW$ is $O(n^{2k} \times T(G,W))$. QED.

So, if W and k are defined as constant, the time complexity of the algorithm $RecoG\text{-}CkW$ is then polynomial since we know a linear time algorithms for checking if the treewidth of a graph is equal to a given integer W [4], and then, $T(G,W) \in O(|G|)$.

Corollary 1. *If W is bounded by a constant, the time complexity of $RecoG\text{-}CkW$ is in $O(n^{2k} \times |G|)$.*

5 Class $\mathcal{C}_W^k$ and Tractability of the Maximum Clique Problem

We now consider the problem of finding a maximum size clique in a graph. We will show that this problem, although known to be NP-hard, can be solved in polynomial time on graph classes of type $\mathcal{C}_W^k$, under the assumption that the base class $\mathcal{C}_W^0$ has a parameter W which is a constant while k is also a constant. To solve this problem, we define the algorithm $MaxCliq\text{-}CkW$. We show its correctness and, then, we evaluate its time complexity. We can then see that, under the assumption that the instance processed as input belongs to a given class $\mathcal{C}_W^k$ for which k and W are constants, then this algorithm has a polynomial time complexity.

At the first step of the algorithm (lines 1–2), if $k=0$, a maximum size clique of G is found, exploiting a tree-decomposition of G whose width is at most W.

For the general case, i.e. $k > 0$, a $\mathcal{C}_W^k$ scheme σ of G is computed with a call to $RecoG\text{-}CkW$. Using this ordering, every vertex x is checked to find a maximum size clique in the associated subgraph $G[N_\sigma^+(x)]$, knowing that this subgraph belongs to $\mathcal{C}_W^{k-1}$. After the **for** loop, we are ensured that K contains the maximum size clique of the graph G.

Theorem 5. *For $k \geq 0$, if $G \in \mathcal{C}_W^k$, the algorithm $MaxCliq$ returns a maximum size clique of G.*

Algorithm 2. MaxCliq

Input: $G = (V, E)$: **graph**; $k \geq 0$, W : **integer**;
Output: K : **set of vertices**;
// K max size clique of $G \in \mathcal{C}_W^k$
1: **if** $k = 0$ **then**
2: $\quad K \leftarrow$ Maximum Size Clique of G;
3: **else**
4: $\quad$ Find a $\mathcal{C}_W^k$ scheme σ using $RecoG\text{-}CkW$;
5: $\quad K \leftarrow \emptyset$;
6: $\quad$ **for** $(i \leftarrow 1; i \leq n; i \leftarrow i + 1)$ **do**
7: $\quad\quad x \leftarrow \sigma[i]$;
8: $\quad\quad K_i \leftarrow \text{MaxCliq}(G[N_\sigma^+(x)], k-1, W)$;
9: $\quad\quad$ **if** $|K_i \cup \{x\}| > |K|$ **then**
10: $\quad\quad\quad K \leftarrow K_i \cup \{x\}$;
11: $\quad\quad$ **end if**
12: $\quad$ **end for**
13: **end if**
14: **return** K;

Proof. We prove this result by induction on k.

For $k = 0$, using a basic algorithm, $MaxCliq$ finds a maximum size clique in the graph G assuming that its treewidth is bounded by W.

Now, consider the graph G and $k > 0$, and assume that $MaxCliq$ is correct when it is called with a parameter k' such that $k' < k$. As $G \in \mathcal{C}_W^k$, and since σ is a $\mathcal{C}_W^k$ scheme, for all i, $1 \leq i \leq n$, each subgraph $G[N_\sigma^+(x_i)]$ belongs to $\mathcal{C}_W^{k-1}$ and therefore, applying the inductive hypothesis about the correction of MaxCliq(), each recursive call MaxCliq$(G[N_\sigma^+(x_i)], k-1, W)$ returns a maximum size clique of $G[N_\sigma^+(x_i)]$ which is assigned to K_i (line 8). Thus, $K_i \cup \{x_i\}$ is a maximum size clique of $G[N_\sigma^+(x_i) \cup \{x_i\}]$. Now, consider K', a maximum size clique of G, and let $x_j \in K'$, be the first vertex of K' with respect to the ordering σ. Necessarily, $K' \subseteq (N_\sigma^+(x_j) \cup \{x_j\})$. Therefore, after the call MaxCliq$(G[N_\sigma^+(x_j)], k-1, W)$, we have $|K' \cup \{x_j\}| = |K_j \cup \{x_j\}|$. Note that here, we consider sizes of cliques rather than cliques themselves because it can exists several (so different) maximum size cliques in a given (sub)graph. So, $MaxCliq$ is correct when it is called with the parameters G, k and W. Finally, for the termination of the algorithm, as for the algorithm $RecoG\text{-}CkW$, we can see that with each recursive call, the value of k decreases strictly, and so the case $k = 0$ will be reached and the execution terminated. QED.

We evaluate the complexity of this algorithm by assuming that the time complexity of computing a clique of maximum size in a graph of treewidth W is given by $TC(G, W)$.

Theorem 6. *For $k \geq 0$, if $G \in \mathcal{C}_W^k$, the time complexity of $MaxCliq$ is $O(k.n^{2k} \times T(G, W) + n^k \times TC(G, W))$.*

Proof. For $k = 0$, assume that the time complexity is bounded by $TC(G, W)$ which is the time needed to compute a clique of maximum size in a graph G whose treewidth is W (line 2).

Now, assume $k \geq 1$ and let $Cliq(k)$ be the time complexity of $MaxCliq$ for a value k. In line 4, the time complexity of the call to $RecoG\text{-}CkW$ is $O(n^{2k} \times T(G, W))$. Next, the instructions in lines 7 to 10 are executed n times. In these lines, it is sufficient to consider the running time of the line 8 which is $Cliq(k-1)$ since the recursive call of $MaxCliq$ considers $k - 1$ as input. So, the time complexity $Cliq(k)$ is given by a recurrence relation based on k:

- For $k = 0$, $Cliq(k) = Cliq(0)$ is $O(TC(G, W))$
- For $k \geq 1$, $Cliq(k) = n^{2k} \times T(G, W) + n.Cliq(k-1)$

The solution for this relation is :

$$Cliq(k) = n^k \times (T(G, W) \times \Sigma_{i=1}^{k} n^i + TC(G, W))$$

which can easily be proved by induction. So, the time complexity of the algorithm $MaxCliq$ can be bounded by $O(k.n^{2k} \times T(G, W) + n^k \times TC(G, W))$. QED.

So, if W and k are defined as constant, the time complexity of the algorithm $MaxCliq$ is then polynomial since we know linear time algorithms for checking the treewidth and looking for its maximal size clique, i.e. $T(G, W)$ and $TC(G, W)$ can be replaced by $|G|$:

Corollary 2. *If W is bounded by a constant, the time complexity of $MaxCliq$ is in $O(k.n^{2k} \times |G|)$.*

We can see that the parameter k plays a fundamental role here, as the practical tractability of the clique problem will be bound in this approach by the value of k. Moreover, it is also well known that difficult instances of the clique problem, but not only of course, have treewidths not bounded by constants. Nevertheless, we show in the next section that such instances can belong to classes of type $\mathcal{C}_W^k$ for which the value of k is small.

6 Classes $\mathcal{C}_W^k$ and Graphs of Unbounded Treewidth

It is well known that numerous hard problems, such as the Clique problem for example, are simple to solve as long as their treewidth is bounded by a constant, while they remain hard for the class of graphs of unbounded treewidth. But we show here that classes of graphs of unbounded treewidth can exist, but for which the parameter k can be equal to 1, and the treewidth W can be equal to a small value constant. This is the case of planar graphs for which we know that their treewidth belongs to $\Theta(\sqrt{n})$ [1]:

Proposition 1. *Planar graphs belong to class $\mathcal{C}_2^1$.*

Proof. To show that planar graphs belong to the class $\mathcal{C}_2^1$, consider a planar graph $G = (V, E)$. We show that for such graphs, there exists an ordering $\sigma = [x_1, x_2, \ldots, x_n]$ of V such that for $i = 1, 2, \ldots, n$, the subgraph $G[N_\sigma^+(x_i)]$ belongs to $\mathcal{C}_3^0$, that is the class of graphs with a treewidth of 2 or less.

It is well known that for any planar graph, there exists a vertex whose degree is at most 5 (a folklore property derived from Euler's formula). So, consider such a vertex x_1 in G. Thus, $|N(x_1)| \leq 5$, $G[N(x_1)]$ cannot contain K_4 because, together with x_1, it would form the clique K_5 which is not possible since G is planar. Moreover, since G is planar and x_1 has 5 neighbors, the densest graph $G[N(x_1)]$ is the one with 7 edges, formed by 1 vertex of degree 4, 2 vertices of degree 3 and 2 vertices of degree 2. This graph has a treewidth equal to 2. Such a graph, to which x1 is added, is isomorphic to the graph given in Fig. 2. Now consider the subgraph of G defined by $G[V \backslash \{x_1\})]$. Since G is planar, $G[V \backslash \{x_1\})]$ is planar too. So G has a vertex x_2, whose degree is at most 5, and the same reasoning as above can be used to show that the subgraph $G[N(x_2) \backslash \{x_1\}]$ belongs to $\mathcal{C}_3^0$. More generally, we can define an ordering $\sigma = [x_1, x_2, \ldots, x_n]$ of V which is a $\mathcal{C}_3^1$ scheme, allowing to prove that G belongs to $\mathcal{C}_3^1$. QED.

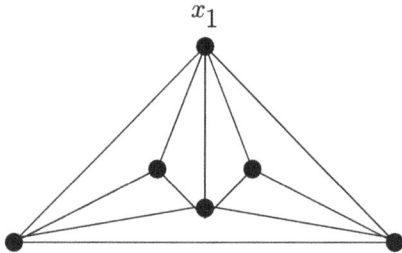

Fig. 2. Densest possible planar graph for which x_1 is of degree 5.

If this result seems significant from a theoretical point of view, it is less relevant when we refer to the problem of the clique of maximum size. Indeed, it is well known that a planar graph cannot possess a clique of size greater than or equal to 5, and thus we then have a trivial polynomial algorithm enumerating all subsets of 4 vertices or less to find a clique of maximum size. So, it is useful to show that there exists other classes of graphs, non-planar for example, for which we have similar properties in terms of parameter k while keeping W equal to a constant. This is the case for the complete bipartite graphs $K_{n,n}$ whose treewidth is n. Indeed, for $K_{n,n}$, it is easy to see that there is a tree-decomposition defined by n bags, each one being associated to one vertex of one on the two sets, and including also the n vertices of the other set. Its treewidth is then n. Moreover, this width is minimal because any optimal triangulation ordering minimizing the size of the largest induced clique (and therefore the width) will have to start with any vertex from $K_{n,n}$ and since these vertices have the same degree, i.e. n, the first vertex will be include in a clique of size $n + 1$, and therefore the width will be at least equal to n. Despite their unbounded treewidth, these graphs belong to the $\mathcal{C}_0^1$ class because the neighborhood of any vertex is a set vertices between which there are no edge. Consequently, the treewidth of this neighborhood is always 0. And this applies regardless of the ordering of the vertices. So every complete bipartite graph on n vertices belongs to $\mathcal{C}_0^1$ while its treewidth is $\frac{n}{2}$.

7 Discussion and Conclusion

As we have seen, the definition of $\mathcal{C}_W^k$ classes makes it possible to design polynomial time algorithms for the Clique problem. This is all the more interesting in that, unlike the use of treewidth, for instances where treewidth is not bounded by a constant, the use of this double parameterization allows some of these instances to be processed in polynomial time. So, a natural question arises. *"Can this definition of a new class of graphs be extended to handle other difficult problems?"* We know that for the case of bounded treewidth, many problems can become tractable because we can obtain efficient approaches for hard problems thanks to FPT algorithms.

Unfortunately, it is not easy to do. Indeed, if we take fairly related graph problems, such as *Independent Set*[1] or *Vertex Cover*[2], the approach used here will not work directly.

In fact, the $\mathcal{C}_W^k$ class seems suited to problems such as Clique due to the definition of the class. Indeed, by constructing a solution as seen in the *MaxCliq-CkW* algorithm, every solution to the problem has a first vertex x_i such that a solution is a set of vertices that belongs to N_σ^+, that is in the neighborhood of this first vertex. And consequently, a global solution is defined in adding x_i to a maximum size clique of $G[N_\sigma^+]$, this subgraph being tractable since it satisfies the right properties. Thus, the existence of a solution is intrinsically linked to the construction of the graph class, which is based on the vertex neighborhood.

On the other hand, if we consider the Independent Set problem, every solution is made up of vertices that are precisely absent from the neighborhood of a first vertex x_i. So, treatment using a similar approach to that based on $\mathcal{C}_W^k$ classes would consist in defining this type of class differently, by considering not the neighborhood of vertices, but the vertices that are not in the neighborhood. But what is possible for a problem like Independent Set is not necessarily so for other problems. For example, if we take Vertex Cover, then we need to be sure not only of the nature of the considered σ ordering, but also of the way in which solutions can be constructed. For a given vertex x_i, which vertices should be related to x_i to ensure that a partial solution in this set of vertices can be extended by adding x_i? While in the case of the Clique problem, this is obvious when considering the neighborhood of x_i, in the case of Vertex Cover, the question remains completely open at this stage.

---

[1] Given a graph G and an integer k, does G have a subgraph of k or more vertices that has no edges.

[2] Given a graph G and an integer k, does G have a subset of k or fewer vertices such that every edge has at least one vertex in that subset.

One possible approach is based on the notion of polynomial transformation, transforming any problem to the Clique problem. However, such an approach would only be possible under the assumption that the structural properties related to the treewidth of the graph are preserved. In another context, this possibility was explored using the notion of "guarded reductions", which are reductions defined by first-order logic formulas showing that guarded reductions preserve bounded treewidth [9].

Acknowledgments. The author would like to thank the anonymous reviewers and Cyril Terrioux for their critical reading of this paper. This work has been funded by the French Agence Nationale de la Recherche, reference Massal'IA ANR-19-CHIA-0013-01.

Disclosure of Interests. The author has no competing interests to declare that are relevant to the content of this article.

References

1. Alon, N., Seymour, P., Thomas, R.: A separator theorem for graphs with an excluded minor and its applications. In: Proceedings of the 22nd Annual ACM Symposium on Theory of Computing, 13–17 May 1990, Baltimore, Maryland, USA, pp. 293–299 (1990)
2. Arnborg, S.: Efficient algorithms for combinatorial problems with bounded decomposability - a survey. BIT **25**(1), 1–23 (1985)
3. Arnborg, S., Corneil, D., Proskuroswk, A.: Complexity of finding embeddings in a k-tree. SIAM J. Discret. Math. **8**(2), 277–284 (1987)
4. Hans, L.: Bodlaender: a linear-time algorithm for finding tree-decompositions of small treewidth. SIAM J. Comput. **25**(6), 12–75 (1996)
5. Courceller, B.: The monadic second-order logic of graphs. I. Recognizable sets of finite graph. Inf. Comput. **85**(1), 1305–1317 (1990)
6. Downey, R.G., Fellows, M.R.: Parameterized Complexity. Monographs in Computer Science. Springer, Heidelberg (1999)
7. Jégou, P.: Decomposition of domains based on the micro-structure of finite constraint satisfaction problems. In: AAAI Proceedings, Washington DC, USA, pp. 731–736 (1993)
8. Johnson, D.S., Trick, M.A.: Cliques, Coloring, and Satisfiability: Second DIMACS Implementation Challenge. DIMACS Series in Discrete Mathematics and Theoretical Computer Science, vol. 26. AMS (1996)
9. Mitchell, D.: Guarded constraint models define treewidth preserving reductions. In: Schiex, T., de Givry, S. (eds.) CP 2019. LNCS, vol. 11802, pp. 350–365. Springer, Cham (2019). https://doi.org/10.1007/978-3-030-30048-7_21
10. Robertson, N., Seymour, P.D.: Graph minors II: algorithmic aspects of treewidth. Algorithms **7**, 309–322 (1986)

Principled Approaches for Learning to Defer with Multiple Experts

Anqi Mao[1], Mehryar Mohri[1,2], and Yutao Zhong[1]

[1] Courant Institute, New York, NY 10012, USA
{aqmao,mohri,yutao}@cims.nyu.edu
[2] Google Research, New York, NY 10011, USA
mohri@google.com

Abstract. We present a study of surrogate losses and algorithms for the general problem of *learning to defer with multiple experts*. We first introduce a new family of surrogate losses specifically tailored for the multiple-expert setting, where the prediction and deferral functions are learned simultaneously. We then prove that these surrogate losses benefit from strong $\mathcal{H}$-consistency bounds. We illustrate the application of our analysis through several examples of practical surrogate losses, for which we give explicit guarantees. These loss functions readily lead to the design of new learning to defer algorithms based on their minimization. While the main focus of this work is a theoretical analysis, we also report the results of several experiments on SVHN and CIFAR-10 datasets.

Keywords: Learning to defer · Learning theory · Consistency

1 Introduction

In many real-world applications, expert decisions can complement or significantly enhance existing models. These experts may consist of humans possessing domain expertise or more sophisticated albeit expensive models. For instance, contemporary language models and dialog-based text generation systems have exhibited susceptibility to generating erroneous information, often referred to as *hallucinations*. Thus, their response quality can be substantially improved by deferring uncertain predictions to more advanced or domain-specific pre-trained models. This particular issue has been recognized as a central challenge for large language models (LLMs) [14,76]. Similar observations apply to other generation systems, including image or video generation, as well as learning models used in various applications such as image classification, image annotation, and speech recognition. Thus, the problem of *learning to defer with multiple experts* has become increasingly critical in applications.

The concept of *learning to defer* can be traced back to the original work on *learning with rejection* or *abstention* based on confidence thresholds [10,18,19,33] (see also [29,62,67,80,81]), rejection or abstention functions [20–22] (see also [15,16,53–55]), or *selective classification* [24,25,27,28,78], and other methods

[1,35,87]. In these studies, either the cost of abstention is not explicit or it is chosen to be a constant.

However, a constant cost does not fully capture all the relevant information in the deferral scenario. It is important to take into account the quality of the expert, whose prediction we rely on. These may be human experts as in several critical applications [8,36,40,71]. To address this gap, [46] incorporated the human expert's decision into the cost and proposed the first *learning to defer (L2D)* framework, which has also been examined in [38,64,65,79]. [58] proposed the first *Bayes-consistent* [9,68,83] surrogate loss for L2D, and subsequent work [44,66] further improved upon it. Another Bayes-consistent surrogate loss in L2D is the one-versus-all loss proposed by [75] that is also studied in [17] as a special case of a general family of loss functions. An additional line of research investigated post-hoc methods [59,63], where [63] proposed an alternative optimization method between the predictor and rejector, and [59] provided a correction to the surrogate losses in [58,75] when they are underfitting. Finally, L2D or its variants have been adopted or studied in various other scenarios [23,26,32,34,44,57,60,69,85].

All the studies mentioned so far mainly focused on learning to defer with a single expert. Most recently, [74] highlighted the significance of *learning to defer with multiple experts* [11,31,37,38,70] and extended the surrogate loss in [58,75] to accommodate the multiple-expert setting, which is currently the only work to propose Bayes-consistent surrogate losses in this scenario. They further showed that a mixture of experts (MoE) approach to multi-expert L2D proposed in [31] is not consistent. More recently, [47] examined a two-stage scenario for learning to defer with multiple experts, which is crucial for various applications. They developed new surrogate losses for this scenario and demonstrated that these are supported by stronger consistency guarantees-specifically, $\mathcal{H}$-consistency bounds as introduced below-implying their Bayes consistency.

Meanwhile, recent work by [4,5] introduced new consistency guarantees, called $\mathcal{H}$-consistency bounds, which they argued are more relevant to learning than Bayes-consistency since they are hypothesis set-specific and non-asymptotic. $\mathcal{H}$-consistency bounds are also stronger guarantees than Bayes-consistency. They established $\mathcal{H}$-consistent bounds for common surrogate losses in standard classification (see also [48,49,86]). This naturally raises the question: can we design deferral surrogate losses that benefit from these more significant consistency guarantees?

Our Contributions. We study the general framework of learning to defer with multiple experts. We first introduce a new family of surrogate losses specifically tailored for the multiple-expert setting, where the prediction and deferral functions are learned simultaneously (Sect. 3). Next, we prove that these surrogate losses benefit from $\mathcal{H}$-consistency bounds (Sect. 4). This implies, in particular, their Bayes-consistency. We illustrate the application of our analysis through several examples of practical surrogate losses, for which we give explicit guarantees. These loss functions readily lead to the design of new learning to defer algorithms based on their minimization. Our $\mathcal{H}$-consistency bounds incorporate

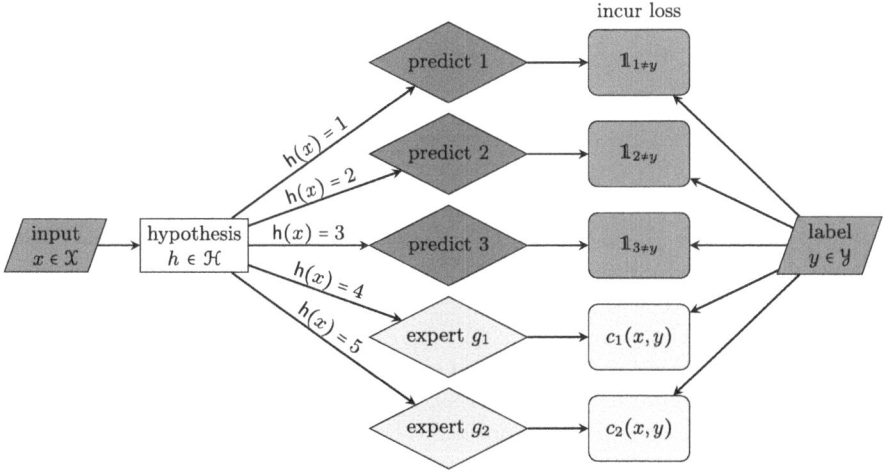

Fig. 1. Illustration of the scenario of learning to defer with multiple experts ($n = 3$ and $n_e = 2$).

a crucial term known as the *minimizability gap*. We show that this makes them more advantageous guarantees than bounds based on the approximation error (Sect. 5). We further demonstrate that our $\mathcal{H}$-consistency bounds can be used to derive generalization bounds for the minimizer of a surrogate loss expressed in terms of the minimizability gaps (Sect. 6). While the main focus of this work is a theoretical analysis, we also report the results of several experiments with SVHN and CIFAR-10 datasets (Sect. 7).

We give a more detailed discussion of related work in Appendix A. We start with the introduction of preliminary definitions and notation needed for our discussion of the problem of learning to defer with multiple experts.

2 Preliminaries

We consider the standard multi-class classification setting with an input space $\mathcal{X}$ and a set of $n \geq 2$ labels $\mathcal{Y} = [n]$, where we use the notation $[n]$ to denote the set $\{1, \ldots, n\}$. We study the scenario of *learning to defer with multiple experts*, where the label set $\mathcal{Y}$ is augmented with n_e additional labels $\{n+1, \ldots, n+n_e\}$ corresponding to n_e pre-defined experts $g_1, \ldots, g_{n_e}$, which are a series of functions mapping from $\mathcal{X} \times \mathcal{Y}$ to $\mathbb{R}$. In this scenario, the learner has the option of returning a label $y \in \mathcal{Y}$, which represents the category predicted, or a label $y = n+j$, $1 \leq j \leq n_e$, in which case it is *deferring* to expert g_j.

We denote by $\overline{\mathcal{Y}} = [n + n_e]$ the augmented label set and consider a hypothesis set $\mathcal{H}$ of functions mapping from $\mathcal{X} \times \overline{\mathcal{Y}}$ to $\mathbb{R}$. The prediction associated by $h \in \mathcal{H}$ to an input $x \in \mathcal{X}$ is denoted by $\mathsf{h}(x)$ and defined as the element in $\overline{\mathcal{Y}}$ with the highest score, $\mathsf{h}(x) = \mathrm{argmax}_{y \in [n+n_e]} h(x, y)$, with an arbitrary but fixed deterministic strategy for breaking ties. We denote by $\mathcal{H}_{\mathrm{all}}$ the family of all measurable functions.

The *deferral loss function* $\mathsf{L}_{\mathrm{def}}$ is defined as follows for any $h \in \mathcal{H}$ and $(x,y) \in \mathcal{X} \times \mathcal{Y}$:

$$\mathsf{L}_{\mathrm{def}}(h,x,y) = \mathbb{1}_{\mathsf{h}(x) \neq y}\mathbb{1}_{\mathsf{h}(x) \in [n]} + \sum_{j=1}^{n_e} c_j(x,y) \mathbb{1}_{\mathsf{h}(x) = n+j} \quad (1)$$

Thus, the loss incurred coincides with the standard zero-one classification loss when $\mathsf{h}(x)$, the label predicted, is in $\mathcal{Y}$. Otherwise, when $\mathsf{h}(x)$ is equal to $n+j$, the loss incurred is $c_j(x,y)$, the cost of deferring to expert g_j. We give an illustration of the scenario of learning to defer with three classes and two experts ($n = 3$ and $n_e = 2$) in Fig. 1. We will denote by $\underline{c}_j \geq 0$ and $\overline{c}_j \leq 1$ finite lower and upper bounds on the cost c_j, that is $c_j(x,y) \in [\underline{c}_j, \overline{c}_j]$ for all $(x,y) \in \mathcal{X} \times \mathcal{Y}$. There are many possible choices for these costs. Our analysis is general and requires no assumption other than their boundedness. One natural choice is to define cost c_j as a function relying on expert g_j's accuracy, for example $c_j(x,y) = \alpha_j \mathbb{1}_{\mathsf{g}_j(x) \neq y} + \beta_j$, with $\alpha_j, \beta_j > 0$, where $\mathsf{g}_j(x) = \operatorname{argmax}_{y \in [n]} g_j(x,y)$ is the prediction made by expert g_j for input x.

Given a distribution $\mathcal{D}$ over $\mathcal{X} \times \mathcal{Y}$, we will denote by $\mathcal{E}_{\mathsf{L}_{\mathrm{def}}}(h)$ the expected deferral loss of a hypothesis $h \in \mathcal{H}$,

$$\mathcal{E}_{\mathsf{L}_{\mathrm{def}}}(h) = \mathbb{E}_{(x,y) \sim \mathcal{D}} [\mathsf{L}_{\mathrm{def}}(h,x,y)], \quad (2)$$

and by $\mathcal{E}^*_{\mathsf{L}_{\mathrm{def}}}(\mathcal{H}) = \inf_{h \in \mathcal{H}} \mathcal{E}_{\mathsf{L}_{\mathrm{def}}}(h)$ its infimum or best-in-class expected loss. We will adopt similar definitions for any surrogate loss function L:

$$\mathcal{E}_{\mathsf{L}}(h) = \mathbb{E}_{(x,y) \sim \mathcal{D}} [\mathsf{L}(h,x,y)], \quad \mathcal{E}^*_{\mathsf{L}}(\mathcal{H}) = \inf_{h \in \mathcal{H}} \mathcal{E}_{\mathsf{L}}(h). \quad (3)$$

3 General Surrogate Losses

In this section, we introduce a new family of surrogate losses specifically tailored for the multiple-expert setting starting from first principles.

The scenario we consider is one where the prediction (first n scores) and deferral functions (last n_e scores) are learned simultaneously. Consider a hypothesis $h \in \mathcal{H}$. Note that, for any $(x,y) \in \mathcal{X} \times \mathcal{Y}$, if the learner chooses to defer to an expert, $\mathsf{h}(x) \in \{n+1, \ldots, n+n_e\}$, then it does not make a prediction of the category, and thus $\mathsf{h}(x) \neq y$. This implies that the following identity holds:

$$\mathbb{1}_{\mathsf{h}(x) \neq y} \mathbb{1}_{\mathsf{h}(x) \in \{n+1, \ldots, n+n_e\}} = \mathbb{1}_{\mathsf{h}(x) \in \{n+1, \ldots, n+n_e\}}.$$

Using this identity and $\mathbb{1}_{\mathsf{h}(x) \in [n]} = 1 - \mathbb{1}_{\mathsf{h}(x) \in \{n+1, \ldots, n+n_e\}}$, we can write the first term of (1) as $\mathbb{1}_{\mathsf{h}(x) \neq y} - \mathbb{1}_{\mathsf{h}(x) \in \{n+1, \ldots, n+n_e\}}$. Note that deferring occurs if and only if one of the experts is selected, that is $\mathbb{1}_{\mathsf{h}(x) \in \{n+1, \ldots, n+n_e\}} = \sum_{j=1}^{n_e} \mathbb{1}_{\mathsf{h}(x) = n+j}$. Therefore, the deferral loss function can be written in the following form for any $h \in \mathcal{H}$ and

$(x, y) \in \mathcal{X} \times \mathcal{Y}$:

$$\mathsf{L}_{\text{def}}(h, x, y)$$

$$= \mathbb{1}_{h(x) \neq y} - \sum_{j=1}^{n_e} \mathbb{1}_{h(x) = n+j} + \sum_{j=1}^{n_e} c_j(x, y) \mathbb{1}_{h(x) = n+j}$$

$$= \mathbb{1}_{h(x) \neq y} + \sum_{j=1}^{n_e} \left(c_j(x, y) - 1 \right) \mathbb{1}_{h(x) = n+j}$$

$$= \mathbb{1}_{h(x) \neq y} + \sum_{j=1}^{n_e} \left(1 - c_j(x, y) \right) \mathbb{1}_{h(x) \neq n+j} + \sum_{j=1}^{n_e} \left(c_j(x, y) - 1 \right).$$

In light of this expression, since the last term $\sum_{j=1}^{n_e} (c_j(x, y) - 1)$ does not depend on h, if ℓ is a surrogate loss for the zero-one multi-class classification loss over the augmented label set $\overline{\mathcal{Y}}$, then L, defined as follows for any $h \in \mathcal{H}$ and $(x, y) \in \mathcal{X} \times \mathcal{Y}$, is a natural surrogate loss for L_{def}:

$$\mathsf{L}(h, x, y) = \ell(h, x, y) + \sum_{j=1}^{n_e} \left(1 - c_j(x, y) \right) \ell(h, x, n + j). \tag{4}$$

We will study the properties of the general family of surrogate losses L thereby defined. Note that in the special case where ℓ is the logistic loss and $n_e = 1$, that is where there is only one pre-defined expert, L coincides with the surrogate loss proposed in [15,54,58]. However, even for that special case, our derivation of the surrogate loss from first principle is new and it is this analysis that enables us to define a surrogate loss for the more general case of multiple experts and other ℓ loss functions. Our formulation also recovers the softmax surrogate loss in [74] when $\ell = \ell_{\log}$ and $c_j(x, y) = 1_{g_j(x) \neq y}$.

4 $\mathcal{H}$-Consistency Bounds for Surrogate Losses

Here, we prove strong consistency guarantees for a surrogate deferral loss L of the form described in the previous section, provided that the loss function ℓ it is based upon admits a similar consistency guarantee with respect to the standard zero-one classification loss.

$\mathcal{H}$-**Consistency Bounds.** To do so, we will adopt the notion of $\mathcal{H}$-*consistency bounds* recently introduced by [4,5] and also studied in [2,3,6,7,48–52,86]. These are guarantees that, unlike Bayes-consistency or excess error bound, take into account the specific hypothesis set $\mathcal{H}$ and do not assume $\mathcal{H}$ to be the family of all measurable functions. Moreover, in contrast with Bayes-consistency, they are not just asymptotic guarantees. In this context, they have the following form: $\mathcal{E}_{\mathsf{L}_{\text{def}}}(h) - \mathcal{E}^*_{\mathsf{L}_{\text{def}}}(\mathcal{H}) \leq f\left(\mathcal{E}_{\mathsf{L}}(h) - \mathcal{E}^*_{\mathsf{L}}(\mathcal{H}) \right)$, where f is a non-decreasing function, typically concave. Thus, when the surrogate estimation loss $\left(\mathcal{E}_{\mathsf{L}}(h) - \mathcal{E}^*_{\mathsf{L}}(\mathcal{H}) \right)$ is reduced to ϵ, the deferral estimation loss $\left(\mathcal{E}_{\mathsf{L}_{\text{def}}}(h) - \mathcal{E}^*_{\mathsf{L}_{\text{def}}}(\mathcal{H}) \right)$ is guaranteed to be at most $f(\epsilon)$.

Minimizability Gaps. A key quantity appearing in these bounds is the *minimizability gap* $\mathcal{M}_\ell(\mathcal{H})$ which, for a loss function ℓ and hypothesis set $\mathcal{H}$, measures the difference of the best-in-class expected loss and the expected pointwise infimum of the loss:

$$\mathcal{M}_\ell(\mathcal{H}) = \mathcal{E}_\ell^*(\mathcal{H}) - \mathbb{E}_x \left[\inf_{h \in \mathcal{H}} \mathbb{E}_{y|x} \left[\ell(h, x, y) \right] \right].$$

By the super-additivity of the infimum, since $\mathcal{E}_\ell^*(\mathcal{H}) = \inf_{h \in \mathcal{H}} \mathbb{E}_x \left[\mathbb{E}_{y|x} \left[\ell(h, x, y) \right] \right]$, the minimizability gap is always non-negative.

When the loss function ℓ only depends on $h(x, \cdot)$ for all h, x, and y, that is $\ell(h, x, y) = \Psi(h(x, 1), \ldots, h(x, n), y)$, for some function Ψ, then it is not hard to show that the minimizability gap vanishes for the family of all measurable functions: $\mathcal{M}_\ell(\mathcal{H}_{\text{all}}) = 0$ [68][lemma 2.5]. It is also null when $\mathcal{E}_\ell^*(\mathcal{H}) = \mathcal{E}_\ell^*(\mathcal{H}_{\text{all}})$, that is when the Bayes-error coincides with the best-in-class error. In general, however, the minimizability gap is non-zero for a restricted hypothesis set $\mathcal{H}$ and is therefore important to analyze. In Sect. 5, we will discuss in more detail minimizability gaps for a relatively broad case and demonstrate that $\mathcal{H}$-consistency bounds with minimizability gaps can often be more favorable than excess error bounds based on the approximation error.

The following theorem is the main result of this section.

Theorem 1 ($\mathcal{H}$-consistency bounds for score-based surrogates). *Assume that ℓ admits an $\mathcal{H}$-consistency bound with respect to the multi-class zero-one classification loss ℓ_{0-1}. Thus, there exists a non-decreasing concave function Γ with $\Gamma(0) = 0$ such that, for any distribution $\mathcal{D}$ and for all $h \in \mathcal{H}$, we have*

$$\mathcal{E}_{\ell_{0-1}}(h) - \mathcal{E}_{\ell_{0-1}}^*(\mathcal{H}) + \mathcal{M}_{\ell_{0-1}}(\mathcal{H}) \leq \Gamma \left(\mathcal{E}_\ell(h) - \mathcal{E}_\ell^*(\mathcal{H}) + \mathcal{M}_\ell(\mathcal{H}) \right).$$

Then, L admits the following $\mathcal{H}$-consistency bound with respect to L_{def}: for all $h \in \mathcal{H}$,

$$\mathcal{E}_{\mathsf{L}_{\text{def}}}(h) - \mathcal{E}_{\mathsf{L}_{\text{def}}}^*(\mathcal{H}) + \mathcal{M}_{\mathsf{L}_{\text{def}}}(\mathcal{H}) \leq \left(n_e + 1 - \sum_{j=1}^{n_e} \underline{c}_j \right) \Gamma \left(\frac{\mathcal{E}_{\mathsf{L}}(h) - \mathcal{E}_{\mathsf{L}}^*(\mathcal{H}) + \mathcal{M}_{\mathsf{L}}(\mathcal{H})}{n_e + 1 - \sum_{j=1}^{n_e} \overline{c}_j} \right). \tag{5}$$

Furthermore, constant factors $\left(n_e + 1 - \sum_{j=1}^{n_e} \underline{c}_j \right)$ and $\frac{1}{n_e + 1 - \sum_{j=1}^{n_e} \overline{c}_j}$ can be removed when Γ is linear.

The proof is given in Appendix C.4. It consists of first analyzing the conditional regret of the deferral loss and that of a surrogate loss. Next, we show how the former can be upper bounded in terms of the latter by leveraging the $\mathcal{H}$-consistency bound of ℓ with respect to the zero-one loss with an appropriate conditional distribution that we construct. This, combined with the results of [5], proves our $\mathcal{H}$-consistency bounds.

Let us emphasize that the theorem is broadly applicable and that there are many choices for the surrogate loss ℓ meeting the assumption of the theorem: [5] showed that a variety of surrogate loss functions ℓ admit an $\mathcal{H}$-consistency bound with respect to the zero-one loss for common hypothesis sets such as linear models and multi-layer neural networks, including *sum losses* [77], *constrained losses* [43], and, as shown more recently by [48] (see also [49,86]), *comp-sum losses*, which include the logistic loss [12,13,72,73], the *sum-exponential loss* and many other loss functions.

Thus, the theorem gives a strong guarantee for a broad family of surrogate losses L based upon such loss functions ℓ. The presence of the minimizability gaps in these bounds is important. In particular, while the minimizability gap can be upper bounded by the approximation error $\mathcal{A}_\ell(\mathcal{H}) = \mathcal{E}_\ell^*(\mathcal{H}) - \mathbb{E}_x\left[\inf_{h \in \mathcal{H}_{\text{all}}} \mathbb{E}_{y|x}\left[\ell(h,x,y)\right]\right] = \mathcal{E}_\ell^*(\mathcal{H}) - \mathcal{E}_\ell^*(\mathcal{H}_{\text{all}})$, it is a finer quantity than the approximation error and can lead to more favorable guarantees.

Note that when the Bayes-error coincides with the best-in-class error, $\mathcal{E}_L^*(\mathcal{H}) = \mathcal{E}_L^*(\mathcal{H}_{\text{all}})$, we have $\mathcal{M}_L(\mathcal{H}) \leq \mathcal{A}_L(\mathcal{H}) = 0$. This leads to the following corollary, using the non-negativity property of the minimizability gap.

Corollary 1. *Assume that ℓ admits an $\mathcal{H}$-consistency bound with respect to the multi-class zero-one classification loss ℓ_{0-1}. Then, for all $h \in \mathcal{H}$ and any distribution such that $\mathcal{E}_L^*(\mathcal{H}) = \mathcal{E}_L^*(\mathcal{H}_{\text{all}})$, the following bound holds:*

$$\mathcal{E}_{L_{\text{def}}}(h) - \mathcal{E}_{L_{\text{def}}}^*(\mathcal{H}) \leq \left(n_e + 1 - \sum_{j=1}^{n_e} \underline{c}_j\right) \Gamma\left(\frac{\mathcal{E}_L(h) - \mathcal{E}_L^*(\mathcal{H})}{n_e + 1 - \sum_{j=1}^{n_e} \overline{c}_j}\right),$$

Furthermore, constant factors $\left(n_e + 1 - \sum_{j=1}^{n_e} \underline{c}_j\right)$ and $\frac{1}{n_e + 1 - \sum_{j=1}^{n_e} \overline{c}_j}$ can be removed when Γ is linear.

Thus, when the estimation error of the surrogate loss, $\mathcal{E}_L(h) - \mathcal{E}_L^*(\mathcal{H})$, is reduced to ϵ, the estimation error of the deferral loss, $\mathcal{E}_{L_{\text{def}}}(h) - \mathcal{E}_{L_{\text{def}}}^*(\mathcal{H})$, is upper bounded by

$$\left(n_e + 1 - \sum_{j=1}^{n_e} \underline{c}_j\right) \Gamma\left(\epsilon \Big/ \left(n_e + 1 - \sum_{j=1}^{n_e} \overline{c}_j\right)\right).$$

Moreover, $\mathcal{H}$-consistency holds since $\mathcal{E}_L(h) - \mathcal{E}_L^*(\mathcal{H}) \to 0$ implies $\mathcal{E}_{L_{\text{def}}}(h) - \mathcal{E}_{L_{\text{def}}}^*(\mathcal{H}) \to 0$.

Table 1 shows several examples of surrogate deferral losses and their corresponding $\mathcal{H}$-consistency bounds, using the multi-class $\mathcal{H}$-consistency bounds known for comp-sum losses ℓ with respect to the zero-one loss [48, Theorem 1]. The bounds have been simplified here using the inequalities $1 \leq n_e + 1 - \sum_{j=1}^{n_e} \overline{c}_j \leq n_e + 1 - \sum_{j=1}^{n_e} \underline{c}_j \leq n_e + 1$. See Appendix D.1 for a more detailed derivation.

Similarly, Table 2 and Table 3 show several examples of surrogate deferral losses with sum losses or constrained losses adopted for ℓ and their corresponding $\mathcal{H}$-consistency bounds, using the multi-class $\mathcal{H}$-consistency bounds in [5, Table 2] and [5, Table 3] respectively. Here too, we present the simplified bounds by using the inequalities $1 \leq n_e + 1 - \sum_{j=1}^{n_c} \overline{c}_j \leq n_e + 1 - \sum_{j=1}^{n_c} \underline{c}_j \leq n_e + 1$. See Appendix D.2 and Appendix D.3 for a more detailed derivation.

5 Benefits of Minimizability Gaps

As already pointed out, the minimizabiliy gap can be upper bounded by the approximation error $\mathcal{A}_\ell(\mathcal{H}) = \mathcal{E}_\ell^*(\mathcal{H}) - \mathbb{E}_x\left[\inf_{h \in \mathcal{H}_{\text{all}}} \mathbb{E}_{y|x}\left[\ell(h,x,y)\right]\right] = \mathcal{E}_\ell^*(\mathcal{H}) -$

Table 1. Examples of the deferral surrogate loss (4) with comp-sum losses adopted for ℓ and their associated $\mathcal{H}$-consistency bounds provided by Corollary 1 (with only the surrogate portion displayed).

ℓ	L	$\mathcal{H}$-consistency bounds
$\ell_{\exp}$	$\sum_{y' \neq y} e^{h(x,y') - h(x,y)} + \sum_{j=1}^{n_e}(1 - c_j(x,y)) \sum_{y' \neq n+j} e^{h(x,y') - h(x,n+j)}$	$\sqrt{2}(n_e + 1)\left(\mathcal{E}_{\mathsf{L}}(h) - \mathcal{E}_{\mathsf{L}}^*(\mathcal{H})\right)^{\frac{1}{2}}$
$\ell_{\log}$	$-\log\left(\frac{e^{h(x,y)}}{\sum_{y' \in \overline{y}} e^{h(x,y')}}\right) - \sum_{j=1}^{n_e}(1 - c_j(x,y)) \log\left(\frac{e^{h(x,n+j)}}{\sum_{y' \in \overline{y}} e^{h(x,y')}}\right)$	$\sqrt{2}(n_e + 1)\left(\mathcal{E}_{\mathsf{L}}(h) - \mathcal{E}_{\mathsf{L}}^*(\mathcal{H})\right)^{\frac{1}{2}}$
ℓ_{gce}	$\frac{1}{\alpha}\left[1 - \left[\frac{e^{h(x,y)}}{\sum_{y' \in \overline{y}} e^{h(x,y')}}\right]^\alpha\right] + \frac{1}{\alpha}\sum_{j=1}^{n_e}(1 - c_j(x,y))\left[1 - \left[\frac{e^{h(x,n+j)}}{\sum_{y' \in \overline{y}} e^{h(x,y')}}\right]^\alpha\right]$	$\sqrt{2n^\alpha}(n_e + 1)\left(\mathcal{E}_{\mathsf{L}}(h) - \mathcal{E}_{\mathsf{L}}^*(\mathcal{H})\right)^{\frac{1}{2}}$
ℓ_{mae}	$1 - \frac{e^{h(x,y)}}{\sum_{y' \in \overline{y}} e^{h(x,y')}} + \sum_{j=1}^{n_e}(1 - c_j(x,y))\left(1 - \frac{e^{h(x,n+j)}}{\sum_{y' \in \overline{y}} e^{h(x,y')}}\right)$	$n\left(\mathcal{E}_{\mathsf{L}}(h) - \mathcal{E}_{\mathsf{L}}^*(\mathcal{H})\right)$

Table 2. Examples of the deferral surrogate loss (4) with sum losses adopted for ℓ and their associated $\mathcal{H}$-consistency bounds provided by Corollary 1 (with only the surrogate portion displayed), where $\Delta_h(x,y,y') = h(x,y) - h(x,y')$, $\Phi_{\text{sq}}(t) = \max\{0, 1-t\}^2$, $\Phi_{\exp}(t) = e^{-t}$, and $\Phi_\rho(t) = \min\{\max\{0, 1 - t/\rho\}, 1\}$.

ℓ	L	$\mathcal{H}$-consistency bounds
$\Phi_{\text{sq}}^{\text{sum}}$	$\sum_{y' \neq y} \Phi_{\text{sq}}(\Delta_h(x,y,y')) + \sum_{j=1}^{n_e}(1 - c_j(x,y)) \sum_{y' \neq n+j} \Phi_{\text{sq}}(\Delta_h(x,n+j,y'))$	$(n_e + 1)\left(\mathcal{E}_{\mathsf{L}}(h) - \mathcal{E}_{\mathsf{L}}^*(\mathcal{H})\right)^{\frac{1}{2}}$
$\Phi_{\exp}^{\text{sum}}$	$\sum_{y' \neq y} \Phi_{\exp}(\Delta_h(x,y,y')) + \sum_{j=1}^{n_e}(1 - c_j(x,y)) \sum_{y' \neq n+j} \Phi_{\exp}(\Delta_h(x,n+j,y'))$	$\sqrt{2}(n_e + 1)\left(\mathcal{E}_{\mathsf{L}}(h) - \mathcal{E}_{\mathsf{L}}^*(\mathcal{H})\right)^{\frac{1}{2}}$
Φ_ρ^{sum}	$\sum_{y' \neq y} \Phi_\rho(\Delta_h(x,y,y')) + \sum_{j=1}^{n_e}(1 - c_j(x,y)) \sum_{y' \neq n+j} \Phi_\rho(\Delta_h(x,n+j,y'))$	$\mathcal{E}_{\mathsf{L}}(h) - \mathcal{E}_{\mathsf{L}}^*(\mathcal{H})$

$\mathcal{E}_\ell^*(\mathcal{H}_{\text{all}})$. It is however a finer quantity than the approximation error and can thus lead to more favorable guarantees. More precisely, as shown by [4,5], for a target loss function ℓ_2 and a surrogate loss function ℓ_1, the excess error bound can be rewritten as

$$\mathcal{E}_{\ell_2}(h) - \mathcal{E}_{\ell_2}^*(\mathcal{H}) + \mathcal{A}_{\ell_2}(\mathcal{H}) \leq \Gamma\left(\mathcal{E}_{\ell_1}(h) - \mathcal{E}_{\ell_1}^*(\mathcal{H}) + \mathcal{A}_{\ell_1}(\mathcal{H})\right),$$

where Γ is typically linear or the square-root function modulo constants. On the other hand, an $\mathcal{H}$-consistency bound can be expressed as follows:

$$\mathcal{E}_{\ell_2}(h) - \mathcal{E}_{\ell_2}^*(\mathcal{H}) + \mathcal{M}_{\ell_2}(\mathcal{H}) \leq \Gamma\left(\mathcal{E}_{\ell_1}(h) - \mathcal{E}_{\ell_1}^*(\mathcal{H}) + \mathcal{M}_{\ell_1}(\mathcal{H})\right).$$

For a target loss function ℓ_2 with discrete outputs, such as the zero-one loss or the deferral loss, we have $\mathbb{E}_x\left[\inf_{h \in \mathcal{H}} \mathbb{E}_{y|x}\left[\ell_2(h,x,y)\right]\right] = \mathbb{E}_x\left[\inf_{h \in \mathcal{H}_{\text{all}}} \mathbb{E}_{y|x}\left[\ell_2(h,x,y)\right]\right]$ when the hypothesis set generates labels that cover all possible outcomes for each input (See [5, Lemma 3], Lemma 1 in Appendix C.1). Consequently, we have $\mathcal{M}_{\ell_2}(\mathcal{H}) = \mathcal{A}_{\ell_2}(\mathcal{H})$. For a surrogate loss function ℓ_1, the minimizability gap is upper bounded by the approximation error, $\mathcal{M}_{\ell_1}(\mathcal{H}) \leq \mathcal{A}_{\ell_1}(\mathcal{H})$, and is generally finer.

Consider a simple binary classification example with the conditional distribution denoted as $\eta(x) = D(Y = 1|X = x)$. Let $\mathcal{H}$ be a family of functions h such that $|h(x)| \leq \Lambda$ for all $x \in \mathcal{X}$, for some $\Lambda > 0$, and such that all values in the range $[-\Lambda, +\Lambda]$ can be achieved. For the exponential-based margin loss, defined as $\ell(h, x, y) = e^{-yh(x)}$, we have

Table 3. Examples of the deferral surrogate loss (4) with constrained losses adopted for ℓ and their associated $\mathcal{H}$-consistency bounds provided by Corollary 1 (with only the surrogate portion displayed), where $\Phi_{\text{hinge}}(t) = \max\{0, 1-t\}$, $\Phi_{\text{sq}}(t) = \max\{0, 1-t\}^2$, $\Phi_{\exp}(t) = e^{-t}$, and $\Phi_\rho(t) = \min\{\max\{0, 1-t/\rho\}, 1\}$ with the constraint that $\sum_{y \in \mathcal{Y}} h(x, y) = 0$.

ℓ	L	$\mathcal{H}$-consistency bounds
$\Phi_{\text{hinge}}^{\text{cstnd}}$	$\sum_{y' \neq y} \Phi_{\text{hinge}}(-h(x,y')) + \sum_{j=1}^{n_e}(1-c_j(x,y))\sum_{y' \neq n+j}\Phi_{\text{hinge}}(-h(x,y'))$	$\mathcal{E}_\mathsf{L}(h) - \mathcal{E}_\mathsf{L}^*(\mathcal{H})$
$\Phi_{\text{sq}}^{\text{cstnd}}$	$\sum_{y' \neq y} \Phi_{\text{sq}}(-h(x,y')) + \sum_{j=1}^{n_e}(1-c_j(x,y))\sum_{y' \neq n+j}\Phi_{\text{sq}}(-h(x,y'))$	$(n_e+1)(\mathcal{E}_\mathsf{L}(h) - \mathcal{E}_\mathsf{L}^*(\mathcal{H}))^{\frac{1}{2}}$
$\Phi_{\exp}^{\text{cstnd}}$	$\sum_{y' \neq y} \Phi_{\exp}(-h(x,y')) + \sum_{j=1}^{n_e}(1-c_j(x,y))\sum_{y' \neq n+j}\Phi_{\exp}(-h(x,y'))$	$\sqrt{2}(n_e+1)(\mathcal{E}_\mathsf{L}(h) - \mathcal{E}_\mathsf{L}^*(\mathcal{H}))^{\frac{1}{2}}$
Φ_ρ^{cstnd}	$\sum_{y' \neq y} \Phi_\rho(-h(x,y')) + \sum_{j=1}^{n_e}(1-c_j(x,y))\sum_{y' \neq n+j}\Phi_\rho(-h(x,y'))$	$\mathcal{E}_\mathsf{L}(h) - \mathcal{E}_\mathsf{L}^*(\mathcal{H})$

$$\mathbb{E}_{y|x}[\ell(h, x, y)] = \eta(x)e^{-h(x)} + (1 - \eta(x))e^{h(x)}.$$

It can be observed that the infimum over all measurable functions can be written as follows, for all x:

$$\inf_{h \in \mathcal{H}_{\text{all}}} \mathbb{E}_{y|x}[\ell(h, x, y)] = 2\sqrt{\eta(x)(1 - \eta(x))},$$

while the infimum over $\mathcal{H}$, $\inf_{h \in \mathcal{H}} \mathbb{E}_{y|x}[\ell(h, x, y)]$, depends on Λ. That infimum over $\mathcal{H}$ is achieved by

$$h(x) = \begin{cases} \min\left\{\frac{1}{2}\log\frac{\eta(x)}{1-\eta(x)}, \Lambda\right\} & \eta(x) \geq 1/2 \\ \max\left\{\frac{1}{2}\log\frac{\eta(x)}{1-\eta(x)}, -\Lambda\right\} & \text{otherwise.} \end{cases}$$

Thus, in the deterministic case, we can explicitly compute the difference between the approximation error and the minimizability gap:

$$\mathcal{A}_\ell(\mathcal{H}) - \mathcal{M}_\ell(\mathcal{H})$$
$$= \mathbb{E}_x\left[\inf_{h \in \mathcal{H}} \mathbb{E}_{y|x}[\ell(h, x, y)] - \inf_{h \in \mathcal{H}_{\text{all}}} \mathbb{E}_{y|x}[\ell(h, x, y)]\right] = e^{-\Lambda}.$$

As the parameter Λ decreases, the hypothesis set $\mathcal{H}$ becomes more restricted and the difference between the approximation error and the minimizability gap increases. In summary, an $\mathcal{H}$-consistency bound can be more favorable than the excess error bound as $\mathcal{M}_{\ell_2}(\mathcal{H}) = \mathcal{A}_{\ell_2}(\mathcal{H})$ when ℓ_2 represents the zero-one loss or deferral loss, and $\mathcal{M}_{\ell_1}(\mathcal{H}) \leq \mathcal{A}_{\ell_1}(\mathcal{H})$. Moreover, we will show in the next section that our $\mathcal{H}$-consistency bounds can lead to learning bounds for the deferral loss and a hypothesis set $\mathcal{H}$ with finite samples.

6 Learning Bounds

For a sample $S = ((x_1, y_1), \ldots, (x_m, y_m))$ drawn from $\mathcal{D}^m$, we will denote by $\widehat{h}_S$ the empirical minimizer of the empirical loss within $\mathcal{H}$ with respect to the surrogate loss function L: $\widehat{h}_S = \operatorname{argmin}_{h \in \mathcal{H}} \frac{1}{m} \sum_{i=1}^{m} \mathsf{L}(h, x_i, y_i)$. Given an $\mathcal{H}$-consistency bound in the form of (5), we can further use it to derive a learning bound for the deferral loss by upper bounding the surrogate estimation error $\mathcal{E}_\mathsf{L}(\widehat{h}_S) - \mathcal{E}_\mathsf{L}^*(\mathcal{H})$ with the complexity (e.g. the Rademacher complexity) of the family of functions associated with L and $\mathcal{H}$: $\mathcal{H}_\mathsf{L} = \{(x, y) \mapsto \mathsf{L}(h, x, y) : h \in \mathcal{H}\}$.

We denote by $\mathfrak{R}_m^\mathsf{L}(\mathcal{H})$ the Rademacher complexity of $\mathcal{H}_\mathsf{L}$ and by B_L an upper bound of the surrogate loss L. Then, we obtain the following learning bound for the deferral loss based on (5).

Theorem 2 (Learning bound). *Under the same assumptions as Theorem 1, for any $\delta > 0$, with probability at least $1 - \delta$ over the draw of an i.i.d sample S of size m, the following deferral loss estimation bound holds for $\widehat{h}_S$:*

$$\mathcal{E}_{\mathsf{L}_{\mathrm{def}}}(\widehat{h}_S) - \mathcal{E}_{\mathsf{L}_{\mathrm{def}}}^*(\mathcal{H}) + \mathcal{M}_\mathsf{L}(\mathcal{H}) \leq \left(n_e + 1 - \sum_{j=1}^{n_e} \underline{c}_j\right) \Gamma\left(\frac{4\mathfrak{R}_m^\mathsf{L}(\mathcal{H}) + 2B_\mathsf{L}\sqrt{\frac{\log \frac{2}{\delta}}{2m}} + \mathcal{M}_\mathsf{L}(\mathcal{H})}{n_e + 1 - \sum_{j=1}^{n_e} \overline{c}_j}\right).$$

The proof is presented in Appendix E. To the best of our knowledge, Theorem 2 provides the first finite-sample guarantee for the estimation error of the minimizer of a surrogate deferral loss L defined for multiple experts. The proof exploits our $\mathcal{H}$-consistency bounds with respect to the deferral loss, as well as standard Rademacher complexity guarantees.

When $\underline{c}_j = 0$ and $\overline{c}_j = 1$ for any $j \in [n_e]$, the right-hand side of the bound admits the following simpler form:

$$(n_e + 1) \, \Gamma\left(4\mathfrak{R}_m^\mathsf{L}(\mathcal{H}) + 2B_\mathsf{L}\sqrt{\frac{\log \frac{2}{\delta}}{2m}} + \mathcal{M}_\mathsf{L}(\mathcal{H})\right).$$

The dependency on the number of experts n_e makes this bound less favorable. There is a trade-off however since, on the other hand, more experts can help us achieve a better accuracy overall and reduce the best-in-class deferral loss. These learning bounds take into account the minimizability gap, which varies as a function of the upper bound Λ on the magnitude of the scoring functions. Thus, both the minimizability gaps and the Rademacher complexity term suggest a regularization controlling the complexity of the hypothesis set and the magnitude of the scores.

Adopting different loss functions ℓ in the definition of our deferral surrogate loss (4) will lead to a different functional form Γ, which can make the bound more or less favorable. For example, a linear form of Γ is in general more favorable than a square-root form modulo a constant. But, the dependency on the number of classes n appearing in Γ (e.g., $\ell = \ell_{\mathrm{gce}}$ or $\ell = \ell_{\mathrm{mae}}$) is also important to take into account since a larger value of n tends to negatively impact the guarantees. We already discussed the dependency on the number of experts n_e in Γ (e.g., $\ell = \ell_{\mathrm{gce}}$ or $\ell = \ell_{\mathrm{exp}}$) and the associated trade-off, which is also important to consider.

Note that the bound of Theorem 2 is expressed in terms of the global complexity of the prediction and deferral scoring functions $\mathcal{H}$. One can however derive a finer bound distinguishing the complexity of the deferral scoring functions and that of the prediction scoring functions following a similar proof and analysis.

Recall that for a surrogate loss L, the minimizability gap $\mathcal{M}_L(\mathcal{H})$ is in general finer than the approximation error $\mathcal{A}_L(\mathcal{H})$, while for the deferral loss, for common hypothesis sets, these two quantities coincide. Thus, our bound can be rewritten as follows for common hypothesis sets:

$$\mathcal{E}_{L_{\text{def}}}(\widehat{h}_S) - \mathcal{E}_{L_{\text{def}}}^*(\mathcal{H}_{\text{all}}) \leq \left(n_e + 1 - \sum_{j=1}^{n_e} \underline{c}_j\right) \Gamma\left(\frac{4\mathfrak{R}_m^L(\mathcal{H}) + 2B_L\sqrt{\frac{\log \frac{2}{\delta}}{2m}} + \mathcal{M}_L(\mathcal{H})}{n_e + 1 - \sum_{j=1}^{n_e} \overline{c}_j}\right).$$

This is more favorable and more relevant than a similar excess loss bound where $\mathcal{M}_L(\mathcal{H})$ is replaced with $\mathcal{A}_L(\mathcal{H})$, which could be derived from a generalization bound for the surrogate loss.

7 Experiments

In this section, we examine the empirical performance of our proposed surrogate loss in the scenario of learning to defer with multiple experts. More specifically, we aim to compare the overall system accuracy for the learned predictor and deferral pairs, considering varying numbers of experts. This comparison provides valuable insights into the performance of our algorithm under different expert configurations. We explore three different scenarios:

- Only a single expert is available, specifically where a larger model than the base model is chosen as the deferral option.
- Two experts are available, consisting of one small model and one large model as the deferral options.
- Three experts are available, including one small model, one medium model, and one large model as the deferral options.

By comparing these scenarios, we evaluate the impact of varying the number and type of experts on the overall system accuracy.

Type of Cost. We carried out experiments with two types of cost functions. For the first type, we selected the cost function to be exactly the misclassification error of the expert: $c_j(x, y) = \mathbb{1}_{\mathsf{g}_j(x) \neq y}$, where $\mathsf{g}_j(x) = \operatorname{argmax}_{y \in [n]} g_j(x, y)$ is the prediction made by expert g_j for input x. In this scenario, the cost incurred for deferring is determined solely based on the expert's accuracy. For the second type, we chose a cost function admitting the form $c_j(x, y) = \mathbb{1}_{\mathsf{g}_j(x) \neq y} + \beta_j$, where an additional non-zero base cost β_j is assigned to each expert. Deferring to a larger model then tends to incur a higher inference cost and hence, the corresponding β_j value for a larger model is higher as well. In addition to the base cost, each expert also incurs a misclassification error, as with the first type. Experimental setup and additional experiments (see Table 6) are included in Appendix B.

Table 4. Overall system accuracy with the first type of cost functions.

	Single expert	Two experts	Three experts
SVHN	92.08 ± 0.15%	93.18 ± 0.18%	93.46 ± 0.12%
CIFAR-10	73.31 ± 0.21%	77.12 ± 0.34%	78.71 ± 0.43%

Table 5. Overall system accuracy with the second type of cost functions.

	Single expert	Two experts	Three experts
SVHN	92.36 ± 0.22%	93.23 ± 0.21%	93.36 ± 0.11%
CIFAR-10	73.70 ± 0.40%	76.29 ± 0.41%	76.43 ± 0.55%

Experimental Results. In Table 4 and Table 5, we report the mean and standard deviation of the system accuracy over three runs with different random seeds. We noticed a positive correlation between the number of experts and the overall system accuracy. Specifically, as the number of experts increases, the performance of the system in terms of accuracy improves. This observation suggests that incorporating multiple experts in the learning to defer framework can lead to better predictions and decision-making. The results also demonstrate the effectiveness of our proposed surrogate loss for deferral with multiple experts.

8 Conclusion

We presented a comprehensive study of surrogate losses for the core challenge of learning to defer with multiple experts. Through our study, we established theoretical guarantees, strongly endorsing the adoption of the loss function family we introduced. This versatile family of loss functions can effectively facilitate the learning to defer algorithms across a wide range of applications. Our analysis offers great flexibility by accommodating diverse cost functions, encouraging exploration and evaluation of various options in real-world scenarios. We encourage further research into the theoretical properties of different choices and their impact on the overall performance to gain deeper insights into their effectiveness.

Disclosure of Interests. The authors have no competing interests to declare that are relevant to the content of this article.

Appendices

A Related Work

The concept of *learning to defer* has its roots in research on abstention, particularly in binary classification scenarios with a constant cost function. Early work by [19] and [18] focused on rejection and set the foundation for subsequent studies on learning with abstention. These studies explored different approaches such as *confidence-based methods* [10,29,33,80], the *predictor-rejector framework* [20–22], or *selective classification* [24,78,81].

[20–22] showed that the confidence-based approach could fail to determine the optimal rejection region when the predictor did not match the Bayes solution. Instead, they proposed a novel *predictor-rejector* framework, for which they gave both Bayes-consistent and *realizable H-consistent* surrogate losses [42,45,82], which achieve state-of-the-art performance in the binary setting.

[24,78] introduced and studied a selective classification based on a predictor and a selector and explored the trade-off between classifier coverage and accuracy, drawing connections to active learning in their analysis.

The confidence-based and predictor-rejector frameworks have been both further analyzed in the context of *multi-class classification.* [1,25,27,62,67] extended the confidence-based method to multi-class settings, while [62] noted that deriving a Bayes-consistent surrogate loss under the *predictor-rejector* framework is quite challenging and left it as an open problem. More recently, [53] presented a series of new theoretical and algorithmic results in this framework, positively resolving this open problem. [55] also examined this framework in the context of learning with a fixed predictor and applied their new algorithms to the task of decontextualization. In response to the challenge in the multi-class setting, [58] formulated a different *score-based* approach to learn the predictor and rejector simultaneously, by introducing an additional scoring function corresponding to rejection. This method has been further explored in a subsequent work [15,54]. The surrogate losses derived under this framework are currently the state-of-the-art [15,54,58].

[28] proposed a new neural network architecture for abstention in the selective classification framework for multi-class classification. They did not derive consistent surrogate losses for this formulation. [87] defined a loss function for the predictor-selector framework based on the doubling rate of gambling that requires almost no modification to the model architecture.

Another line of research studied multi-class abstention using an *implicit criterion* [1,16,25,35], by directly modeling regions with high confidence.

However, a constant cost does not fully capture all the relevant information in the deferral scenario. It is important to take into account the quality of the expert, whose prediction we rely on. These may be human experts as in several critical applications [8,36,40,71]. To address this gap, [46] incorporated the human expert's decision into the cost and proposed the first *learning to defer (L2D)* framework, which has also been examined in [38,64,65,79]. [58] proposed the first *Bayes-consistent* [9,68,83] surrogate loss for L2D, and subsequent work [44,66] further improved upon it. Another Bayes-consistent surrogate loss in L2D is the one-versus-all loss proposed by [75] that is also studied in [17] as a special case of a general family of loss functions. An additional line of research investigated post-hoc methods [59,63], where [63] proposed an alternative optimization method between the predictor and rejector, and [59] provided a correction to the surrogate losses in [58,75] when they are underfitting. Finally,

L2D or its variants have been adopted or studied in various other scenarios [23,26,32,34,44,57,60,69,85].

All the studies mentioned so far mainly focused on learning to defer with a single expert. Most recently, [74] highlighted the significance of *learning to defer with multiple experts* [11,31,37,38,70] and extended the surrogate loss in [58,75] to accommodate the multiple-expert setting, which is currently the only work to propose Bayes-consistent surrogate losses in this scenario. They further showed that a mixture of experts (MoE) approach to multi-expert L2D proposed in [31] is not consistent. More recently, [47] examined a two-stage scenario for learning to defer with multiple experts, which is crucial for various applications. They developed new surrogate losses for this scenario and demonstrated that these are supported by stronger consistency guarantees-specifically, $\mathcal{H}$-consistency bounds as introduced below-implying their Bayes consistency.

Meanwhile, recent work by [4,5] introduced new consistency guarantees, called $\mathcal{H}$-consistency bounds, which they argued are more relevant to learning than Bayes-consistency since they are hypothesis set-specific and non-asymptotic. $\mathcal{H}$-consistency bounds are also stronger guarantees than Bayes-consistency. They established $\mathcal{H}$-consistent bounds for common surrogate losses in standard classification (see also [48,49,86]).

In this work, we study the general framework of learning to defer with multiple experts. Furthermore, we design deferral surrogate losses that benefit from these more significant consistency guarantees, namely, $\mathcal{H}$-consistency bounds, in the general multiple-expert setting.

B Experimental Details

Experimental setup. For our experiments, we used two popular datasets: CIFAR-10 [41] and SVHN (Street View House Numbers) [61]. CIFAR-10 consists of 60,000 color images in 10 different classes, with 6,000 images per class. The dataset is split into 50,000 training images and 10,000 test images. SVHN contains images of house numbers captured from Google Street View. It consists of 73,257 images for training and 26,032 images for testing. We trained for 50 epochs on CIFAR-10 and 15 epochs on SVHN without any data augmentation.

In our experiments, we adopted the ResNet [30] architecture for the base model and selected various sizes of ResNet models as experts in each scenario. Throughout all three scenarios, we used ResNet-4 for both the predictor and the deferral models. In the first scenario, we chose ResNet-10 as the expert model. In the second scenario, we included ResNet-10 and ResNet-16 as expert models. The third scenario involves ResNet-10, ResNet-16, and ResNet-28 as expert models with increasing complexity. The expert models are pre-trained on the training data of SVHN and CIFAR-10 respectively.

During the training process, we simultaneously trained the predictor ResNet-4 and the deferral model ResNet-4. We adopted the Adam optimizer [39] with a batch size of 128 and a weight decay of 1×10^{-4}. We used our proposed deferral surrogate loss (4) with the generalized cross-entropy loss being adopted for ℓ. As suggested by [84], we set the parameter α to 0.7.

Table 6. Comparison of our proposed deferral surrogate loss with the one-vs-all (OvA) surrogate loss in an intriguing setting where multiple experts are available and each of them has a clear domain of expertise.

Method	System accuracy (%)	Ratio of deferral (%)										
		all the classes			classes 0 to 2			classes 3 to 5			classes 6 to 9		
		predictor	expert 1	expert 2	predictor	expert 1	expert 2	predictor	expert 1	expert 2	predictor	expert 1	expert 2
Ours	92.19	61.43	17.38	21.19	46.77	49.67	3.57	33.60	3.33	63.07	92.88	3.43	3.70
OvA	91.39	59.72	16.78	23.50	48.63	47.67	3.70	27.87	2.47	69.67	92.73	3.50	3.78

For the second type of cost functions, we set the base costs as follows: $\beta_1 = 0.1$, $\beta_2 = 0.12$ and $\beta_3 = 0.14$ for the SVHN dataset and $\beta_1 = 0.3$, $\beta_2 = 0.32$, $\beta_3 = 0.34$ for the CIFAR-10 dataset, where β_1 corresponds to the cost associated with the smallest expert model, ResNet-10, β_2 to that of the medium model, ResNet-16, and β_3 to that of the largest expert model, ResNet-28. A base cost value that is not too far from the misclassification error of expert models encourages in practice a reasonable amount of input instances to be deferred. We observed that the performance remains close for other neighboring values of base costs.

Additional Experiments. Here, we share additional experimental results in an intriguing setting where multiple experts are available and each of them has a clear domain of expertise. We report below the empirical results of our proposed deferral surrogate loss and the one-vs-all (OvA) surrogate loss proposed in recent work [74], which is the state-of-the-art surrogate loss for learning to defer with multiple experts, on CIFAR-10. In this setting, the two experts have a clear domain of expertise. The expert 1 is always correct on the first three classes, 0 to 2, and predicts uniformly at random for other classes; the expert 2 is always correct on the next three classes, 3 to 5, and generates random predictions otherwise. We train a ResNet-16 for the predictor/deferral model.

As shown in Table 6, our method achieves comparable system accuracy with OvA. Among the images in classes 0 to 2, only 3.57% is deferred to expert 2 which predicts uniformly at random. Similarly, among the images in classes 3 to 5, only 3.33% is deferred to expert 1. For the rest of the images in classes 6 to 9, the predictor decides to learn to classify them by itself and actually makes 92.88% of the final predictions. This illustrates that our proposed surrogate loss is effective and comparable to the baseline.

C Proof of $\mathcal{H}$-Consistency Bounds for Deferral Surrogate Losses

To prove $\mathcal{H}$-consistency bounds for our deferral surrogate loss functions, we will show how the *conditional regret* of the deferral loss can be upper bounded in terms of the *conditional regret* of the surrogate loss. The general theorems proven by [5, Theorem 4, Theorem 5] then guarantee our $\mathcal{H}$-consistency bounds.

For any $x \in \mathcal{X}$ and $y \in \mathcal{Y}$, let $p(x,y)$ denote the conditional probability of $Y = y$ given $X = x$ for any $y \in \mathcal{Y}$. Then, for any $x \in \mathcal{X}$, the *conditional $\mathsf{L}_{\mathrm{def}}$-loss* $\mathcal{C}_{\mathsf{L}_{\mathrm{def}}}(h,x)$ and *conditional regret (or calibration gap)* $\Delta \mathcal{C}_{\mathsf{L}_{\mathrm{def}}}(h,x)$ of a hypothesis $h \in \mathcal{H}$ are defined by

$$\mathcal{C}_{\mathsf{L}_{\mathrm{def}}}(h,x) = \mathbb{E}_{y|x}[\mathsf{L}_{\mathrm{def}}(h,x,y)] = \sum_{y \in \mathcal{Y}} p(x,y)\mathsf{L}_{\mathrm{def}}(h,x,y)$$

$$\Delta \mathcal{C}_{\mathsf{L}_{\mathrm{def}}}(h,x) = \mathcal{C}_{\mathsf{L}_{\mathrm{def}}}(h,x) - \mathcal{C}^*_{\mathsf{L}_{\mathrm{def}}}(\mathcal{H},x),$$

where $\mathcal{C}^*_{\mathsf{L}_{\mathrm{def}}}(\mathcal{H},x) = \inf_{h \in \mathcal{H}} \mathcal{C}_{\mathsf{L}_{\mathrm{def}}}(h,x)$. Similar definitions hold for the surrogate loss L. To bound $\Delta \mathcal{C}_{\mathsf{L}_{\mathrm{def}}}(h,x)$ in terms of $\Delta \mathcal{C}_{\mathsf{L}}(h,x)$, we first give more explicit expressions for these conditional regrets.

To do so, it will be convenient to use the following definition for any $x \in \mathcal{X}$ and $y \in [n + n_e]$:

$$q(x,y) = \begin{cases} p(x,y) & y \in \mathcal{Y} \\ 1 - \sum_{y \in \mathcal{Y}} p(x,y) c_j(x,y) & n+1 \leq y \leq n + n_e. \end{cases}$$

Note that $q(x,y)$ is non-negative but, in general, these quantities do not sum to one. We denote by $\overline{q}(x,y) = \frac{q(x,y)}{Q}$ their normalized counterparts which represent probabilities, where $Q = \sum_{y \in [n+n_e]} q(x,y)$.

For any $x \in \mathcal{X}$, we will denote by $\mathsf{H}(x)$ the set of labels generated by hypotheses in $\mathcal{H}$: $\mathsf{H}(x) = \{\mathsf{h}(x) : h \in \mathcal{H}\}$. We denote by $y_{\max} \in [n + n_e]$ the label associated by q to an input $x \in \mathcal{X}$, defined as $y_{\max} = \mathrm{argmax}_{y \in [n+n_e]} q(x,y)$, with the same deterministic strategy for breaking ties as that of $\mathsf{h}(x)$.

C.1 Conditional Regret of the Deferral Loss

With these definitions, we can now express the conditional loss and regret of the deferral loss.

Lemma 1. *For any $x \in \mathcal{X}$, the minimal conditional $\mathsf{L}_{\mathrm{def}}$-loss and the calibration gap for $\mathsf{L}_{\mathrm{def}}$ can be expressed as follows:*

$$\mathcal{C}^*_{\mathsf{L}_{\mathrm{def}}}(\mathcal{H},x) = 1 - \max_{y \in \mathsf{H}(x)} q(x,y)$$

$$\Delta \mathcal{C}_{\mathsf{L}_{\mathrm{def}},\mathcal{H}}(h,x) = \max_{y \in \mathsf{H}(x)} q(x,y) - q(x,\mathsf{h}(x)).$$

Proof. The conditional L_{def}-risk of h can be expressed as follows:

$$\begin{aligned}
\mathcal{C}_{\mathsf{L}_{\text{def}}}(h,x) &= \mathop{\mathbb{E}}_{y|x}\left[\mathsf{L}_{\text{def}}(h,x,y)\right] \\
&= \mathop{\mathbb{E}}_{y|x}\left[\mathbb{1}_{\mathsf{h}(x)\neq y}\right]\mathbb{1}_{\mathsf{h}(x)\in[n]} + \sum_{j=1}^{n_e}\mathop{\mathbb{E}}_{y|x}\left[c_j(x,y)\right]\mathbb{1}_{\mathsf{h}(x)=n+j} \\
&= \sum_{y\in\mathcal{Y}} q(x,y)\mathbb{1}_{\mathsf{h}(x)\neq y}\mathbb{1}_{\mathsf{h}(x)\in[n]} + \sum_{j=1}^{n_e}(1-q(x,n+j))\mathbb{1}_{\mathsf{h}(x)=n+j} \\
&= (1-q(x,\mathsf{h}(x)))\mathbb{1}_{\mathsf{h}(x)\in[n]} + \sum_{j=1}^{n_e}(1-q(x,\mathsf{h}(x)))\mathbb{1}_{\mathsf{h}(x)=n+j} \\
&= 1 - q(x,\mathsf{h}(x)).
\end{aligned}$$

Then, the minimal conditional L_{def}-risk is given by

$$\mathcal{C}^*_{\mathsf{L}_{\text{def}}}(\mathcal{H},x) = 1 - \max_{y\in\mathsf{H}(x)} q(x,y),$$

and the calibration gap can be expressed as follows:

$$\begin{aligned}
\Delta\mathcal{C}_{\mathsf{L}_{\text{def}},\mathcal{H}}(h,x) &= \mathcal{C}_{\mathsf{L}_{\text{def}}}(h,x) - \mathcal{C}^*_{\mathsf{L}_{\text{def}}}(\mathcal{H},x) \\
&= \max_{y\in\mathsf{H}(x)} q(x,y) - q(x,\mathsf{h}(x)),
\end{aligned}$$

which completes the proof.

C.2 Conditional Regret of a Surrogate Deferral Loss

Lemma 2. *For any $x \in \mathcal{X}$, the conditional surrogate L-loss and regret can be expressed as follows:*

$$\mathcal{C}_{\mathsf{L}}(h,x) = \sum_{y\in[n+n_e]} q(x,y)\ell(h,x,y)$$

$$\Delta\mathcal{C}_{\mathsf{L}}(h,x) = \sum_{y\in[n+n_e]} q(x,y)\ell(h,x,y)$$
$$- \inf_{h\in\mathcal{H}} \sum_{y\in[n+n_e]} q(x,y)\ell(h,x,y).$$

Proof. By definition, $\mathcal{C}_L(h,x)$ is the conditional-L loss can be expressed as follows:

$$\begin{aligned}
&\mathcal{C}_L(h,x)\\
&= \mathbb{E}_y\left[L(h,x,y)\right]\\
&= \mathbb{E}_y\left[\ell(h,x,y)\right] + \sum_{j=1}^{n_e} \mathbb{E}_{y|x}\left[(1-c_j(x,y))\right]\ell(h,x,n+j) \qquad (6)\\
&= \sum_{y\in\mathcal{Y}} q(x,y)\ell(h,x,y) + \sum_{j=1}^{n_e} q(x,n+j)\ell(h,x,n+j)\\
&= \sum_{y\in[n+n_e]} q(x,y)\ell(h,x,y),
\end{aligned}$$

which ends the proof.

C.3 Conditional Regret of Zero-One Loss

We will also make use of the following result for the zero-one loss $\ell_{0-1}(h,x,y) = \mathbb{1}_{h(x)\neq y}$ with label space $[n+n_e]$ and the conditional probability vector $\overline{q}(x,\cdot)$, which characterizes the minimal conditional ℓ_{0-1}-loss and the corresponding calibration gap [5, Lemma 3].

Lemma 3. *For any $x \in \mathcal{X}$, the minimal conditional ℓ_{0-1}-loss and the calibration gap for ℓ_{0-1} can be expressed as follows:*

$$\mathcal{C}^*_{\ell_{0-1}}(x) = 1 - \max_{y\in\mathsf{H}(x)} \overline{q}(x,y)$$

$$\Delta\mathcal{C}_{\ell_{0-1}}(h,x) = \max_{y\in\mathsf{H}(x)} \overline{q}(x,y) - \overline{q}(x,\mathsf{h}(x)).$$

C.4 Proof of $\mathcal{H}$-Consistency Bounds for Deferral Surrogate Losses (Theorem 1)

Theorem 1 ($\mathcal{H}$-consistency bounds for score-based surrogates). *Assume that ℓ admits an $\mathcal{H}$-consistency bound with respect to the multi-class zero-one classification loss ℓ_{0-1}. Thus, there exists a non-decreasing concave function Γ with $\Gamma(0) = 0$ such that, for any distribution $\mathcal{D}$ and for all $h \in \mathcal{H}$, we have*

$$\mathcal{E}_{\ell_{0-1}}(h) - \mathcal{E}^*_{\ell_{0-1}}(\mathcal{H}) + \mathcal{M}_{\ell_{0-1}}(\mathcal{H}) \leq \Gamma\left(\mathcal{E}_\ell(h) - \mathcal{E}^*_\ell(\mathcal{H}) + \mathcal{M}_\ell(\mathcal{H})\right).$$

Then, L admits the following $\mathcal{H}$-consistency bound with respect to L_{def}: for all $h\in\mathcal{H}$,

$$\mathcal{E}_{L_{def}}(h) - \mathcal{E}^*_{L_{def}}(\mathcal{H}) + \mathcal{M}_{L_{def}}(\mathcal{H}) \leq \left(n_e + 1 - \sum_{j=1}^{n_e} \underline{c}_j\right)\Gamma\left(\frac{\mathcal{E}_L(h) - \mathcal{E}^*_L(\mathcal{H}) + \mathcal{M}_L(\mathcal{H})}{n_e + 1 - \sum_{j=1}^{n_e} \overline{c}_j}\right). \qquad (5)$$

Furthermore, constant factors $\left(n_e + 1 - \sum_{j=1}^{n_c} \underline{c}_j\right)$ and $\frac{1}{n_e+1-\sum_{j=1}^{n_e}\overline{c}_j}$ can be removed when Γ is linear.

Proof. We denote the normalization factor as $Q = \sum_{y \in [n+n_e]} q(x,y) = n_e + 1 - \mathbb{E}_y[c_j(x,y)]$, which is a constant that ensures the sum of $\overline{q}(x,y) = \frac{q(x,y)}{Q}$ is equal to 1. By Lemma 1, the calibration gap of L_{def} can be expressed and upper-bounded as follows:

$$\Delta \mathcal{C}_{\mathsf{L}_{\text{def}}}(h,x)$$
$$= \max_{y \in \mathsf{H}(x)} q(x,y) - q(x, \mathsf{h}(x)) \qquad \text{(Lemma 1)}$$
$$= Q \left(\max_{y \in \mathsf{H}(x)} \overline{q}(x,y) - \overline{q}(x, \mathsf{h}(x)) \right)$$
$$= Q \Delta \mathcal{C}_{\ell_{0-1}}(h,x) \qquad \text{(Lemma 3)}$$
$$\leq Q \Gamma \left(\Delta \mathcal{C}_{\ell, \mathcal{H}}(h,x) \right) \qquad (\mathcal{H}\text{-consistency bound of } \ell)$$
$$= Q \Gamma \Bigg(\sum_{y \in [n+n_e]} \overline{q}(x,y) \ell(h,x,y)$$
$$\qquad - \inf_{h \in \mathcal{H}} \sum_{y \in [n+n_e]} \overline{q}(x,y) \ell(h,x,y) \Bigg)$$
$$= Q \Gamma \Bigg(\sum_{y \in [n+n_e]} \frac{q(x,y)}{Q} \ell(h,x,y)$$
$$\qquad - \inf_{h \in \mathcal{H}} \sum_{y \in [n+n_e]} \frac{q(x,y)}{Q} \ell(h,x,y) \Bigg)$$
$$= Q \Gamma \left(\frac{1}{Q} \Delta \mathcal{C}_{\mathsf{L}}(h,x) \right). \qquad \text{(Lemma 2)}$$

Thus, taking expectations gives:

$$\mathcal{E}_{\mathsf{L}_{\text{def}}}(h) - \mathcal{E}^*_{\mathsf{L}_{\text{def}}}(\mathcal{H}) + \mathcal{M}_{\mathsf{L}_{\text{def}}}(\mathcal{H})$$
$$= \mathbb{E}_X [\Delta \mathcal{C}_{\mathsf{L}_{\text{def}}}(h,x)]$$
$$\leq \mathbb{E}_X \left[Q \Gamma \left(\frac{1}{Q} \Delta \mathcal{C}_{\mathsf{L}}(h,x) \right) \right]$$
$$\leq Q \Gamma \left(\frac{1}{Q} \mathbb{E}_X [\Delta \mathcal{C}_{\mathsf{L}}(h,x)] \right) \qquad \text{(concavity of } \Gamma \text{ and Jensen's ineq.)}$$
$$= Q \Gamma \left(\frac{\mathcal{E}_{\mathsf{L}}(h) - \mathcal{E}^*_{\mathsf{L}}(\mathcal{H}) + \mathcal{M}_{\mathsf{L}}(\mathcal{H})}{Q} \right)$$
$$= \left(n_e + 1 - \mathbb{E}_y [c_j(x,y)] \right) \Gamma \left(\frac{\mathcal{E}_{\mathsf{L}}(h) - \mathcal{E}^*_{\mathsf{L}}(\mathcal{H}) + \mathcal{M}_{\mathsf{L}}(\mathcal{H})}{n_e + 1 - \mathbb{E}_y [c_j(x,y)]} \right)$$
$$\leq \left(n_e + 1 - \sum_{j=1}^{n_e} \underline{c}_j \right) \Gamma \left(\frac{\mathcal{E}_{\mathsf{L}}(h) - \mathcal{E}^*_{\mathsf{L}}(\mathcal{H}) + \mathcal{M}_{\mathsf{L}}(\mathcal{H})}{n_e + 1 - \sum_{j=1}^{n_e} \overline{c}_j} \right) \quad (\underline{c}_j \leq c_j(x,y) \leq \overline{c}_j, \forall j \in [n_e])$$

and $\mathcal{E}_{\mathsf{L}_{\mathrm{def}}}(h) - \mathcal{E}^*_{\mathsf{L}_{\mathrm{def}}}(\mathcal{H}) + \mathcal{M}_{\mathsf{L}_{\mathrm{def}}}(\mathcal{H}) \leq \Gamma\left(\mathcal{E}_{\mathsf{L}}(h) - \mathcal{E}^*_{\mathsf{L}}(\mathcal{H}) + \mathcal{M}_{\mathsf{L}}(\mathcal{H})\right)$ when Γ is linear, which completes the proof.

D Examples of Deferral Surrogate Losses and Their $\mathcal{H}$-Consistency Bounds

D.1 ℓ Being Adopted as Comp-Sum Losses

Example: $\ell = \ell_{\exp}$. Plug in $\ell = \ell_{\exp} = \sum_{y' \neq y} e^{h(x,y') - h(x,y)}$ in (4), we obtain

$$\mathsf{L} = \sum_{y' \neq y} e^{h(x,y') - h(x,y)} + \sum_{j=1}^{n_e} (1 - c_j(x,y)) \sum_{y' \neq n+j} e^{h(x,y') - h(x,n+j)}.$$

By [48, Theorem 1], $\ell_{\exp}$ admits an $\mathcal{H}$-consistency bound with respect to ℓ_{0-1} with $\Gamma(t) = \sqrt{2t}$, using Corollary 1, we obtain

$$\mathcal{E}_{\mathsf{L}_{\mathrm{def}}}(h) - \mathcal{E}^*_{\mathsf{L}_{\mathrm{def}}}(\mathcal{H}) \leq \sqrt{2}\left(n_e + 1 - \sum_{j=1}^{n_e} c_j\right) \left(\frac{\mathcal{E}_{\mathsf{L}}(h) - \mathcal{E}^*_{\mathsf{L}}(\mathcal{H})}{n_e + 1 - \sum_{j=1}^{n_e} \overline{c}_j}\right)^{\frac{1}{2}}.$$

Since $1 \leq n_e + 1 - \sum_{j=1}^{n_e} \overline{c}_j \leq n_e + 1 - \sum_{j=1}^{n_e} c_j \leq n_e + 1$, the bound can be simplified as

$$\mathcal{E}_{\mathsf{L}_{\mathrm{def}}}(h) - \mathcal{E}^*_{\mathsf{L}_{\mathrm{def}}}(\mathcal{H}) \leq \sqrt{2}(n_e + 1) \left(\mathcal{E}_{\mathsf{L}}(h) - \mathcal{E}^*_{\mathsf{L}}(\mathcal{H})\right)^{\frac{1}{2}}.$$

Example: $\ell = \ell_{\log}$. Plug in $\ell = \ell_{\log} = -\log\left[\frac{e^{h(x,y)}}{\sum_{y' \in \overline{y}} e^{h(x,y')}}\right]$ in (4), we obtain

$$\mathsf{L} = -\log\left(\frac{e^{h(x,y)}}{\sum_{y' \in \overline{y}} e^{h(x,y')}}\right) - \sum_{j=1}^{n_e} (1 - c_j(x,y)) \log\left(\frac{e^{h(x,n+j)}}{\sum_{y' \in \overline{y}} e^{h(x,y')}}\right).$$

By [48, Theorem 1], $\ell_{\log}$ admits an $\mathcal{H}$-consistency bound with respect to ℓ_{0-1} with $\Gamma(t) = \sqrt{2t}$, using Corollary 1, we obtain

$$\mathcal{E}_{\mathsf{L}_{\mathrm{def}}}(h) - \mathcal{E}^*_{\mathsf{L}_{\mathrm{def}}}(\mathcal{H}) \leq \sqrt{2}\left(n_e + 1 - \sum_{j=1}^{n_e} c_j\right) \left(\frac{\mathcal{E}_{\mathsf{L}}(h) - \mathcal{E}^*_{\mathsf{L}}(\mathcal{H})}{n_e + 1 - \sum_{j=1}^{n_e} \overline{c}_j}\right)^{\frac{1}{2}}.$$

Since $1 \leq n_e + 1 - \sum_{j=1}^{n_e} \overline{c}_j \leq n_e + 1 - \sum_{j=1}^{n_e} c_j \leq n_e + 1$, the bound can be simplified as

$$\mathcal{E}_{\mathsf{L}_{\mathrm{def}}}(h) - \mathcal{E}^*_{\mathsf{L}_{\mathrm{def}}}(\mathcal{H}) \leq \sqrt{2}(n_e + 1) \left(\mathcal{E}_{\mathsf{L}}(h) - \mathcal{E}^*_{\mathsf{L}}(\mathcal{H})\right)^{\frac{1}{2}}.$$

Example: $\ell = \ell_{\text{gce}}$. Plug in $\ell = \ell_{\text{gce}} = \frac{1}{\alpha}\left[1 - \left[\frac{e^{h(x,y)}}{\sum_{y'\in \overline{y}} e^{h(x,y')}}\right]^\alpha\right]$ in (4), we obtain

$$\mathsf{L} = \frac{1}{\alpha}\left[1 - \left[\frac{e^{h(x,y)}}{\sum_{y'\in \overline{y}} e^{h(x,y')}}\right]^\alpha\right]$$
$$+ \frac{1}{\alpha}\sum_{j=1}^{n_e}(1 - c_j(x,y))\left[1 - \left[\frac{e^{h(x,n+j)}}{\sum_{y'\in \overline{y}} e^{h(x,y')}}\right]^\alpha\right].$$

By [48, Theorem 1], ℓ_{gce} admits an $\mathcal{H}$-consistency bound with respect to ℓ_{0-1} with $\Gamma(t) = \sqrt{2n^\alpha t}$, using Corollary 1, we obtain

$$\mathcal{E}_{\mathsf{L}_{\text{def}}}(h) - \mathcal{E}_{\mathsf{L}_{\text{def}}}^*(\mathcal{H}) \le \sqrt{2n^\alpha}\left(n_e + 1 - \sum_{j=1}^{n_e}\underline{c}_j\right)\left(\frac{\mathcal{E}_{\mathsf{L}}(h) - \mathcal{E}_{\mathsf{L}}^*(\mathcal{H})}{n_e + 1 - \sum_{j=1}^{n_e}\overline{c}_j}\right)^{\frac{1}{2}}.$$

Since $1 \le n_e + 1 - \sum_{j=1}^{n_e}\overline{c}_j \le n_e + 1 - \sum_{j=1}^{n_e}\underline{c}_j \le n_e + 1$, the bound can be simplified as

$$\mathcal{E}_{\mathsf{L}_{\text{def}}}(h) - \mathcal{E}_{\mathsf{L}_{\text{def}}}^*(\mathcal{H}) \le \sqrt{2n^\alpha}(n_e + 1)\left(\mathcal{E}_{\mathsf{L}}(h) - \mathcal{E}_{\mathsf{L}}^*(\mathcal{H})\right)^{\frac{1}{2}}.$$

Example: $\ell = \ell_{\text{mae}}$. Plug in $\ell = \ell_{\text{mae}} = 1 - \frac{e^{h(x,y)}}{\sum_{y'\in \overline{y}} e^{h(x,y')}}$ in (4), we obtain

$$\mathsf{L} = 1 - \frac{e^{h(x,y)}}{\sum_{y'\in \overline{y}} e^{h(x,y')}}$$
$$+ \sum_{j=1}^{n_e}(1 - c_j(x,y))\left(1 - \frac{e^{h(x,n+j)}}{\sum_{y'\in \overline{y}} e^{h(x,y')}}\right).$$

By [48, Theorem 1], ℓ_{mae} admits an $\mathcal{H}$-consistency bound with respect to ℓ_{0-1} with $\Gamma(t) = nt$, using Corollary 1, we obtain

$$\mathcal{E}_{\mathsf{L}_{\text{def}}}(h) - \mathcal{E}_{\mathsf{L}_{\text{def}}}^*(\mathcal{H}) \le n\left(\mathcal{E}_{\mathsf{L}}(h) - \mathcal{E}_{\mathsf{L}}^*(\mathcal{H})\right).$$

D.2 ℓ Being Adopted as Sum Losses

Example: $\ell = \Phi_{\text{sq}}^{\text{sum}}$. Plug in $\ell = \Phi_{\text{sq}}^{\text{sum}} = \sum_{y'\ne y}\Phi_{\text{sq}}(h(x,y) - h(x,y'))$ in (4), we obtain

$$\mathsf{L} = \sum_{y'\ne y}\Phi_{\text{sq}}\left(\Delta_h(x,y,y')\right)$$
$$+ \sum_{j=1}^{n_e}(1 - c_j(x,y))\sum_{y'\ne n+j}\Phi_{\text{sq}}\left(\Delta_h(x,n+j,y')\right),$$

where $\Delta_h(x,y,y') = h(x,y) - h(x,y')$ and $\Phi_{\text{sq}}(t) = \max\{0, 1-t\}^2$. By [5, Table 2], $\Phi_{\text{sq}}^{\text{sum}}$ admits an $\mathcal{H}$-consistency bound with respect to ℓ_{0-1} with $\Gamma(t) = \sqrt{t}$, using

Corollary 1, we obtain

$$\mathcal{E}_{\mathsf{L}_{\mathrm{def}}}(h) - \mathcal{E}^*_{\mathsf{L}_{\mathrm{def}}}(\mathcal{H}) \leq \left(n_e + 1 - \sum_{j=1}^{n_e} \underline{c}_j\right) \left(\frac{\mathcal{E}_\mathsf{L}(h) - \mathcal{E}^*_\mathsf{L}(\mathcal{H})}{n_e + 1 - \sum_{j=1}^{n_e} \overline{c}_j}\right)^{\frac{1}{2}}.$$

Since $1 \leq n_e + 1 - \sum_{j=1}^{n_e} \overline{c}_j \leq n_e + 1 - \sum_{j=1}^{n_e} \underline{c}_j \leq n_e + 1$, the bound can be simplified as

$$\mathcal{E}_{\mathsf{L}_{\mathrm{def}}}(h) - \mathcal{E}^*_{\mathsf{L}_{\mathrm{def}}}(\mathcal{H}) \leq (n_e + 1)\left(\mathcal{E}_\mathsf{L}(h) - \mathcal{E}^*_\mathsf{L}(\mathcal{H})\right)^{\frac{1}{2}}.$$

Example: $\ell = \Phi^{\mathrm{sum}}_{\mathrm{exp}}$. Plug in $\ell = \Phi^{\mathrm{sum}}_{\mathrm{exp}} = \sum_{y' \neq y} \Phi_{\mathrm{exp}}(h(x,y) - h(x,y'))$ in (4), we obtain

$$\mathsf{L} = \sum_{y' \neq y} \Phi_{\mathrm{exp}}(\Delta_h(x,y,y'))$$

$$+ \sum_{j=1}^{n_e}(1 - c_j(x,y)) \sum_{y' \neq n+j} \Phi_{\mathrm{exp}}(\Delta_h(x, n+j, y')),$$

where $\Delta_h(x,y,y') = h(x,y) - h(x,y')$ and $\Phi_{\mathrm{exp}}(t) = e^{-t}$. By [5, Table 2], $\Phi^{\mathrm{sum}}_{\mathrm{exp}}$ admits an $\mathcal{H}$-consistency bound with respect to ℓ_{0-1} with $\Gamma(t) = \sqrt{2t}$, using Corollary 1, we obtain

$$\mathcal{E}_{\mathsf{L}_{\mathrm{def}}}(h) - \mathcal{E}^*_{\mathsf{L}_{\mathrm{def}}}(\mathcal{H}) \leq \sqrt{2}\left(n_e + 1 - \sum_{j=1}^{n_e} \underline{c}_j\right) \left(\frac{\mathcal{E}_\mathsf{L}(h) - \mathcal{E}^*_\mathsf{L}(\mathcal{H})}{n_e + 1 - \sum_{j=1}^{n_e} \overline{c}_j}\right)^{\frac{1}{2}}.$$

Since $1 \leq n_e + 1 - \sum_{j=1}^{n_e} \overline{c}_j \leq n_e + 1 - \sum_{j=1}^{n_e} \underline{c}_j \leq n_e + 1$, the bound can be simplified as

$$\mathcal{E}_{\mathsf{L}_{\mathrm{def}}}(h) - \mathcal{E}^*_{\mathsf{L}_{\mathrm{def}}}(\mathcal{H}) \leq \sqrt{2}(n_e + 1)\left(\mathcal{E}_\mathsf{L}(h) - \mathcal{E}^*_\mathsf{L}(\mathcal{H})\right)^{\frac{1}{2}}.$$

Example: $\ell = \Phi^{\mathrm{sum}}_\rho$. Plug in $\ell = \Phi^{\mathrm{sum}}_\rho = \sum_{y' \neq y} \Phi_\rho(h(x,y) - h(x,y'))$ in (4), we obtain

$$\mathsf{L} = \sum_{y' \neq y} \Phi_\rho(\Delta_h(x,y,y'))$$

$$+ \sum_{j=1}^{n_e}(1 - c_j(x,y)) \sum_{y' \neq n+j} \Phi_\rho(\Delta_h(x, n+j, y')),$$

where $\Delta_h(x,y,y') = h(x,y) - h(x,y')$ and $\Phi_\rho(t) = \min\{\max\{0, 1 - t/\rho\}, 1\}$. By [5, Table 2], Φ^{sum}_ρ admits an $\mathcal{H}$-consistency bound with respect to ℓ_{0-1} with $\Gamma(t) = t$, using Corollary 1, we obtain

$$\mathcal{E}_{\mathsf{L}_{\mathrm{def}}}(h) - \mathcal{E}^*_{\mathsf{L}_{\mathrm{def}}}(\mathcal{H}) \leq \mathcal{E}_\mathsf{L}(h) - \mathcal{E}^*_\mathsf{L}(\mathcal{H}).$$

D.3 ℓ Being Adopted as Constrained Losses

Example: $\ell = \Phi_{\text{hinge}}^{\text{cstnd}}$. Plug in $\ell = \Phi_{\text{hinge}}^{\text{cstnd}} = \sum_{y' \neq y} \Phi_{\text{hinge}}(-h(x, y'))$ in (4), we obtain

$$\mathsf{L} = \sum_{y' \neq y} \Phi_{\text{hinge}}(-h(x, y')) + \sum_{j=1}^{n_e}(1 - c_j(x, y)) \sum_{y' \neq n+j} \Phi_{\text{hinge}}(-h(x, y')),$$

where $\Phi_{\text{hinge}}(t) = \max\{0, 1 - t\}$ with the constraint that $\sum_{y \in \mathcal{Y}} h(x, y) = 0$. By [5, Table 3], $\Phi_{\text{hinge}}^{\text{cstnd}}$ admits an $\mathcal{H}$-consistency bound with respect to ℓ_{0-1} with $\Gamma(t) = t$, using Corollary 1, we obtain

$$\mathcal{E}_{\mathsf{L}_{\text{def}}}(h) - \mathcal{E}_{\mathsf{L}_{\text{def}}}^*(\mathcal{H}) \leq \mathcal{E}_{\mathsf{L}}(h) - \mathcal{E}_{\mathsf{L}}^*(\mathcal{H}).$$

Example: $\ell = \Phi_{\text{sq}}^{\text{cstnd}}$. Plug in $\ell = \Phi_{\text{sq}}^{\text{cstnd}} = \sum_{y' \neq y} \Phi_{\text{sq}}(-h(x, y'))$ in (4), we obtain

$$\mathsf{L} = \sum_{y' \neq y} \Phi_{\text{sq}}(-h(x, y')) + \sum_{j=1}^{n_e}(1 - c_j(x, y)) \sum_{y' \neq n+j} \Phi_{\text{sq}}(-h(x, y')),$$

where $\Phi_{\text{sq}}(t) = \max\{0, 1 - t\}^2$ with the constraint that $\sum_{y \in \mathcal{Y}} h(x, y) = 0$. By [5, Table 3], $\Phi_{\text{sq}}^{\text{cstnd}}$ admits an $\mathcal{H}$-consistency bound with respect to ℓ_{0-1} with $\Gamma(t) = \sqrt{t}$, using Corollary 1, we obtain

$$\mathcal{E}_{\mathsf{L}_{\text{def}}}(h) - \mathcal{E}_{\mathsf{L}_{\text{def}}}^*(\mathcal{H}) \leq \left(n_e + 1 - \sum_{j=1}^{n_e} \underline{c}_j\right) \left(\frac{\mathcal{E}_{\mathsf{L}}(h) - \mathcal{E}_{\mathsf{L}}^*(\mathcal{H})}{n_e + 1 - \sum_{j=1}^{n_e} \overline{c}_j}\right)^{\frac{1}{2}}.$$

Since $1 \leq n_e + 1 - \sum_{j=1}^{n_e} \overline{c}_j \leq n_e + 1 - \sum_{j=1}^{n_e} \underline{c}_j \leq n_e + 1$, the bound can be simplified as

$$\mathcal{E}_{\mathsf{L}_{\text{def}}}(h) - \mathcal{E}_{\mathsf{L}_{\text{def}}}^*(\mathcal{H}) \leq (n_e + 1)\left(\mathcal{E}_{\mathsf{L}}(h) - \mathcal{E}_{\mathsf{L}}^*(\mathcal{H})\right)^{\frac{1}{2}}.$$

Example: $\ell = \Phi_{\exp}^{\text{cstnd}}$. Plug in $\ell = \Phi_{\exp}^{\text{cstnd}} = \sum_{y' \neq y} \Phi_{\exp}(-h(x, y'))$ in (4), we obtain

$$\mathsf{L} = \sum_{y' \neq y} \Phi_{\exp}(-h(x, y')) + \sum_{j=1}^{n_e}(1 - c_j(x, y)) \sum_{y' \neq n+j} \Phi_{\exp}(-h(x, y')),$$

where $\Phi_{\exp}(t) = e^{-t}$ with the constraint that $\sum_{y \in \mathcal{Y}} h(x, y) = 0$. By [5, Table 3], $\Phi_{\exp}^{\text{cstnd}}$ admits an $\mathcal{H}$-consistency bound with respect to ℓ_{0-1} with $\Gamma(t) = \sqrt{2t}$, using Corollary 1, we obtain

$$\mathcal{E}_{\mathsf{L}_{\text{def}}}(h) - \mathcal{E}_{\mathsf{L}_{\text{def}}}^*(\mathcal{H}) \leq \sqrt{2}\left(n_e + 1 - \sum_{j=1}^{n_e} \underline{c}_j\right) \left(\frac{\mathcal{E}_{\mathsf{L}}(h) - \mathcal{E}_{\mathsf{L}}^*(\mathcal{H})}{n_e + 1 - \sum_{j=1}^{n_e} \overline{c}_j}\right)^{\frac{1}{2}}.$$

Since $1 \le n_e + 1 - \sum_{j=1}^{n_c} \overline{c}_j \le n_e + 1 - \sum_{j=1}^{n_c} \underline{c}_j \le n_e + 1$, the bound can be simplified as
$$\mathcal{E}_{\mathsf{L}_{\mathrm{def}}}(h) - \mathcal{E}^*_{\mathsf{L}_{\mathrm{def}}}(\mathcal{H}) \le \sqrt{2}(n_e + 1)\left(\mathcal{E}_{\mathsf{L}}(h) - \mathcal{E}^*_{\mathsf{L}}(\mathcal{H})\right)^{\frac{1}{2}}.$$

Example: $\ell = \Phi_\rho^{\mathrm{cstnd}}$. Plug in $\ell = \Phi_\rho^{\mathrm{cstnd}} = \sum_{y' \ne y} \Phi_\rho(-h(x, y'))$ in (4), we obtain
$$\mathsf{L} = \sum_{y' \ne y} \Phi_\rho(-h(x, y'))$$
$$+ \sum_{j=1}^{n_c} (1 - c_j(x, y)) \sum_{y' \ne n+j} \Phi_\rho(-h(x, y')),$$

where $\Phi_\rho(t) = \min\{\max\{0, 1 - t/\rho\}, 1\}$ with the constraint that $\sum_{y \in \mathcal{Y}} h(x, y) = 0$. By [5, Table 3], $\Phi_\rho^{\mathrm{cstnd}}$ admits an $\mathcal{H}$-consistency bound with respect to ℓ_{0-1} with $\Gamma(t) = t$, using Corollary 1, we obtain
$$\mathcal{E}_{\mathsf{L}_{\mathrm{def}}}(h) - \mathcal{E}^*_{\mathsf{L}_{\mathrm{def}}}(\mathcal{H}) \le \mathcal{E}_{\mathsf{L}}(h) - \mathcal{E}^*_{\mathsf{L}}(\mathcal{H}).$$

E Proof of Learning Bounds for Deferral Surrogate Losses (Theorem 2)

Theorem 2 (Learning bound). *Under the same assumptions as Theorem 1, for any $\delta > 0$, with probability at least $1 - \delta$ over the draw of an i.i.d sample S of size m, the following deferral loss estimation bound holds for $\widehat{h}_S$:*

$$\mathcal{E}_{\mathsf{L}_{\mathrm{def}}}(\widehat{h}_S) - \mathcal{E}^*_{\mathsf{L}_{\mathrm{def}}}(\mathcal{H}) + \mathcal{M}_{\mathsf{L}}(\mathcal{H}) \le \left(n_e + 1 - \sum_{j=1}^{n_c} \overline{c}_j\right)\Gamma\left(\frac{4\mathfrak{R}^{\mathsf{L}}_m(\mathcal{H}) + 2B_{\mathsf{L}}\sqrt{\frac{\log \frac{2}{\delta}}{2m}} + \mathcal{M}_{\mathsf{L}}(\mathcal{H})}{n_e + 1 - \sum_{j=1}^{n_c} \overline{c}_j}\right).$$

Proof. By using the standard Rademacher complexity bounds [56], for any $\delta > 0$, with probability at least $1 - \delta$, the following holds for all $h \in \mathcal{H}$:
$$\left|\mathcal{E}_{\mathsf{L}}(h) - \widehat{\mathcal{E}}_{\mathsf{L},S}(h)\right| \le 2\mathfrak{R}^{\mathsf{L}}_m(\mathcal{H}) + B_{\mathsf{L}}\sqrt{\frac{\log(2/\delta)}{2m}}.$$

Fix $\epsilon > 0$. By the definition of the infimum, there exists $h^* \in \mathcal{H}$ such that $\mathcal{E}_{\mathsf{L}}(h^*) \le \mathcal{E}^*_{\mathsf{L}}(\mathcal{H}) + \epsilon$. By definition of $\widehat{h}_S$, we have
$$\mathcal{E}_{\mathsf{L}}(\widehat{h}_S) - \mathcal{E}^*_{\mathsf{L}}(\mathcal{H})$$
$$= \mathcal{E}_{\mathsf{L}}(\widehat{h}_S) - \widehat{\mathcal{E}}_{\mathsf{L},S}(\widehat{h}_S) + \widehat{\mathcal{E}}_{\mathsf{L},S}(\widehat{h}_S) - \mathcal{E}^*_{\mathsf{L}}(\mathcal{H})$$
$$\le \mathcal{E}_{\mathsf{L}}(\widehat{h}_S) - \widehat{\mathcal{E}}_{\mathsf{L},S}(\widehat{h}_S) + \widehat{\mathcal{E}}_{\mathsf{L},S}(h^*) - \mathcal{E}^*_{\mathsf{L}}(\mathcal{H})$$
$$\le \mathcal{E}_{\mathsf{L}}(\widehat{h}_S) - \widehat{\mathcal{E}}_{\mathsf{L},S}(\widehat{h}_S) + \widehat{\mathcal{E}}_{\mathsf{L},S}(h^*) - \mathcal{E}^*_{\mathsf{L}}(h^*) + \epsilon$$
$$\le 2\left[2\mathfrak{R}^{\mathsf{L}}_m(\mathcal{H}) + B_{\mathsf{L}}\sqrt{\frac{\log(2/\delta)}{2m}}\right] + \epsilon.$$

Since the inequality holds for all $\epsilon > 0$, it implies:
$$\mathcal{E}_{\mathsf{L}}(\widehat{h}_S) - \mathcal{E}^*_{\mathsf{L}}(\mathcal{H}) \le 4\mathfrak{R}^{\mathsf{L}}_m(\mathcal{H}) + 2B_{\mathsf{L}}\sqrt{\frac{\log(2/\delta)}{2m}}.$$

Plugging in this inequality in the bound (5) completes the proof.

References

1. Acar, D.A.E., Gangrade, A., Saligrama, V.: Budget learning via bracketing. In: International Conference on Artificial Intelligence and Statistics, pp. 4109–4119 (2020)
2. Awasthi, P., Frank, N., Mao, A., Mohri, M., Zhong, Y.: Calibration and consistency of adversarial surrogate losses. Adv. Neural Inf. Process. Syst. **34**, 9804–9815 (2021)
3. Awasthi, P., Mao, A., Mohri, M., Zhong, Y.: A finer calibration analysis for adversarial robustness. arXiv preprint arXiv:2105.01550 (2021)
4. Awasthi, P., Mao, A., Mohri, M., Zhong, Y.: $\mathscr{H}$-consistency bounds for surrogate loss minimizers. In: International Conference on Machine Learning (2022)
5. Awasthi, P., Mao, A., Mohri, M., Zhong, Y.: Multi-class $\mathscr{H}$-consistency bounds. Adv. Neural Inf. Process. Syst. **35**, 782–795 (2022)
6. Awasthi, P., Mao, A., Mohri, M., Zhong, Y.: Theoretically grounded loss functions and algorithms for adversarial robustness. In: International Conference on Artificial Intelligence and Statistics, pp. 10077–10094 (2023)
7. Awasthi, P., Mao, A., Mohri, M., Zhong, Y.: DC-programming for neural network optimizations. J. Glob. Optim. 1–17 (2024)
8. Bansal, G., Nushi, B., Kamar, E., Horvitz, E., Weld, D.S.: Is the most accurate AI the best teammate? optimizing AI for teamwork. In: Proceedings of the AAAI Conference on Artificial Intelligence, pp. 11405–11414 (2021)
9. Bartlett, P.L., Jordan, M.I., McAuliffe, J.D.: Convexity, classification, and risk bounds. J. Am. Stat. Assoc. **101**(473), 138–156 (2006)
10. Bartlett, P.L., Wegkamp, M.H.: Classification with a reject option using a hinge loss. J. Mach. Learn. Res. **9**(8) (2008)
11. Benz, N.L.C., Rodriguez, M.G.: Counterfactual inference of second opinions. In: Uncertainty in Artificial Intelligence, pp. 453–463. PMLR (2022)
12. Berkson, J.: Application of the logistic function to bio-assay. J. Am. Stat. Assoc. **39**, 357–365 (1944)
13. Berkson, J.: Why I prefer logits to probits. Biometrics **7**(4), 327–339 (1951)
14. Bubeck, S., et al.: Sparks of artificial general intelligence: early experiments with gpt-4. arXiv preprint arXiv:2303.12712 (2023)
15. Cao, Y., et al.: Generalizing consistent multi-class classification with rejection to be compatible with arbitrary losses. Adv. Neural Inf. Process. Syst. **35**, 521–534 (2022)
16. Charoenphakdee, N., Cui, Z., Zhang, Y., Sugiyama, M.: Classification with rejection based on cost-sensitive classification. In: International Conference on Machine Learning, pp. 1507–1517 (2021)
17. Charusaie, M.A., Mozannar, H., Sontag, D., Samadi, S.: Sample efficient learning of predictors that complement humans. In: International Conference on Machine Learning, pp. 2972–3005 (2022)
18. Chow, C.: On optimum recognition error and reject tradeoff. IEEE Trans. Inf. Theory **16**(1), 41–46 (1970)
19. Chow, C.K.: An optimum character recognition system using decision functions. IRE Trans. Electron. Comput. **4**, 247–254 (1957)
20. Cortes, C., DeSalvo, G., Mohri, M.: Boosting with abstention. Adv. Neural Inf. Process. Syst. **29**, 1660–1668 (2016)
21. Cortes, C., DeSalvo, G., Mohri, M.: Learning with rejection. In: International Conference on Algorithmic Learning Theory, pp. 67–82 (2016)

22. Cortes, C., DeSalvo, G., Mohri, M.: Theory and algorithms for learning with rejection in binary classification. In: Annals of Mathematics and Artificial Intelligence, pp. 1–39 (2023)
23. De, A., Koley, P., Ganguly, N., Gomez-Rodriguez, M.: Regression under human assistance. In: Proceedings of the AAAI Conference on Artificial Intelligence, pp. 2611–2620 (2020)
24. El-Yaniv, R., et al.: On the foundations of noise-free selective classification. J.0 Mach. Learn. Res. **11**(5) (2010)
25. Gangrade, A., Kag, A., Saligrama, V.: Selective classification via one-sided prediction. In: International Conference on Artificial Intelligence and Statistics, pp. 2179–2187 (2021)
26. Gao, R., Saar-Tsechansky, M., De-Arteaga, M., Han, L., Lee, M.K., Lease, M.: Human-AI collaboration with bandit feedback. arXiv preprint arXiv:2105.10614 (2021)
27. Geifman, Y., El-Yaniv, R.: Selective classification for deep neural networks. Adv. Neural Inf. Process. Syst. **30**, 1–10 (2017)
28. Geifman, Y., El-Yaniv, R.: Selectivenet: a deep neural network with an integrated reject option. In: International Conference on Machine Learning, pp. 2151–2159 (2019)
29. Grandvalet, Y., Rakotomamonjy, A., Keshet, J., Canu, S.: Support vector machines with a reject option. Adv. Neural Inf. Process. Syst. **21**, 1–8 (2008)
30. He, K., Zhang, X., Ren, S., Sun, J.: Deep residual learning for image recognition. In: Proceedings of the IEEE Conference on Computer Vision and Pattern Recognition, pp. 770–778 (2016)
31. Hemmer, P., Schellhammer, S., Vössing, M., Jakubik, J., Satzger, G.: Forming effective human-AI teams: building machine learning models that complement the capabilities of multiple experts. arXiv preprint arXiv:2206.07948 (2022)
32. Hemmer, P., Thede, L., Vössing, M., Jakubik, J., Kühl, N.: Learning to defer with limited expert predictions. arXiv preprint arXiv:2304.07306 (2023)
33. Herbei, R., Wegkamp, M.: Classification with reject option. Can. J. Stat. (2005)
34. Joshi, S., Parbhoo, S., Doshi-Velez, F.: Pre-emptive learning-to-defer for sequential medical decision-making under uncertainty. arXiv preprint arXiv:2109.06312 (2021)
35. Kalai, A.T., Kanade, V., Mansour, Y.: Reliable agnostic learning. J. Comput. Syst. Sci. **78**(5), 1481–1495 (2012)
36. Kamar, E., Hacker, S., Horvitz, E.: Combining human and machine intelligence in large-scale crowdsourcing. In: AAMAS, pp. 467–474 (2012)
37. Kerrigan, G., Smyth, P., Steyvers, M.: Combining human predictions with model probabilities via confusion matrices and calibration. Adv. Neural. Inf. Process. Syst. **34**, 4421–4434 (2021)
38. Keswani, V., Lease, M., Kenthapadi, K.: Towards unbiased and accurate deferral to multiple experts. In: Proceedings of the 2021 AAAI/ACM Conference on AI, Ethics, and Society, pp. 154–165 (2021)
39. Kingma, D.P., Ba, J.: Adam: a method for stochastic optimization. arXiv preprint arXiv:1412.6980 (2014)
40. Kleinberg, J., Lakkaraju, H., Leskovec, J., Ludwig, J., Mullainathan, S.: Human decisions and machine predictions. Q. J. Econ. **133**(1), 237–293 (2018)
41. Krizhevsky, A.: Learning multiple layers of features from tiny images. Toronto University, Technical report (2009)
42. Kuznetsov, V., Mohri, M., Syed, U.: Multi-class deep boosting. Adv. Neural Inf. Process. Syst. **27**, 2501–2509 (2014)

43. Lee, Y., Lin, Y., Wahba, G.: Multicategory support vector machines: theory and application to the classification of microarray data and satellite radiance data. J. Am. Stat. Assoc. **99**(465), 67–81 (2004)
44. Liu, J., Gallego, B., Barbieri, S.: Incorporating uncertainty in learning to defer algorithms for safe computer-aided diagnosis. Sci. Rep. **12**(1), 1762 (2022)
45. Long, P., Servedio, R.: Consistency versus realizable H-consistency for multiclass classification. In: International Conference on Machine Learning, pp. 801–809 (2013)
46. Madras, D., Creager, E., Pitassi, T., Zemel, R.: Learning adversarially fair and transferable representations. arXiv preprint arXiv:1802.06309 (2018)
47. Mao, A., Mohri, C., Mohri, M., Zhong, Y.: Two-stage learning to defer with multiple experts. Adv. Neural Inf. Process. Syst. **36**, 1–29 (2023)
48. Mao, A., Mohri, M., Zhong, Y.: Cross-entropy loss functions: theoretical analysis and applications. In: International Conference on Machine Learning (2023)
49. Mao, A., Mohri, M., Zhong, Y.: H-consistency bounds: characterization and extensions. Adv. Neural Inf. Process. Syst. **36**, 1–39 (2023)
50. Mao, A., Mohri, M., Zhong, Y.: H-consistency bounds for pairwise misranking loss surrogates. In: International Conference on Machine Learning (2023)
51. Mao, A., Mohri, M., Zhong, Y.: Ranking with abstention. In: ICML 2023 Workshop the Many Facets of Preference-Based Learning (2023)
52. Mao, A., Mohri, M., Zhong, Y.: Structured prediction with stronger consistency guarantees. Adv. Neural Inf. Process. Syst. **36**, 46903–46937 (2023)
53. Mao, A., Mohri, M., Zhong, Y.: Predictor-rejector multi-class abstention: theoretical analysis and algorithms. In: International Conference on Algorithmic Learning Theory (2024)
54. Mao, A., Mohri, M., Zhong, Y.: Theoretically grounded loss functions and algorithms for score-based multi-class abstention. In: International Conference on Artificial Intelligence and Statistics (2024)
55. Mohri, C., Andor, D., Choi, E., Collins, M., Mao, A., Zhong, Y.: Learning to reject with a fixed predictor: application to decontextualization. In: International Conference on Learning Representations (2024)
56. Mohri, M., Rostamizadeh, A., Talwalkar, A.: Foundations of Machine Learning, 2nd edn. MIT Press, Cambridge (2018)
57. Mozannar, H., Satyanarayan, A., Sontag, D.: Teaching humans when to defer to a classifier via exemplars. In: Proceedings of the AAAI Conference on Artificial Intelligence, pp. 5323–5331 (2022)
58. Mozannar, H., Sontag, D.: Consistent estimators for learning to defer to an expert. In: International Conference on Machine Learning, pp. 7076–7087 (2020)
59. Narasimhan, H., Jitkrittum, W., Menon, A.K., Rawat, A.S., Kumar, S.: Post-hoc estimators for learning to defer to an expert. Adv. Neural Inf. Process. Syst. **35**, 29292–29304 (2022)
60. Narasimhan, H., Menon, A.K., Jitkrittum, W., Kumar, S.: Learning to reject meets ood detection: are all abstentions created equal? arXiv preprint arXiv:2301.12386 (2023)
61. Netzer, Y., Wang, T., Coates, A., Bissacco, A., Wu, B., Ng, A.Y.: Reading digits in natural images with unsupervised feature learning. Adv. Neural Inf. Process. Syst. (2011)
62. Ni, C., Charoenphakdee, N., Honda, J., Sugiyama, M.: On the calibration of multiclass classification with rejection. Adv. Neural Inf. Process. Syst. **32**, 2582–2592 (2019)

63. Okati, N., De, A., Rodriguez, M.: Differentiable learning under triage. Adv. Neural. Inf. Process. Syst. **34**, 9140–9151 (2021)
64. Pradier, M.F., Zazo, J., Parbhoo, S., Perlis, R.H., Zazzi, M., Doshi-Velez, F.: Preferential mixture-of-experts: interpretable models that rely on human expertise as much as possible. AMIA Summits Transl. Sci. Proc. **2021**, 525 (2021)
65. Raghu, M., Blumer, K., Corrado, G., Kleinberg, J., Obermeyer, Z., Mullainathan, S.: The algorithmic automation problem: prediction, triage, and human effort. arXiv preprint arXiv:1903.12220 (2019)
66. Raman, N., Yee, M.: Improving learning-to-defer algorithms through fine-tuning. arXiv preprint arXiv:2112.10768 (2021)
67. Ramaswamy, H.G., Tewari, A., Agarwal, S.: Consistent algorithms for multiclass classification with an abstain option. Electron. J. Stat. **12**(1), 530–554 (2018)
68. Steinwart, I.: How to compare different loss functions and their risks. Constr. Approx. **26**(2), 225–287 (2007)
69. Straitouri, E., Singla, A., Meresht, V.B., Gomez-Rodriguez, M.: Reinforcement learning under algorithmic triage. arXiv preprint arXiv:2109.11328 (2021)
70. Straitouri, E., Wang, L., Okati, N., Rodriguez, M.G.: Provably improving expert predictions with conformal prediction. arXiv preprint arXiv:2201.12006 (2022)
71. Tan, S., Adebayo, J., Inkpen, K., Kamar, E.: Investigating human+ machine complementarity for recidivism predictions. arXiv preprint arXiv:1808.09123 (2018)
72. Verhulst, P.F.: Notice sur la loi que la population suit dans son accroissement. Correspondance mathématique et physique **10**, 113–121 (1838)
73. Verhulst, P.F.: Recherches mathématiques sur la loi d'accroissement de la population. Nouveaux Mémoires de l'Académie Royale des Sciences et Belles-Lettres de Bruxelles **18**, 1–42 (1845)
74. Verma, R., Barrejón, D., Nalisnick, E.: Learning to defer to multiple experts: consistent surrogate losses, confidence calibration, and conformal ensembles. In: International Conference on Artificial Intelligence and Statistics, pp. 11415–11434 (2023)
75. Verma, R., Nalisnick, E.: Calibrated learning to defer with one-vs-all classifiers. In: International Conference on Machine Learning, pp. 22184–22202 (2022)
76. Wei, J., et al.: Emergent abilities of large language models. CoRR arxiv:2206.07682 (2022)
77. Weston, J., Watkins, C.: Multi-class support vector machines. Technical report, Citeseer (1998)
78. Wiener, Y., El-Yaniv, R.: Agnostic selective classification. Adv. Neural Inf. Process. Syst. **24**, 1–9 (2011)
79. Wilder, B., Horvitz, E., Kamar, E.: Learning to complement humans. In: International Joint Conferences on Artificial Intelligence, pp. 1526–1533 (2021)
80. Yuan, M., Wegkamp, M.: Classification methods with reject option based on convex risk minimization. J. Mach. Learn. Res. **11**(1) (2010)
81. Yuan, M., Wegkamp, M.: SVMs with a reject option. In: Bernoulli (2011)
82. Zhang, M., Agarwal, S.: Bayes consistency vs. H-consistency: the interplay between surrogate loss functions and the scoring function class. Adv. Neural Inf. Process. Syst. (2020)
83. Zhang, T.: Statistical behavior and consistency of classification methods based on convex risk minimization. Ann. Stat. **32**(1), 56–85 (2004)
84. Zhang, Z., Sabuncu, M.: Generalized cross entropy loss for training deep neural networks with noisy labels. Adv. Neural Inf. Process. Syst. (2018)

85. Zhao, J., Agrawal, M., Razavi, P., Sontag, D.: Directing human attention in event localization for clinical timeline creation. In: Machine Learning for Healthcare Conference, pp. 80–102 (2021)
86. Zheng, C., Wu, G., Bao, F., Cao, Y., Li, C., Zhu, J.: Revisiting discriminative vs. generative classifiers: theory and implications. arXiv preprint arXiv:2302.02334 (2023)
87. Ziyin, L., Wang, Z., Liang, P.P., Salakhutdinov, R., Morency, L.P., Ueda, M.: Deep gamblers: learning to abstain with portfolio theory. arXiv preprint arXiv:1907.00208 (2019)

On Sample Reuse Methods for Answering k-wise Statistical Queries

Lev Reyzin and Duan Tu[✉]

University of Illinois at Chicago, Chicago, IL 60607, USA
{lreyzin,dtu4}@uic.edu

Abstract. This paper examines the computational and sample complexity of answering k-wise statistical queries, which were introduced by Felman and Ghazi [9] as a generalization to the standard statistical query model of Kearns [11]. In particular, our paper studies two sample reuse schemes: (1) reusing independent "pseudo-samples" for adaptive queries and (2) reusing dependent k-wise samples for non-adaptive queries. Comparing to a baseline non-reuse strategy, we show that the first reuse method offers a trade-off between k, the arity of the query, and M, the total number of queries to be answered. We also show that the second reuse method performs no worse than the baseline, and possibly better, from the perspective of variance reduction.

Keywords: Statistical query learning · Sample reuse · Differential privacy

1 Introduction

In this paper we study the sample complexity of answering M different k-wise statistical queries. k-wise statistical queries are a generalization of the statistical query model introduced by Kearns [11] and widely studied thereafter [3,4]. While unary statistical queries look at the expectation of a function $q: X \to \{0,1\}$ from one data point onto a binary range, k-wise queries $q: X^k \to \{0,1\}$ use samples of size $k \geq 1$. The importance of being able to answer k-wise queries for larger values of k is illustrated by Felman and Ghazi [9], who showed that as k increases, strictly more problems can be solved using k-wise queries.

Known methods for answering statistical queries (SQs) include strategies ranging from straightforward sampling methods to more involved approaches involving principled sample reuse from the perspective of adaptive data analysis [5]. In this paper we analyze these varying approaches for the more general k-wise case and find a trade-off in which method is the best depending on the relative values of k, the arity of the query, and M, the total number of queries to be answered. We also give a different view for a known strategy for sample reuse, and show that it performs no worse than the original non-reuse strategy, and possibly better.

There are two natural ways to improve beyond the straightforward sampling approach: (1) reuse independent *pseudo-samples* for adaptive queries and (2) reuse dependent k-wise samples for non-adaptive queries. In the first method, each pseudo-sample is composed by drawing k i.i.d points from the sample domain, so pseudo-samples are mutually independent and identically distributed according to the product distribution. Intuitively, the first method reuses the same set of i.i.d. pseudo-samples among all queries, while the second method draws a new set of points for each query and takes all possible size k subsets to create dependent k-wise samples. It is worth noting that we are only concerned with the case of adaptive queries for method (1), since results for non-adaptive queries are already given by VC theory.

In the remaining parts of the paper, we shall provide rigorous statements of definitions and useful technical tools in Preliminaries, introduce the naive sampling approach in Baseline simulation of an SQ oracle, and then discuss results of the two sample reuse methods in the last two sections.

2 Preliminaries

We first give the definition of a k-wise SQ oracle.

Definition 1 (Feldman and Ghazi [9]). *Let $\mathcal{D}$ be a distribution over a domain X and $\tau > 0$. A k-wise statistical query oracle $\mathrm{STAT}_{\mathcal{D}}^{(k)}(\phi, \tau)$ is an oracle that given as input any query function $\phi : X^k \to \{0,1\}$ and a value τ, returns some value v such that $|v - \mathbb{E}_{\mathbf{x} \sim \mathcal{D}^k}[\phi(\mathbf{x})]| \leq \tau$.*

The goal of statistical query learning, as originally defined by Kearns [11], is to learn a target class of functions efficiently, achieving PAC guarantees while using the SQ oracle instead of labeled examples. In this case, to efficiently SQ-learn a function class, one wants to make a polynomial number of calls to the $\mathrm{STAT}_{\mathcal{D}}^{(k)}$ oracle, using tolerances τ such that $\frac{1}{\tau}$ is polynomially bounded away from 0, and using query functions ϕ evaluable in polynomial time. For detailed definitions and more about the SQ model, see Reyzin [12].

Next, we are interested in the sensitivity of query functions that are fed to the SQ oracle. The ℓ_1-sensitivity of a query measures the magnitude by which perturbing a single data point can change the query output in the worst case. It is an important parameter in determining the algorithm's required accuracy when answering queries.

Definition 2 (Dwork and Roth [7]). *The ℓ_1-sensitivity of a function $f : \mathbb{N}^{|X|} \to \mathbb{R}$ is*

$$\Delta f = \max_{\substack{x,y \in \mathbb{N}^{|X|} \\ \|x-y\|_1 = 1}} \|f(x) - f(y)\|_1$$

$$= \max_{\substack{x,y \in \mathbb{N}^{|X|} \\ \|x-y\|_1 = 1}} \sum_{i=1}^{|X|} |f(x_i) - f(y_i)|.$$

One main technique we use to study reusing pseudo-samples among adaptive queries is the Transfer Theorem developed in Bassily et al. [2]. The theorem says a differentially private learner that is accurate with respect to its samples generalizes to the population from which the samples were drawn. Bassily et al. [2] uses the term "max-KL stability" to refer to the differential privacy model of Dwork et al. [6], emphasizing it as one of the various notions of stability in machine learning. We state the definition of a differentially private learner and the Transfer Theorem as follows.

Definition 3 (Dwork and Roth [7]). *A randomized algorithm $\mathcal{M}$ with domain $\mathbb{N}^{|\mathcal{X}|}$ is (ε, δ)-differentially private if for all $\mathcal{S} \subseteq \mathrm{Range}(\mathcal{M})$ and for all $x, y \in \mathbb{N}^{|\mathcal{X}|}$ such that $\|x - y\|_1 \leq 1$:*

$$\mathbb{P}[\mathcal{M}(x) \in \mathcal{S}] \leq \exp(\varepsilon) \Pr[\mathcal{M}(y) \in \mathcal{S}] + \delta,$$

where the probability space is over the randomness of $\mathcal{M}$.

Before introducing the Transfer Theorem (Lemma 1), let us define what it means for an algorithm to be accurate with respect to a collection of samples and with respect to a population.

Definition 4 (Bassily et al. [2]). *A mechanism $\mathcal{M}$ is (α, β)-accurate with respect to samples of size n from $\mathcal{X}$ for M adaptively chosen queries from Φ, if for every adversary $\mathcal{A}$, which gives output $(a_1, \cdots, a_M)$,*

$$\mathbb{P}\left[\max_{j \in [M]} \left|a_j - \frac{1}{n} \sum_{i \in [n]} \phi_j(\mathbf{x}_i)\right| \leq \alpha \right] \geq 1 - \beta.$$

A mechanism $\mathcal{M}$ is (α, β)-accurate with respect to the population for M adaptively chosen queries from Φ given n samples $\mathbf{x} \in \mathcal{X}$, if for every adversary $\mathcal{A}$, which gives output $(a_1, \cdots, a_M)$,

$$\mathbb{P}\left[\max_{j \in [M]} \left|a_j - \mathbb{E}\, \phi_j(\mathbf{x})\right| \leq \alpha \right] \geq 1 - \beta.$$

Now we are ready to state the Transfer Theorem.

Lemma 1 (Transfer Theorem, by Bassily et al. [2]). *Let Φ be a family of Δ-sensitive queries on X^k. Assume that for some $\alpha, \beta \in (0, 0.1)$, an algorithm $\mathcal{A}$ is*

1. *$(\varepsilon' = \alpha/64\Delta n, \delta' = \alpha\beta/32\Delta n)$-max-KL stable for M adaptively chosen queries from Φ and*
2. *$(\alpha' = \alpha/8, \beta' = \alpha\beta/16\Delta n)$-accurate with respect to its n samples from X^k for M adaptively chosen queries from Φ.*

Then $\mathcal{A}$ is (α, β)-accurate with respect to the population for M adaptively chosen queries from Φ given n samples from X^k.

One can achieve the privacy requirement (via max-KL stability) in the Transfer Theorem through the *Laplace mechanism*. Recall that Δf denotes the ℓ_1-sensitivity of function f. Recall that the Laplace Distribution centered at 0 with scale parameter b has probability density function

$$\mathrm{Lap}(b) = \frac{1}{2b} \exp\left(-\frac{|x|}{b}\right).$$

Definition 5 (Dwork and Roth [7]). *Given any function $f : \mathbb{N}^{|\mathcal{X}|} \to \mathbb{R}^k$, the Laplace mechanism is defined as:*

$$\mathcal{M}_L(x, f(\cdot), \varepsilon) = f(x) + (Y_1, \cdots, Y_k),$$

where Y_i are i.i.d. random variables drawn from the Laplace Distribution of scale parameter $\Delta f/\varepsilon$, denoted as $\mathrm{Lap}(\Delta f/\varepsilon)$.

Let $\mathcal{A}$ be an algorithm that calculates the average of a function $\phi : X^k \to \{0, 1\}$ over n samples. Suppose ϕ has ℓ_1-sensitivity Δ. After computing the true average value a, the Laplace mechanism outputs $v = a + y$ where $y \sim \mathrm{Lap}(\Delta/\varepsilon)$ is drawn from the Laplace distribution with scale parameter Δ/ε.

Lemma 2 (Dwork and Roth [7]). *For any function $f : \mathbb{N}^{|\mathcal{X}|} \to \mathbb{R}^k$, the Laplace mechanism guarantees $(\varepsilon, 0)$-differential privacy.*

It is easy to come up with a high probability bound on the amount of noise added by the Laplace mechanism.

Lemma 3 (Dwork and Roth [7]). *Let $f : \mathbb{N}^{|\mathcal{X}|} \to \mathbb{R}^k$. Let the output of the Laplace mechanism be $y = \mathcal{M}_L(x, f(\cdot), \varepsilon)$. Then $\forall \delta \in (0, 1]$:*

$$\mathbb{P}\left[\|f(x) - y\|_\infty \geq \ln\left(\frac{k}{\delta}\right) \cdot \left(\frac{\Delta f}{\varepsilon}\right)\right] \leq \delta.$$

3 Baseline Simulation of an SQ Oracle

In this section we discuss the sample complexity of learning with k-wise SQs without sample reuse. An algorithm simulates a k-wise SQ oracle by taking empirical averages. This simulation is extended from the one given by Kearns [11] for unary SQ oracles. In the k-wise case, to each query function ϕ_i the learner feeds it a fresh batch of $\tilde{n}$ i.i.d. pseudo-samples $S_i = \{\mathbf{x}_1, \cdots, \mathbf{x}_{\tilde{n}}\}$, where each pseudo-sample $\mathbf{x}_j = (x_{j_1}, \cdots x_{j_k})$ consists of k sample points. Then the learner computes the empirical average of $\phi_i(\mathbf{x})$ over the set of pseudo-sample S_i. With high probability, the empirical average will fall within the amount of tolerance allowed by the SQ oracle from the true expectation of ϕ_i, thanks to concentration inequalities. Proposition 1 provides the quantitative result of this baseline approach.

Proposition 1. *Suppose there exists an SQ learner that makes M k-wise statistical queries of tolerance τ to learn over a class $\mathcal{C}$, then there exists a simulation algorithm, which does not reuse any samples, for which a set of i.i.d. samples of size*

$$n = O\left(k\frac{M}{\tau^2}\log\left(\frac{M}{\delta}\right)\right)$$

is sufficient to PAC learn $\mathcal{C}$ with error bounded by ε and probability of failure bounded by δ.

Proof. We take the specified number of samples and partition them into

$$\tilde{n} = \frac{n}{Mk} = O\left(\frac{1}{\tau^2}\log\left(\frac{M}{\delta}\right)\right)$$

i.i.d. pseudo-samples for each query function ϕ_i. The Hoeffding bound guarantees that the empirical average over $\tilde{n}$ pseudo-samples falls within $\pm\tau$ from $\mathbb{E}[\phi_i(\mathbf{x})]$ with probability $\geq 1 - \frac{\delta}{M}$. Then apply the union bound and we obtain that with probability $\geq 1 - \delta$, the empirical average falls within $\pm\tau$ from the true expectation for all queries. Hence we have successfully simulated the k-wise SQ learner with high probability, fulfilling the PAC requirements. □

4 Independent Pseudo-Samples for Adaptive Queries

In this section we discuss the reuse of independent pseudo-samples for adaptive queries. Suppose there exists a k-wise SQ learner that efficiently SQ-learns a function class by asking M adaptive k-wise queries $\phi_1, \cdots, \phi_M$. Similar to the baseline case, our algorithm (Algorithm 1) simulates a k-wise SQ oracle through taking empirical averages. However, what is different from the baseline case is that Algorithm 1 partitions the set of n samples $x \sim \mathcal{D}$ into $\tilde{n} = n/k$ parts to create $\tilde{n}$ i.i.d. pseudo-samples $\mathbf{x} = (x_1, \cdots, x_k) \sim \mathcal{D}^k$. It then reuses the same set of pseudo-samples among all queries when taking their empirical averages.

Now we state our main sample complexity result. Theorem 1 provides the optimal sample complexity for an algorithm that reuses the same set of independent pseudo-samples while answering adaptive queries.

Theorem 1. *Suppose there exists an SQ learner that makes M k-wise statistical queries of tolerance τ to learn a class $\mathcal{C}$, then there exists a simulation algorithm, which reuses independent pseudo-samples among the M queries, for which a set of i.i.d. samples of size*

$$n = O\left(\frac{k^2\sqrt{M}}{\tau^2}\log\left(\max\left\{M,\frac{k}{\tau}\right\}\frac{1}{\delta}\right)\right)$$

$$= O\left(\frac{k^2\sqrt{M}}{\tau^2}\log\left(\frac{Mk}{\tau\delta}\right)\right)$$

is sufficient to PAC-learn $\mathcal{C}$ with error bounded by ε and probability of failure bounded by δ.

Comparing Theorem 1 to the naive bound in Proposition 1, we observe an interesting trade-off between the arity of the query k, and the total number of queries M. The trade-off suggests that only when a learner uses a large amount of short queries ($k < \sqrt{M}$) is it worth to reuse pseudo-samples.

It is worth noting that Algorithm 1 is specific to k-wise statistical queries and it differs from approaches that work for low-sensitivity queries in general. In addition to having low sensitivity, statistical queries and their k-wise generalizations have the additional property that they can be evaluated on k points at a time, and are therefore amenable to sampling techniques, which can produce potential speedups (see Fish, Reyzin, and Rubinstein [10]). This allows us to evaluate our queries on pseudo-samples, each of which consists of k sample points.

Algorithm 1. Reusing Independent Pseudo-samples for Adaptive Queries

Inputs. Sample points $x \in X$ and k-wise Statistical Queries $\phi_1,\cdots,\phi_M$, where $\phi_i : X^k \to \{0,1\}$ for all $i \in [M]$.
Outputs. $v \in \mathbb{R}^M$.

1: Draw $O\left(\frac{k^2\sqrt{M}}{\tau^2}\log(\frac{Mk}{\tau\delta})\right)$ i.i.d. sample points $x \sim \mathcal{D}$. Create $\tilde{n} = \frac{n}{k}$ copies of i.i.d. pseudo-samples $\mathbf{x}_j = (x_{j_1},\cdots,x_{j_k}) \sim \mathcal{D}^k$, where $j = 1,\cdots,\tilde{n}$.
2: **for** $i = 1,\cdots,M$ **do**
3:     **for** $j = 1,\cdots,\tilde{n}$ **do**
4:         $a_{ij} \leftarrow \phi_i(\mathbf{x}_j)$
5:     **end for**
6:     Draw Laplace noise $y \sim \text{Lap}\left(\frac{128k^2\sqrt{M}}{\tau n}\right)$.
7:     $v_i \leftarrow \left(\frac{1}{\tilde{n}}\sum_{j=1}^{\tilde{n}} a_{ij}\right) + y$
8: **end for**
9: $v \leftarrow (v_1,\cdots,v_M)$

The sample complexity achieved by Algorithm 1 is no worse than the bound $\tilde{O}\left(\frac{\sqrt{M}}{\tau^2}\right)$ known for general low-sensitivity queries in Bassily et al. [2]. Their approach for general low-sensitivity queries takes time $\text{poly}(n, \log|X|)$ per oracle call, whereas our Algorithm 1 runs in $\text{poly}(k, \log|X|)$ time per call to the oracle (assuming polynomial-time evaluability of the respective queries). This can potentially create a dramatic improvement in running time, since the straightforward non-sampling approach for exactly evaluating a k-wise query on a sample of n points would be to evaluate it on all k-point subsets, which is indeed polynomial in n but exponential in k. In fact, we go on to analyze that particular approach towards the end of the paper.

In the remaining parts of this section, we first discuss a couple technical tools used to prove Theorem 1 and then we give the proof itself.

4.1 Privacy Composition

To ensure that the simulation generalizes to the sample distribution, we apply Lemma 1 (Transfer Theorem), which demands the algorithm be differentially private. The algorithm composes multiple query functions, so in order to achieve the required level of privacy overall, we need to use results on privacy composition to figure out what level of privacy is required for each individual query.

There are two well-known bounds on the privacy of query composition: simple composition and advanced composition. Simple composition provides the elementary bound that, when a learner uses independent queries, its privacy equals to the sum of privacy of all queries. Advanced composition deals with the more complicated situation, one where the learner poses adaptive queries to the same database repeatedly. We shall see that under appropriate choice of parameters, advanced composition offers tighter privacy bound than simple composition (by a factor of $\sqrt{M}$). The exact statements of the two composition results are provided by Lemma 4 and Lemma 5.

Lemma 4 (Simple Composition, as presented in Dwork and Roth [7]). *Let $\mathcal{A}_i : X^k \to \{0,1\}$ be an $(\varepsilon_i, \delta_i)$-differentially private algorithm for $i = 1, \cdots, M$. Then $\mathcal{A} = (\mathcal{A}_1, \cdots, \mathcal{A}_M)$ is $\left(\sum_{i=1}^{M} \varepsilon_i, \sum_{i=1}^{M} \delta_i\right)$-differentially private.*

Lemma 5 (Advanced Composition, by Dwork, Rothblum, and Vadhan [8]). *For all $\varepsilon, \delta, \delta' \geq 0$, the class of (ε, δ)-deferentially private mechanisms satisfies $(\varepsilon', M\delta + \delta')$-differential privacy under M-fold adaptive composition for*

$$\varepsilon' = \varepsilon\sqrt{2M\ln(1/\delta')} + M\varepsilon(e^\varepsilon - 1).$$

Observe that

$$\varepsilon' \leq \varepsilon\sqrt{2M\ln(1/\delta')} + M\varepsilon^2 = O\left(\varepsilon\sqrt{M\ln(1/\delta')}\right)$$

when ε is small. By choosing δ' small, say $\delta' = 1/e$, it can be shown that the M-fold adaptive composition satisfies $(\varepsilon\sqrt{M}, \delta M)$ - differential privacy [2].

Theorem 1 uses advanced composition of privacy. It is important to mention that advanced composition is necessary when analyzing pseudo-sample reuse. Since the algorithm uses adaptive queries, it needs to be strict when budgeting the privacy level for each query. Otherwise, an excess amount of Laplace noise would need to be added, which will overturn the effect of sample reuse. As shown in Theorem 2, if the algorithm composed privacy of the queries as if they were independent, the resulting sample complexity is actually worse than the baseline bound.

Theorem 2. *Under the setting of Theorem 1, except that suppose the simulation algorithm treats the M queries as if they were independent and calculates their overall privacy through simple composition, a set of i.i.d. samples of size*

$$n = O\left(\frac{k^2 M}{\tau^2} \log\left(\max\left\{M, \frac{k}{\tau}\right\} \frac{1}{\delta}\right)\right)$$
$$= O\left(\frac{k^2 M}{\tau^2} \log\left(\frac{Mk}{\tau\delta}\right)\right)$$

is sufficient to PAC learn $\mathcal{C}$ with error bounded by ε and probability of failure bounded by δ.

We omit the proof for Theorem 2 since it closely resembles that of Theorem 1, with the only difference being the privacy composition calculations.

4.2 Laplace Mechanism

Now that we know to use advanced composition, let us consider how to achieve the desired level of privacy for each query function. As suggested by Lemma 2, we adopt Laplace mechanism, the standard technique that offers privacy guarantee for algorithms.

For each ϕ_i, Algorithm 1 outputs $v_i = a_i + y$, where a_i is the empirical average of ϕ_i over a large set of pseudo-samples and y is a small Laplace noise parameter. There are two key considerations when choosing the parameters. First, the sample set needs to be large enough so that the empirical average is close to the true expectation with high probability. Second, the Laplace noise needs to be small enough so that it does not steer the empirical average away from the expected average too far, but in the meantime still large enough to maintain privacy.

Using Lemma 2, we choose $y \sim \text{Lap}(\Delta \cdot \frac{128k}{\tau})$, which preserves $(\frac{\tau}{128k}, 0)$-differential privacy for each query, surpassing the requirement of the Transfer Theorem (Lemma 1). Here Δ is the ℓ_1-sensitivity of the empirical average of ϕ over all pseudo-samples. In an attempt to simplify the writing, we abuse notation and use $\Delta\phi$ to represent the aforementioned ℓ_1-sensitivity.

Proposition 2. *The ℓ_1-sensitivity of the empirical average of $\phi: X^k \to \{0,1\}$ is $\Delta\phi \leq \frac{k}{n}$.*

Proof. Among all pseudo-samples $\mathbf{x}_i \in S$ and $\mathbf{x}'_i \in S'$ where $i = 1, \cdots, \tilde{n}$, exactly one pair is different $\mathbf{x}_j \neq \mathbf{x}'_j$. Then $|\phi(x_i) - \phi(x'_i)| = 0$ for all $i \neq j$, while $|\phi(x_j) - \phi(x'_j)| \leq 1$. Therefore,

$$\Delta\phi = \max_{\substack{S,S' \subseteq X^k \\ \text{s.t. } \|S-S'\|_1 = 1}} \left\| \frac{1}{\tilde{n}} \sum_{i=1}^{\tilde{n}} \left(\phi(\mathbf{x}_i) - \phi(\mathbf{x}'_i) \right) \right\|_1,$$

which can trivially be bounded as $\Delta\phi \leq 1/\tilde{n} = k/n$. □

Proof of Theorem 1

Given an efficient k-wise SQ learner that learns $\mathcal{C}$ approximately correct (to an error ε), the empirical average simulation wishes to mimic the learner's query outputs with high probability. In the language of the Transfer Theorem (Lemma 1), that is to say the simulator needs to be (τ, δ)-accurate with respect to the population. We prove Theorem 1 using the Transfer Theorem.

Proof (Proof of Theorem 1). Given the total allowed error of τ, we allocate $\tau/2$ to the empirical average and $\tau/2$ to the added Laplace noise. We first analyze the empirical average. To achieve $(\tau/2, \delta)$-accuracy with respect to the population for M adaptively chosen queries, the Transfer Theorem demands

(i) the simulation is $\left(\frac{\tau}{128k}, \frac{\tau\delta}{64k}\right)$-differentially private for M adaptive queries,
(ii) the simulation is $\left(\frac{\tau}{16}, \frac{\tau\delta}{32k}\right)$-accurate with respect to n samples for M adaptive queries.

To satisfy (i), we adopt advanced composition. According to Lemma 5, each of the M queries needs to be $\left(\frac{\tau}{128k\sqrt{M}}, \frac{\tau\delta}{64kM}\right)$-differentially private to obtain the composed privacy stated in (i). We know each query has ℓ_1-sensitivity k/n through Lemma 2. Then following the standard technique stated in Lemma 2, we add Laplace noise of scale $\frac{128k^2\sqrt{M}}{\tau n}$ to each query average, which achieves $\left(\frac{\tau}{128k\sqrt{M}}, 0\right)$-differential privacy, surpassing the needed amount. Lemma 3 verifies that the added Laplace noise is bounded above by $\frac{128k^2\sqrt{M}}{\tau n} \log \frac{2M}{\delta}$ with probability $\geq 1 - \frac{\delta}{2M}$. In order to restrict the amount of Laplace noise within $\tau/2$ with high probability, we ask that

$$\frac{128k^2\sqrt{M}}{\tau n} \log \frac{2M}{\delta} \leq \frac{\tau}{2},$$

which implies

$$n = O\left(\frac{k^2\sqrt{M}}{\tau^2} \log \frac{M}{\delta}\right) \tag{1}$$

is sufficient. Now let us consider (ii). It suffices to show that for all queries ϕ_i, the simulator's output a_i satisfies

$$\mathbb{P}\left[|\text{err}_\mathbf{x}(\phi_i, a_i)| \leq \frac{\tau}{16}\right] \geq 1 - \frac{\tau\delta}{32k}.$$

We know that
$$a_i = \frac{1}{\tilde{n}} \sum_{j=1}^{\tilde{n}} \phi_i(\mathbf{x}_j) + \text{Lap}\left(\frac{128k\sqrt{M}}{\tau \tilde{n}}\right),$$

so for all i,
$$|\text{err}_\mathbf{x}(\phi_i, a_i)| = |a_i - \phi_i(\mathbf{x})|$$
$$= \left|a_i - \frac{1}{\tilde{n}} \sum_{j=1}^{\tilde{n}} \phi_i(x_j)\right|$$
$$= \text{Lap}\left(\frac{128k\sqrt{M}}{\tau \tilde{n}}\right).$$

According to Lemma 3, it is easy to verify that with probability $\geq 1 - \frac{\tau\delta}{32k}$, the Laplace noise of scale $\frac{128k\sqrt{M}}{\tau\tilde{n}}$ is $O(\frac{k^2\sqrt{M}}{\tau n} \log \frac{k}{\tau\delta})$. To satisfy (ii), we ask that $\frac{128k^2\sqrt{M}}{\tau n} \log \frac{32k}{\tau\delta} \leq \frac{\tau}{16}$, which implies

$$n = O\left(\frac{k^2\sqrt{M}}{\tau^2} \log \frac{k}{\tau\delta}\right) \tag{2}$$

is sufficient. Combining inequalities (1) and (2), we get

$$n = O\left(\max\left\{\frac{k^2\sqrt{M}}{\tau^2} \log \frac{M}{\delta}, \frac{k^2\sqrt{M}}{\tau^2} \log \frac{k}{\tau\delta}\right\}\right)$$
$$= O\left(\frac{k^2\sqrt{M}}{\tau^2} \log\left(\max\left\{M, \frac{k}{\tau}\right\} \frac{1}{\delta}\right)\right)$$
$$= O\left(\frac{k^2\sqrt{M}}{\tau^2} \log\left(\frac{Mk}{\tau\delta}\right)\right),$$

which completes the proof. □

5 Dependent k-wise Samples for Non-adaptive Queries

Now we examine the second reuse method. Algorithm 2 draws n i.i.d. sample points $x \sim X$ and partitions them into M equal parts, $S_1, \cdots, S_M$, to be used by M queries. Denote the size of each part as $|S_i| = \hat{n}$, so the total number of samples is $n = M\hat{n}$. For each query, the algorithm calculates its empirical average over $\binom{\hat{n}}{k}$ k-wise samples, which are generated by taking all size k subsets of S_i.

Algorithm 2. Dependent k-wise Samples for Non-adaptive Queries

Inputs. Sample points $x \in X$ and k-wise Statistical Queries $\phi_i : X^k \to \{0, 1\}$, where $i = 1, \cdots, M$.
Outputs. $v = (v_1, \cdots, v_M) \in \mathbb{R}^M$.
1: Draw n i.i.d. sample points $x \sim \mathcal{D}$ and partition them into M equal parts $S_1, \cdots, S_M$, where $|S_i| = \hat{n}$.
2: **for** $i = 1, \cdots, M$ **do**
3:     Take all size k subsets of S_i to create k-wise samples $\mathbf{x}_j = (x_{j_1}, \cdots, x_{j_k})$, where $j = 1, \cdots, \binom{\hat{n}}{k}$.
4:     Compute the empirical average of ϕ_i
$$v_i \leftarrow \frac{1}{\binom{\hat{n}}{k}} \sum_{j=1}^{\binom{\hat{n}}{k}} \phi_i(\mathbf{x}_j).$$
5: **end for**
6: $v \leftarrow (v_1, \cdots, v_M)$.

In contrast to creating independent pseudo-samples, Algorithm 2 uses all k-subsets of the provided sample set, yielding additional k-wise samples, although it fails to maintain their independence since each point contributes to $(k-1)$ samples.

We can analyze these dependent k-wise samples from the perspective of a hypergraph. In the language of hypergraphs, we can think of each sample point as a vertex and each k-wise sample as a k-hyperedge. The learner is given K_n^k, a complete hypergraph on n vertices, whose hyperedges contain k vertices (assuming k divides n). The learner uses k-hyperedges as inputs to the queries. Notice that the hyperedges are not independent with each other. Fortunately, we can bypass the hyperedge dependency through Baranyai's Theorem.

Theorem 3 (Baranyai [1]). *Every K_n^k hypergraph decomposes into a disjoint collection of 1-factors.*

Recall that a 1-factor is a set of hyperedges that touch each vertex in K_n^k exactly once. Intuitively, we can think of a 1-factor as a perfect matching. With the guarantee of decomposition given by Baranyai's Theorem, we are able to interpret the collection of dependent hyperedges as a set of perfect matchings. Although these matchings are dependent on one another, they each contain independent hyperedges within themselves. Figure 1 gives an example of when $n = 6$ and $k = 2$. As shown by Fig. 1, K_6^2 can be decomposed into a disjoint union of 1-factors, each of which consists of three mutually independent edges.

How well do dependent k-wise samples perform when we use them to estimate the expected value through empirical average? In each 1-factor, the independent hyperedges act like pseudo-samples introduced in Algorithm 1. Accordingly, in Theorem 4 we provide accuracy bounds of dependent k-wise samples by comparing its variance to that of independent pseudo-samples.

To set up Theorem 4, let $\phi : X^k \to \{0, 1\}$ be a k-wise statistical query, S be a set of samples $x \sim \mathcal{D}$, and suppose $|S| = n$, where k divides n. Let Y_p, Y_a be random variables that represent the empirical average of ϕ under the two

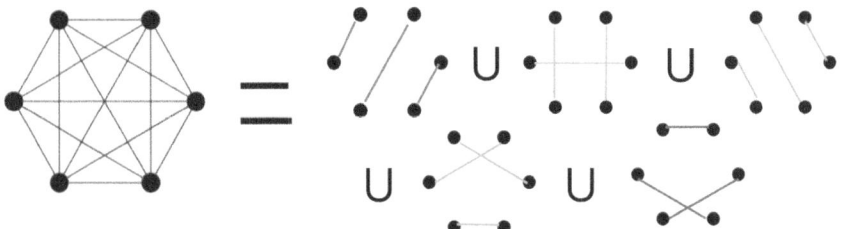

Fig. 1. An illustration of a decomposition of K_6^2 into five disjoint perfect matchings

sampling schemes respectively: creating n/k independent pseudo-samples and taking all $\binom{n}{k}$ k-subsets of S. The expected value of Y_p and Y_a both equal to $\mathbb{E}(\phi)$.

Theorem 4. *The variance of Y_p and Y_a satisfy*

$$\frac{1}{\binom{n-1}{k-1}}\mathrm{Var}(Y_p) \leq \mathrm{Var}(Y_a) \leq \mathrm{Var}(Y_p).$$

Proof. We first study the upper bound. Construct a complete hypergraph K_n^k with the given n sample points. With guarantee from Baranyai's theorem, we can decompose K_n^k into 1-factors $G_1, \cdots, G_m$, where $m = \binom{n-1}{k-1}$. Each G_i contains n/k i.i.d. hyperedges of length k. The vertices in these i.i.d. hyperedges form i.i.d. pseudo-samples used in Algorithm 1. Let Y_{G_i} be a random variable that represents the empirical average of ϕ over pseudo-samples in G_i. By previous analysis, we know for all $i = 1, \cdots, \binom{n-1}{k-1}$,

$$\mathrm{Var}(Y_p) = \mathrm{Var}(Y_{G_i}).$$

Observe that taking an empirical average over all $\binom{n}{k}$ hyperedges in K_n^k is equivalent to taking an average of all the empirical averages over $G_1, \cdots, G_m$. Therefore,

$$\mathrm{Var}(Y_a) = \mathrm{Var}\left(\frac{1}{m}\sum_{i=1}^{m} Y_{G_i}\right).$$

Since Y_{G_i} are i.i.d. random variables, we can denote $\mathrm{Var}(Y_{G_i}) = \sigma^2$ for all $i = 1, \cdots, m$. Then we can prove the upper bound

$$\mathrm{Var}(Y_a) = \mathrm{Var}\left(\frac{1}{m}\sum_{i=1}^{m} Y_{G_i}\right)$$
$$= \frac{1}{m^2}\left(\sum_i \mathrm{Var}(Y_{G_i}) + \sum_{i \neq j} \mathrm{Cov}(Y_{G_i}, Y_{G_j})\right)$$
$$\leq \frac{1}{m^2}\left(m\sigma^2 + (m^2 - m)\sqrt{\sigma^2\sigma^2}\right)$$
$$= \sigma^2.$$

The inequality uses the well-known fact that for any two random variables X_i, X_j,
$$\mathrm{Cov}(X_i, X_j) \le \sqrt{\mathrm{Var}(X_i)\mathrm{Var}(X_j)}.$$
The lower bound follows similar reasoning.

$$\mathrm{Var}(Y_a) = \frac{1}{m^2}\left(\sum_i \mathrm{Var}(Y_{G_i}) + \sum_{i \ne j} \mathrm{Cov}(Y_{G_i}, Y_{G_j})\right)$$
$$\ge \frac{1}{m^2}\sum_i^m \sigma^2$$
$$= \frac{\sigma^2}{m}.$$

This completes the proof. □

Therefore, we find that while Algorithm 2 may take longer to run than baseline sampling (due to its exponential dependence on k), the variance in its estimates will never be worse, which should lead to an improved (or at least not degraded) sample complexity.

Our analysis in Algorithm 2 corresponds to exact evaluation of k-wise statistical queries. If we added, e.g. Laplace noise, to add stability to Algorithm 2, this would be closer to the approach of Bassily et al. [2] for adaptive data reuse. As it turns out, we can achieve the same bound of $\tilde{O}(\frac{k^2\sqrt{M}}{\tau^2})$ in Algorithm 1 but at lower computational cost. This gives an improvement over the work of Bassily et al. [2] for the case of k-wise statistical queries. Their approach, however, is more general.

Acknowledgments. This work was supported in part by National Science Foundation grant ECCS-2217023. We thank an anonymous reviewer of this paper for detailed and helpful comments.

References

1. Baranyai, Z.: On the factrization of the complete uniform hypergraphs. Infinite Finite Sets 1 (1974). https://cir.nii.ac.jp/crid/1571417125147391744
2. Bassily, R., Nissim, K., Smith, A., Steinke, T., Stemmer, U., Ullman, J.: Algorithmic stability for adaptive data analysis. SIAM J. Comput. **50**(3), STOC16-377–STOC16-405 (2021). https://doi.org/10.1137/16M1103646
3. Blum, A., Furst, M.L., Jackson, J.C., Kearns, M.J., Mansour, Y., Rudich, S.: Weakly learning DNF and characterizing statistical query learning using fourier analysis. In: Leighton, F.T., Goodrich, M.T. (eds.) Proceedings of the Twenty-Sixth Annual ACM Symposium on Theory of Computing, 23–25 May 1994, Montréal, Québec, Canada, pp. 253–262. ACM (1994). https://doi.org/10.1145/195058.195147
4. Blum, A., Kalai, A., Wasserman, H.: Noise-tolerant learning, the parity problem, and the statistical query model. J. ACM **50**(4), 506–519 (2003)

5. Dwork, C., Feldman, V., Hardt, M., Pitassi, T., Reingold, O., Roth, A.: The reusable holdout: preserving validity in adaptive data analysis. Science **349**(6248), 636–638 (2015)
6. Dwork, C., McSherry, F., Nissim, K., Smith, A.D.: Calibrating noise to sensitivity in private data analysis. J. Priv. Confid. **7**(3), 17–51 (2016). https://doi.org/10.29012/jpc.v7i3.405
7. Dwork, C., Roth, A.: The algorithmic foundations of differential privacy. Found. Trends Theor. Comput. Sci. **9**(3-4), 211–407 (2014). https://doi.org/10.1561/0400000042
8. Dwork, C., Rothblum, G.N., Vadhan, S.P.: Boosting and differential privacy. In: 51th Annual IEEE Symposium on Foundations of Computer Science, FOCS 2010, Las Vegas, Nevada, USA, pp. 51–60. IEEE Computer Society (2010)
9. Feldman, V., Ghazi, B.: On the power of learning from k-wise queries. In: Papadimitriou, C.H. (ed.) 8th Innovations in Theoretical Computer Science Conference, ITCS 2017, 9–11 January 2017, Berkeley, CA, USA. LIPIcs, vol. 67, pp. 41:1–41:32. Schloss Dagstuhl - Leibniz-Zentrum für Informatik (2017)
10. Fish, B., Reyzin, L., Rubinstein, B.I.P.: Sampling without compromising accuracy in adaptive data analysis. In: Kontorovich, A., Neu, G. (eds.) Algorithmic Learning Theory, ALT 2020, 8–11 February 2020, San Diego, CA, USA. Proceedings of Machine Learning Research, vol. 117, pp. 297–318. PMLR (2020)
11. Kearns, M.J.: Efficient noise-tolerant learning from statistical queries. J. ACM **45**(6), 983–1006 (1998)
12. Reyzin, L.: Statistical queries and statistical algorithms: foundations and applications. CoRR abs/2004.00557 (2020). https://arxiv.org/abs/2004.00557

Neural Diffusion Graph Convolutional Network for Predicting Heat Transfer in Selective Laser Melting

Benjamin Uhrich[1,2(✉)], Tim Häntschel[1,2], Martin Schäfer[3], and Erhard Rahm[1,2]

[1] Center for Scalable Data Analytics and Artificial Intelligence Dresden/Leipzig, Leipzig, Germany
uhrich@informatik.uni-leipzig.de
[2] Leipzig University, Leipzig, Germany
[3] SIEMENS AG, Berlin, Germany
http://www.scads.ai/

Abstract. The quality of components produced through additive manufacturing processes, such as selective laser melting (SLM), is significantly influenced by heat transfer phenomena. Numerical simulations have emerged as valuable tools for gaining a deeper understanding of these processes. Deep investigation is made possible by a large amount of sensor data in this area. Both offers the potential to reduce the cost and time associated with empirical experimentation. Physics-informed neural networks (PINNs) combine the data-driven capabilities of deep neural networks with the mathematical formulations of physical laws, such as heat diffusion. In particular, the gap between numerical simulations and data observations can be bridged. In this paper, we present a novel neural diffusion graph convolutional network (NDGCN) designed to reveal physically interpretable parameters and accurately predict heat transfer dynamics during the SLM process. Our methodology involves representing the fabricated part as a graph model, constructed from high-dimensional data. This facilitates the integration of complex geometries and thermal properties into our predictive framework.

Keywords: Graph convolutional network · Neural diffusion · 3D printing

1 Introduction

SLM has emerged as a transformative technology in additive manufacturing, enabling the fabrication of intricate components with unprecedented precision. However, the quality of SLM-produced parts is profoundly influenced by the dynamic interplay of circulating heating and cooling processes. Undesirable thermomechanical distortions, porosity, and cracks are often caused by local and global overheating or rapid cooling. To this end, a thorough understanding of heat transfer phenomena is required to optimize manufacturing results. In the pursuit of such understanding, mathematical foundations have played a central role. A partial differential equation (PDE) is used for modeling heat diffusion. These equation provide a rigorous framework for comprehending the underlying physics of heat transfer. Nonetheless, there exists a challenge: the disparities between numerical analysis based on these equation and real-world observations obtained from sensors. Furthermore, the collection of comprehensive data of an entire component in the complicated SLM process proves to be impossible. Traditional thermal imagers can only capture surface temperatures, leaving the dynamics of the building's interior largely undetected. To address these challenges, PINNs have emerged as a powerful tool by harmonizing data-driven modeling with the diffusion equation. This fusion of data-driven and physics-based approaches not only expedites the training process but also extends our ability to predict heat transfer behaviors within SLM, even in areas devoid of sensor data. Building on the success of neural networks, Graph Convolutional Neural Networks (GCNs) have demonstrated remarkable predictive capabilities in various applications involving large-scale data relationships. Intriguingly, the mathematical description of GCNs functionality closely aligns with the structure of the explicit Euler method, offering a promising avenue for knowledge transfer [4]. In this paper, we propose a tailored NDGCN. It is designed to model discretized temperature fields within a built part as node features, facilitating predictions of heat transfer during additive manufacturing. Additionally, we explore the incorporation of different regularization techniques based on the knowledge of heat transfer properties to enhance training efficiency and accuracy. To the best of our knowledge, this work represents the first application of a NDGCN in the context of SLM for heat transfer prediction. We make an important contribution to this as follows:

- We introduce a NDGCN for the discovery of physically interpretable parameters and the prediction of temperature fields using real measurement data.
- Our approach is benchmarked against numerical simulations employing the finite volume method (FVM) and a PINN solution.
- We show that the stability of the learning algorithm can be improved using heat transport properties in the form of control loss functions.

This paper is organized in the following manner. Related work that serves as a foundation for this paper is presented in Sect. 2. Section 3 is a benchmark of the approach in terms of learning physical parameters and in comparison to other methods. This includes a FVM solution and a PINN solution. Section 4 describes the data integration. A summary and discussion is given in Sect. 5.

2 Related Work

2.1 Numerical Analysis and PINNs

Numerical analysis are indispensable in scientific and engineering research. These methods are particularly valuable when analytical solutions are either infeasible or non-existent due to the complexity of the underlying physics, such as the heat transfer in SLM. The state of the art for approximately solve PDEs are the finite difference method, FVM, and the finite element method. Mukherjee *et al.* presented a pioneering work for simulating heat transfer and fluid flow in AM using the FVM [15,16]. PINNs represent a recent innovation at the intersection of machine learning and numerical simulations. For SLM processes involving highly physical interactions, obtaining sufficient data to train accurate machine learning models can be prohibitively expensive, time-consuming, or even impossible. Traditional data-driven methods such as deep learning often struggle with data scarcity because they rely on large data sets for effective generalization. PINNs offer an elegant solution to this problem by seamlessly incorporating domain-specific knowledge of heat transfer into neural network architectures. Raissi et al. has laid the foundation for this area of research and for much of the successful work of the past few years [17]. A comprehensive overview of PINNs in heat transfer problems is given by Cai et al. [3]. Uhrich et al. are given a heat forecasting on a built part in AM using simplified 2-dimensional PINN models [21]. A multi-model, physics-informed machine learning approach based on thermal and grayscale image analysis is presented by Bauer et al. [2].

2.2 Differential Equation Inspired Neural Networks

Deep Neural Networks are often seen as complex "black boxes," making it challenging to understand their inner workings, despite their primary goal of learning patterns in data. This lack of transparency makes it difficult to assess a neural network's suitability for a specific task, often requiring trial and error. Additionally, the unpredictability of trained models in real-world situations adds complexity, causing concerns about their stability, robustness, and reliability. For this purpose, Weinan was the first author to introduce the bridge between deep residual networks (Resnets) [10] and ODEs [7]. The use of ODE-inspired network design for single image super-resolution is given by He et al.. The authors proposed several network architectures based on Runge-Kutta methods [11]. An ODE transformer network is recently presented by Khoshsirat and Kambhamettu [13]. Ruthotto and Haber are given more specific DNN architectures that are motivated by PDEs. The foundation for the network design are the different classes of PDEs [18]. Alt et al. has also presented the transfer of the rich set of numerical foundations from PDEs to DNNs [1]. Shen et al. extend the explicit forward Euler method with the implicit backward counterpart and present a neural network for single image dehazing inspired by implicit Euler ODEs in their work [19]. Uhrich et al. showed how to predict valve failure using a neural network inspired by mathematical formulations of the operation of an electrodynamic valve [20].

2.3 Neural Diffusion Graph Networks

A novel approach that views deep learning on graphs as a continuous diffusion process and interprets Graph Neural Networks as discrete approximations of an underlying PDE is introduced by Chamberlain et al.. They addresses challenges commonly faced by graph learning models, such as depth, oversmoothing, and bottlenecks, while ensuring stability with respect to data perturbations in both implicit and explicit discretization schemes. The authors develop linear and nonlinear versions of GRAND, which demonstrate competitive performance on various standard graph benchmarks [4]. Two interesting applications of such approaches are climate modeling and stream water temperature prediction, presented by Choi et al. and Jia et al., respectively [6, 12]. We follow the work of Chamberlain et al., which serves as a foundation for our approach.

3 Methodology and Benchmarking

3.1 Methodology NDGCN

A Graph data structure are a collection of vertices and edges $G = (V, E)$, $V = \{1, \ldots, N\}$, $E \subseteq V \times V$. Vertex distances d are described by the euclidean distance of vertex positions $p(v) \in \mathbb{R}^3$. For simplicity, a cubic mesh graph with equidistant vertices is used for benchmarking, as shown in Fig. 1. Heat diffusion processes in SLM can be described by the heat equation:

$$\dot{T}(x,t) = \frac{\kappa(T(x,t))}{c(T(x,t))\rho} \Delta T(x,t) + Q(x,t), \tag{1}$$

$$(x_1, x_2, x_3) \in \Omega, \ t \in [0, T]$$

with the thermal conductivity κ, the specific heat c, the heat generated per volume Q and the density ρ. To approximate a given initial-boundary value problem on a graph model, a NDGCN is described in the following explicit scheme using forward time difference:

$$T_{n+1} = T_n + \delta t \frac{\kappa(T_n)}{c(T_n)\rho} L T_n + \delta t \cdot Q_n \tag{2}$$

The vertices of the graph represents temperature features $T : V \times [0, \tau) \to \mathbb{R}$ and the edges are weighted $\omega(v1, v2, t) : V \times V \times [0, \tau)$, where $w = \frac{\kappa}{\rho c}$. L represents the laplacian matrix, which is a discretized form of the laplace operator:

$$L = A - D \tag{3}$$

where A is the adjacency matrix, containing the connectivity between adjacent vertices, and D is the diagonal degree matrix, i.e. the matrix that has as the i-th diagonal element, the sum of the entries in the i-th row of A.

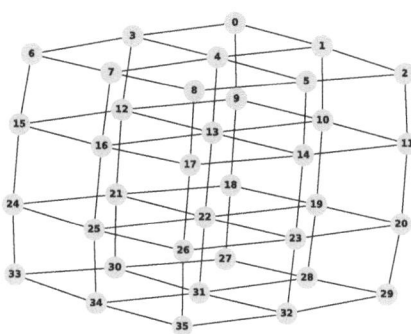

Fig. 1. Potential Cubic Mesh - Vertices and edges are equidistant

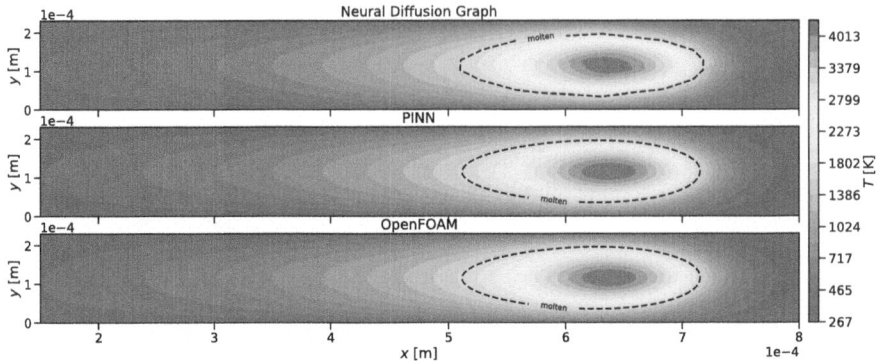

Fig. 2. Heat Transfer Prediction - The predicted solution of the NDGCN is compared with the FVM and a PINN solution as a baseline [22]. The NDGCN approach is able to reproduce the PINN and FVM solution

Due to the equal distribution of all vertices and edges, the trainable weights can be equated with the physical parameters $\kappa(T)$ and $c(T)$. The timesteps are equivalent to the number of layers in the NDGCN. The initial condition and Dirichlet boundary is formulized in the following manner:

$$T(t_0, x) = T_0, \quad (x_1, x_2, x_3) \in \Omega \tag{4}$$

$$T(t, x) = T_\mathcal{B}, \quad (x_1, x_2, x_3) \in \Omega, \quad t \in [0, T] \tag{5}$$

The Robin boundary condition is defined by the gaussian heat flux q of the energy source across the top surface:

$$\kappa \frac{\partial T}{\partial n} = q(x, t), \quad (x_1, x_2) \in \Omega_S \tag{6}$$

$$q = \frac{2AP}{\pi r_b^2} \exp\left(\frac{2(\mathbf{r_0} - \mathbf{v}t)^2}{r_b^2}\right) \tag{7}$$

with $\mathbf{r_0} = (\frac{x_{1b}}{4}, \frac{x_{2b}}{2})$
where P is the laser power, x_{1b} and x_{2b} are the distance from the laser beam axis, r_b is the radius of the laser, and A is the absorption of the laser energy. v is the scanning velocity.

3.2 Benchmark

The performance of the approach to identify the physical parameters of 316L stainless steel is evaluated to benchmark the NDGCN model. The approach relies on utilizing the numerical solution of the boundaries as training data, in particular for the upper boundary. The training of the NDGCN model is guided by a loss function designed to incorporate Dirichlet boundary conditions, a crucial aspect in accurately modeling

Table 1. Thermal and mechanical properties

Properties	
Laser Power P (W)	250
Ambient Temperature T_0 (K)	293.15
Density ρ (kg/m^3)	$0.6 \cdot 7800$
Thermal conductivity κ (W/(m K))	$7.092 + 0.636 \times 10^{-2}$ T
Identified thermal conductivity κ (W/(m K))	$8.436 + 0.219 \times 10^{-2}$ T
Specific heat c (J/(kg K))	$330.9 + 0.563T - 4.015 \times 10^{-4}T^2$ $+9.465 \times 10^{-8}T^3$
Identified specific heat c (J(kg K))	$319.15 + 0.676T - 2.285 \times 10^{-3}T^2$
Absorption of the laser energy A	9×10^{-2}
Laser Beam Radius r (m)	1.4×10^{-4}

physical systems. This loss function serves as the driving force for training the model, ensuring it captures the essential physics of the problem:

$$\mathcal{L} = \frac{1}{N_v} \sum_{i=0}^{N_v-1} (T_\mathcal{B}(t^i, x^i) - T_{pred}(t^i, x^i))^2 \tag{8}$$

where N_v are the number of vertices on the boundaries. Figure 2 visually demonstrates the efficacy of the NDGCN approach in reproducing the solutions obtained through FVM approximation and PINN solution. The successful reproduction of these solutions highlights the robustness and accuracy of the model. Figure 3 provides further insight into the capabilities of the NDGCN approach. It showcases that the model can predict heat transfer patterns within the constructed part without the need for additional data.

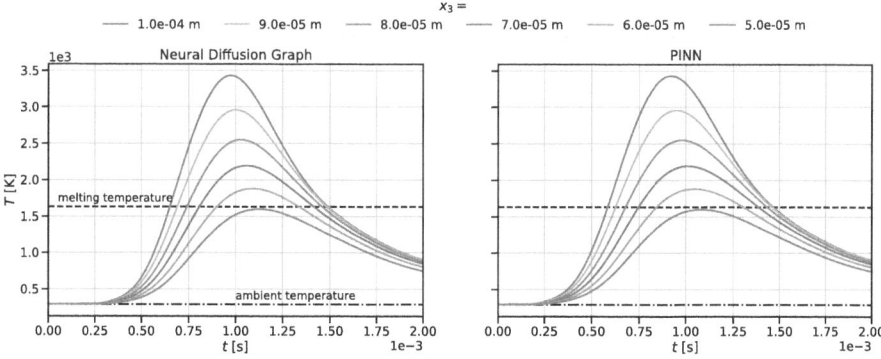

Fig. 3. Prediction of inside the built - The NDGCN model is able to predict the heat transfer in the inside of the built. The results are compared against the PINN solution

The trainable parameters of the NDGCN model are physically interpretable, as shown in Table 1. The trainable weights are modelled as parameters of the specific heat and thermal conductivity.

4 Data Integration

An arbitrary spatial structure can be approximately described using a simplicial 3-complex. For each layer in a printing job, the objective is to represent the partial object printed up to that point, by a simplicial complex, such that a) the shape of the complex closely resembles the shape of the part, and b) the dynamics of the heat distribution in the part can be modeled by a diffusion process on the underlying graph, i.e. the graph whose vertices and edges are given by the 0- and 1-simplices in the complex. Thermal images of the surface are taken periodically during the printing process. The shape of each layer of the part can be detected from these images. Stacking this information for all layers, we obtain a 3-dimensional model of the object.

4.1 Graph Construction from Thermal Images

We associate the pixels of a thermal image with points on a plane in $\mathbb{R}^3$, that is parallel to the (x, y)-plane and contains the current layer of the printing job. A point is assumed to be part of the printed object, if the temperature value of the associated pixel is above an empirical threshold. If this is the case, the point is added to a point cloud in $\mathbb{R}^3$, giving the pixel coordinates a third component for the height of the layer. A representative subset needs to be selected from the composite point cloud to build the simplicial complex. To this end, we make use of the pruning method described in [14], that iteratively removes the point with the highest scale-invariant density (SID).

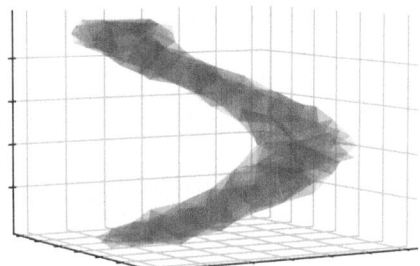

Fig. 4. Simplicial Complex - created from a helix-shaped point cloud, using SID-based pruning, Delaunay triangulation and restriction to the interior of the alpha shape

Since the data comes from $\mathbb{R}^3$ instead of a plane, we adapt the SID, by replacing the r-density with the 3-dimensional analogue:

$$d_r(v_i) = \frac{\#\{\|v_j - v_i\| < r \,:\, 1 \leq j \leq N,\, i \neq j\}}{\frac{4}{3}\pi r^3} \tag{9}$$

Which is the number of data points in the r-ball around v_i, divided by the volume of the ball. The scale-invariant density is defined as the integral over all r-densities:

$$d(v_i) = \int_0^\infty d_r(v_i) dr \tag{10}$$

and by a similar calculation as in [14], we see that:

$$d(v_i) = \left(\frac{8}{3}\pi\right)^{-1} \sum_{j \neq i} \|v_i - v_j\|^{-2} \tag{11}$$

In each step, the point with the highest spatial redundancy (measured by the SID) is removed. The resulting subset is relatively homogeneously distributed over the interior of the point cloud, but contains many boundary points, as their surrounding is partially void, resulting in a lower SID. This is desirable, as the boundary points define the shape of the printed object. For building the graphs representing the partially printed object, we iterate over the number of layers n, ranging from 1 to the total number N of layers in the printing job. For each n, the pruned set of points from the previous step is considered together with the points from the n-th layer. We restrict pruning to points from the top k layers ($k \ll N$), thus the representation for the first $n - k$ layers is inherited. A simplicial complex is constructed from the set of pruned points, using *Delaunay triangulation* [5]. The printed part must not necessarily be convex, but the shape produced by the Delaunay triangulation always is. To extract only the simplices that are within the boundary of the printed part, we use an *alpha shape* [8] to determine the hull of the point cloud. Removing the simplices that are not encased by the alpha shape, we end up with a simplicial 3-complex that resembles the shape of our printed object (see Fig. 4). The vertices and edges of this complex are used to define a graph.

The vertices are categorized based on spatial position:

1. vertices in the lowest layer are assigned to the **bottom boundary** class
2. vertices in the surface layer are assigned to the **top boundary** class
3. vertices that are part of a surface of the alpha shape, but neither in the top- nor bottom boundary are assigned to the **side boundary** class
4. vertices that are not part of either of these classes are assigned to the **interior** class

$C_i = C(v_i)$ denotes the class of the vertex v_i, $i = 1, \ldots N$.

4.2 Modelling Internal Heat Using Graph Diffusion

Due to non-equidistant vertices in the graph and a random laser trajectory in the 3D printing process, we propose a more complex model to predict the heat transfer based on real measurement data. Revisiting (1), we simplify notation by expressing the conductivity term $\frac{\kappa}{c\rho}$ as a single parameter α:

$$\dot{T}(x,t) = \alpha\big(T(x,t)\big) \Delta T(x,t) - Q\big(x, T(x,t)\big) \tag{12}$$

For a discrete approximation of the heat process on a graph, we replace the Laplacian Δ with the Graph Laplacian L (3).

$$(LT)(v_i,t) = \sum_{v_j \sim v_i} a_{ij} T(v_j,t) - \Big(\sum_{v_j \sim v_i} a_{ij}\Big) T(v_i,t)$$
$$= \sum_{v_j \sim v_i} a_{ij}\big(T(v_j,t) - T(v_i,t)\big) \tag{13}$$

As a result, we obtain the diffusion equation on the graph:

$$\dot{T}(t) = \alpha\big(T(v_i,t)\big)(LT)(t) - Q\big(v_i, T(v_i,t)\big)$$
$$= \sum_{v_j \sim v_i} \Big(\alpha(T(v_i,t)) a_{ij}\big(T(v_j,t) - T(v_i,t)\big)\Big) - Q\big(v_i, T(v_i,t)\big) \tag{14}$$

This equation can be rewritten using a state-dependent adjacency $A(T) = \big(\alpha(T(v_i,t)) \cdot a_{ij}\big)_{i,j}$, and $L(T)$ accordingly, to get:

$$\dot{T} = L(T)T - Q(T) \tag{15}$$

Here T is a vector of temperature values and Q is a vector-valued function.

Let $c_{ij}(T)$ denote the entries of $A(T)$. We assume that c_{ij} only depends on the temperature of the vertices it connects, i.e. $c_{ij}(T) = c_{ij}(T_i, T_j)$. Furthermore, c_{ij} should not depend on the specific vertices v_i, v_j, but only on their local properties, i.e.

1. their distance $\varrho_{ij} = \|v_i - v_j\|_2$
2. the vertex classes $C(v_i), C(v_j)$
3. the scale-invariant densities $d(v_i), d(v_j)$

Under these assumptions, there is some function φ, modelling the connectivity of adjacent nodes:

$$c_{ij}(T) = \varphi(\varrho_{ij}, T_i, T_j, C_i, C_j, d_i, d_j) \tag{16}$$

In a similar fashion, the i-th entry of the dissipation vector $Q(T) = \big(Q_i(T)\big)_i$ should only depend on the local properties of v_i, i.e. T_i, C_i, d_i. This implies that there is some function ψ, s.t.:

$$Q(T) = \psi(T_i, C_i, d_i) \tag{17}$$

For the neural diffusion model, the functions φ, ψ are approximated by single-hidden-layer networks, and L, Q are composed by the evaluations of these models.

We propose a NDGCN model for predicting the temperature change over discrete time-steps for arbitrary graphs, defined by the equation:

$$T_{n+1} = T_n + \Big(\sum_{j=1}^{k} \Theta_j \big(\delta t L(T_n)\big)^j \Big(T_n - \frac{\delta t}{2} Q(T_n)\Big)\Big) - \delta t \cdot Q(T_n) \tag{18}$$

This model contains the trainable submodels φ, ψ determining L and Q, as well as the trainable parameters Θ_j, which define a graph convolution. The model Eq. (18) is motivated by the Taylor series for a solution of the homogeneous heat equation:

$$T(t+\delta t) = \sum_{j=0}^{\infty} \frac{1}{j!} \left(\left(\frac{\partial}{\partial t}\right)^j T \right)(t)(\delta t)^j$$

$$= T(t) + \sum_{j=1}^{\infty} \frac{1}{j!} \left(\left(\frac{\partial}{\partial t}\right)^j T \right)(t)(\delta t)^j \qquad (19)$$

Note that for a solution of the homogeneous heat equation, its derivative w.r.t. t is again a solution of the heat equation, since $0 = \frac{\partial}{\partial t}(\dot{T} - \alpha \Delta T) = \dot{T} - \alpha \Delta \dot{T}$. Hence, we can replace $\left(\frac{\partial}{\partial t}\right)^j$ by $(\alpha \Delta)^j$ in the Taylor series. Approximating $\alpha \Delta$ by $L(T)$, we obtain:

$$T(t+\delta t) \approx T(t) + \sum_{j=1}^{\infty} \frac{1}{j!} \left(L(T)^j T \right)(t)(\delta t)^j \qquad (20)$$

The model Eq. (18) is derived from this by evaluating the first k terms of the series, accounting for the inhomogeneity due to dissipation, and relaxing the fixed coefficients $1/j!$ to trainable parameters.

Training this kind of model entails several practical challenges: First, using the model to predict changes in surface temperature, requires knowing the initial state T_0 for all vertices, including those that cannot be observed. Additionally, the limited observability of vertices likely leads to an underdetermined optimization problem, which raises doubts about the model's ability to accurately represent the internal heat state. Finally, the model's locality means that vertices beyond a certain distance from the surface (namely, k steps in the graph) do not directly influence the observations, and thus are not tracked by the learning algorithm.

To address the first problem, the initial state T_0 is constructed by iterative model predictions for each layer, starting from the first layer, where all vertices can be observed. However, this requires a relatively well-trained model in the first place. To resolve this circular dependency, we start by training the model on just the first two layers, and then gradually increase the number of layers when the model has learned a good prediction for the current depth. For controlling the other problems, we rely on regularizing loss functions, that are motivated by known properties of heat conduction and enforce those properties on the model predictions.

First, we ensure that the relative magnitude of the convolution parameters is consistent with the derivation from the Taylor series, using the loss term:

$$\mathcal{L}_\Theta = \sum_{j=1}^{k} \left(\frac{\Theta_j j!}{\Theta_1} - 1 \right)^2 \qquad (21)$$

For the connectivity model φ, it is known from the discretization of the continuous Laplace, that the connectivity of two adjacent vertices should be proportional to the inverse square of their distance, ϱ_{ij}^{-2}. Assuming $\varphi(\varrho_{ij}) = \tau \varrho_{ij}^{-2}$ (keeping all other parameters constant), it follows $\varphi'(\varrho_{ij}) = -2\tau \varrho_{ij}^{-3}$, and thus $\frac{\varphi'(\varrho_{ij})}{\varphi(\varrho_{ij})} = -2\varrho_{ij}^{-1}$, which is independent of the scale τ. Therefore, the second regularizing loss term is given by:

$$\mathcal{L}_\varphi = \sum_{i,j: i \sim j} \left(\frac{\varphi'(\varrho_{ij})}{\varphi(\varrho_{ij})} - \left(-2\varrho_{ij}^{-1} \right) \right)^2 \tag{22}$$

which ensures consistency of the edge weights with the distances of the connected vertices. Furthermore, dissipation can only occur at the boundary, so $\psi = 0$ is required for interior points, motivating the loss:

$$\mathcal{L}_\psi = \sum_{i:\, C_i = \text{int.}} \psi(T_i, C_i, d_i)^2 \tag{23}$$

Using knowledge from the theoretical study of PDEs, it is also possible to make statements about the temporal evolution of the heat state. First, the total thermal energy in the body can only change because of dissipation, thus $\sum_i T_{n+1}(i) = \sum_i \left(T_n(i) - Q_n(i) \right)$. Therefore, a loss function is proposed:

$$\mathcal{L}_{\text{heat}} = \left(\sum_i \left(T_{n+1}(i) - T_n(i) + Q_n(i) \right) \right)^2 \tag{24}$$

Another well known property of the evolution of heat distribution is the maximum principle (see for example [9], §2.3.). For our purpose, it suggests that the temperature at a vertex is within the range given by the minimum and maximum over its previous temperature and the temperatures of connected vertices. This is expressed by the regularizing loss terms:

$$\mathcal{L}_{\max} = \sum_i \max\left(0, T_{n+1}(i) - \max(M)\right)^2 \tag{25}$$

$$\mathcal{L}_{\min} = \sum_i \max\left(0, \min(M) - T_{n+1}(i)\right)^2 \tag{26}$$

$$M = \{T_n(i);\, T_{n+1}(j),\, j \sim i\}$$

Furthermore, a potential energy for the heat distribution can be defined. It is given by:

$$E(T,t) = \int_U \left(T(x,t) - \overline{T}(t) \right)^2 dx \tag{27}$$

For the time-differential of this energy, one can compute:

$$\dot{E}(T,t) = 2 \int_U \left(T(x,t) - \overline{T}(t) \right) \dot{T}(x,t) dx$$

$$= 2\alpha \int_U \left(T(x,t) - \overline{T}(t) \right) \Delta T(x,t) dx$$

$$= 2\alpha \Bigg(\underbrace{\int_{\partial U} \left(T(x,t) - \overline{T}(t) \right) (\nabla T(x,t) \cdot \nu) dS}_{\text{energy from dissipation}} - \underbrace{\int_U |\nabla T(x,t)|^2 dx}_{\geq 0} \Bigg) \tag{28}$$

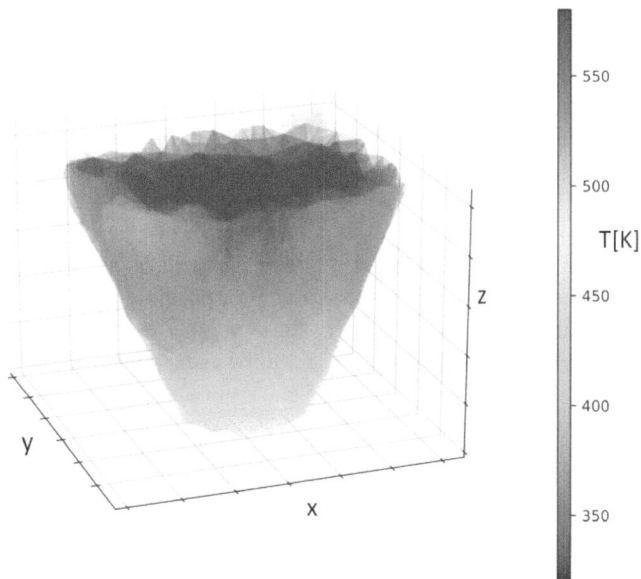

Fig. 5. Predicting heat transfer of a 350-layer printed inverted pyramid - Heat transfer can be predicted over a long period of time by integrating real measured data into a graph model

Assuming dissipation is relatively small, it should roughly hold that $\dot{E} \leq 0$, so $E(T_{n+1}) \leq E(T_n)$. Thus, we introduce another loss term:

$$\mathcal{L}_{\text{energy}} = \max\left(0, E(T_{n+1}) - E(T_n)\right) \tag{29}$$

For training the model, the regularizing loss functions $\mathcal{L}_\Theta, \mathcal{L}_\varphi, \mathcal{L}_\psi, \mathcal{L}_{\text{heat}}, \mathcal{L}_{\min}, \mathcal{L}_{\max}$, and $\mathcal{L}_{\text{energy}}$, as well as the prediction loss

$$\mathcal{L}_{\text{data}} = \sum_{i:\ C_i=\text{top}} \left(T_{n+1}(i) - T_{n+1}^{(data)}(i)\right)^2 \tag{30}$$

are added using appropriate weights, and the descent algorithm seeks to find a minimum for the sum, i.e. a model state that fits the data without violating known properties of heat diffusion. This restricts the admissible set of solutions, thus mitigating the problem of underdetermination, and the regularizing loss functions also regard vertices that are disconnected from the surface and therefore not captured by $\mathcal{L}_{\text{data}}$.

4.3 Results

The model is restricted for learning a consistent diffusion process, both with the laws of physics and the observations.

To generate the initial states for the training steps, the model training starts only for the first layer (where we can observe the complete state), and then gradually increase the number of layers used for training.

The model is trained on the first 100 layers of a print job for an inverted pyramid frustum to evaluate predictions for all 500 layers. The optimisation was performed with the ADAM optimizer. We used a learning rate of $\eta = 1 \times 10^{-5}$ and decay rates $\beta_1 = 0.5$ and $\beta_2 = 0.99$, which estimate the first and second moments of the gradient less than usual. Our model architecture contains the same number of layers as the thermal images are generated. The frame rate has been limited to 3 Hz in order to be in control of the amount of data. The internal state of the model after printing up to 350 layers is shown in Fig. 5. For now, it is not possible to verify these predictions, as the lower layers cannot be observed. However, the distributions seem to be consistent with the heat process in an object.

5 Conclusion and Discussion

In this paper, we have proposed an NDGCN approach for predicting heat transfer in SLM processes. First, we have given a benchmark for our model to reproduce synthetic data and discover physically interpretable parameters modeled as trainable weights of the network. This foundational step underscores the potential of our approach in capturing underlying physical processes. In the second part, a generated graph data structure is used to integrate thermal imaging data. In addition, we extend our model with several regularization principles to predict heat transfer using real measurement data. While this work represents an essential proof of concept with promising initial results, several avenues for future research and refinement are evident. First, the second part of the paper lacks a comprehensive evaluation of our results in comparison to baseline models, including other machine learning approaches. This comparative analysis will provide valuable insights into the effectiveness of our NDGCN in relation to existing methodologies. Additionally, we anticipate expanding our dataset to encompass a broader range of built geometries and scenarios, thereby enhancing the robustness and applicability of our model. This broader dataset will not only facilitate a more thorough evaluation but also enable us to explore diverse industrial applications. Furthermore, we acknowledge the importance of addressing the computational complexity of our NDGCN model relative to other techniques like PINNs or numerical analysis. Such an evaluation will help in understanding the computational trade-offs associated with our approach and guide its optimization for practical use cases.

Acknowledgments. The authors acknowledge the financial support by the Federal Ministry of Education and Research of Germany and by the Sächsische Staatsministerium für Wissenschaft Kultur und Tourismus in the program Center of Excellence for AI-research "Center for Scalable Data Analytics and Artificial Intelligence Dresden/Leipzig", project identification number: ScaDS.AI.

References

1. Alt, T., Schrader, K., Augustin, M., Peter, P., Weickert, J.: Connections between numerical algorithms for PDEs and neural networks. J. Math. Imaging Vision **65**(1), 185–208 (2023). https://doi.org/10.1007/s10851-022-01106-x
2. Bauer, M., Uhrich, B., Schäfer, M., Theile, O., Augenstein, C., Rahm, E.: Multi-modal artificial intelligence in additive manufacturing: combining thermal and camera images for 3D-print quality monitoring. In: Proceedings of the 25th International Conference on Enterprise Information Systems, pp. 539–546. SCITEPRESS - Science and Technology Publications (2023). https://doi.org/10.5220/0011967500003467
3. Cai, S., Wang, Z., Wang, S., Perdikaris, P., Karniadakis, G.E.: Physics-informed neural networks for heat transfer problems. J. Heat Transfer **143**(6) (2021). https://doi.org/10.1115/1.4050542
4. Chamberlain, B., Rowbottom, J., Gorinova, M.I., Bronstein, M., Webb, S., Rossi, E.: Grand: graph neural diffusion. In: Meila, M., Zhang, T. (eds.) Proceedings of the 38th International Conference on Machine Learning. Proceedings of Machine Learning Research, vol. 139, pp. 1407–1418. PMLR (2021). https://proceedings.mlr.press/v139/chamberlain21a.html
5. Chen, L., Xu, J.C.: Optimal delaunay triangulations. J. Comput. Math. 299–308 (2004)
6. Choi, H., Choi, J., Hwang, J., Lee, K., Lee, D., Park, N.: Climate modeling with neural advection-diffusion equation. Knowl. Inf. Syst. **65**(6), 2403–2427 (2023). https://doi.org/10.1007/s10115-023-01829-2
7. E, W.: A proposal on machine learning via dynamical systems. Commun. Math. Stat. **5**(1), 1–11 (2017). https://doi.org/10.1007/s40304-017-0103-z
8. Edelsbrunner, H., Mücke, E.P.: Three-dimensional alpha shapes. ACM Trans. Graph. (TOG) **13**(1), 43–72 (1994)
9. Evans, L.C.: Partial Differential Equations, vol. 19. American Mathematical Society (2022)
10. He, K., Zhang, X., Ren, S., Sun, J.: Deep residual learning for image recognition. In: 2016 IEEE Conference on Computer Vision and Pattern Recognition (CVPR), pp. 770–778 (2016). https://doi.org/10.1109/CVPR.2016.90
11. He, X., Mo, Z., Wang, P., Liu, Y., Yang, M., Cheng, J.: Ode-inspired network design for single image super-resolution. In: 2019 IEEE/CVF Conference on Computer Vision and Pattern Recognition (CVPR), pp. 1732–1741. IEEE (2019). https://doi.org/10.1109/CVPR.2019.00183
12. Jia, X., Chen, S., Zheng, C., Xie, Y., Jiang, Z., Kalanat, N.: Physics-guided graph diffusion network for combining heterogeneous simulated data: an application in predicting stream water temperature. In: Shekhar, S., Zhou, Z.H., Chiang, Y.Y., Stiglic, G. (eds.) Proceedings of the 2023 SIAM International Conference on Data Mining (SDM), pp. 361–369. Society for Industrial and Applied Mathematics, Philadelphia (2023). https://doi.org/10.1137/1.9781611977653.ch41
13. Khoshsirat, S., Kambhamettu, C.: A transformer-based neural ode for dense prediction. Mach. Vis. Appl. **34**(6) (2023). https://doi.org/10.1007/s00138-023-01465-4
14. Kurz, G., Holoch, M., Biber, P.: Geometry-based graph pruning for lifelong slam. In: 2021 IEEE/RSJ International Conference on Intelligent Robots and Systems (IROS), pp. 3313–3320. IEEE (2021)
15. Mukherjee, T., Wei, H.L., De, A., DebRoy, T.: Heat and fluid flow in additive manufacturing – part ii: powder bed fusion of stainless steel, and titanium, nickel and aluminum base alloys. Comput. Mater. Sci. **150**, 369–380 (2018). https://doi.org/10.1016/j.commatsci.2018.04.027
16. Mukherjee, T., Wei, H.L., De, A., DebRoy, T.: Heat and fluid flow in additive manufacturing– part i: modeling of powder bed fusion. Comput. Mater. Sci. **150**, 304–313 (2018). https://doi.org/10.1016/j.commatsci.2018.04.022

17. Raissi, M., Perdikaris, P., Karniadakis, G.E.: Physics-informed neural networks: a deep learning framework for solving forward and inverse problems involving nonlinear partial differential equations. J. Comput. Phys. **378**, 686–707 (2019). https://doi.org/10.1016/j.jcp.2018.10.045
18. Ruthotto, L., Haber, E.: Deep neural networks motivated by partial differential equations. J. Math. Imaging Vision **62**(3), 352–364 (2020). https://doi.org/10.1007/s10851-019-00903-1
19. Shen, J., Li, Z., Yu, L., Xia, G.S., Yang, W.: Implicit euler ode networks for single-image dehazing. In: 2020 IEEE/CVF Conference on Computer Vision and Pattern Recognition Workshops (CVPRW), pp. 877–886. IEEE (2020). https://doi.org/10.1109/CVPRW50498.2020.00117
20. Uhrich, B., Hlubek, N., Häntschel, T., Rahm, E.: Using differential equation inspired machine learning for valve faults prediction. In: 2023 IEEE 21st International Conference on Industrial Informatics (INDIN), pp. 1–8. IEEE (2023). https://doi.org/10.1109/INDIN51400.2023.10217897
21. Uhrich, B., Schäfer, M., Theile, O., Rahm, E.: Using physics-informed machine learning to optimize 3D printing processes. In: Correia Vasco, J.O., et al. (eds.) ProDPM 2021. Springer Tracts in Additive Manufacturing, pp. 206–221. Springer, Cham (2023). https://doi.org/10.1007/978-3-031-33890-8_18
22. Uhrich, B., Pfeifer, N., Schäfer, M., et al.: Physics-informed deep learning to quantify anomalies for real-time fault mitigation in 3D printing. Appl. Intell. **54**(6), 4736–4755 (2024). https://doi.org/10.1007/s10489-024-05402-4. Springer

New Proportion Measures of Discrimination Based on Natural Direct and Indirect Effects

Ryusei Shingaki(✉) and Manabu Kuroki

Graduate School of Engineering Science, Yokohama National University, Yokohama, Japan
shingaki.ryusei@gmail.com, kuroki-manabu-zm@ynu.ac.jp

Abstract. Discrimination-aware data mining is expected to play an important role in data-driven decision making, as "BIG data" can be obtained from the actual society. To build the appropriate decision making system, AI researchers and practitioners have proposed various discrimination measures. However, most of the existing discrimination measures cannot be interpreted as the "proportion" and thus may not provide the comparable evaluation of the discrimination level. To evaluate how much of discrimination is based on a sensitive feature directly, indirectly, or totally, we propose three proportion measures of discrimination using natural direct and indirect effects [12]. The effectiveness of the proposed discrimination measures is confirmed on Adult Census Data [2].

Keywords: Natural direct effect · Natural indirect effect · Proportion measures of discrimination

1 Introduction

Current society is complicated and increasingly relies on decision-making systems based on observed data. In this situation, discrimination-aware data mining is expected to play an important role in data-driven decision making, as "BIG data" can be obtained from the society [16,20]. In recent years, AI researchers and practitioners have developed a number of discrimination-aware data mining algorithms, using various measures to evaluate how much of discrimination is based on a sensitive (or protected) feature [20]. Such measures are called discrimination measures in this paper.

Regarding the definition of discrimination, [20] stated that

> In the context of data mining and machine learning, non-discrimination can be defined as follows: (1) people that are similar in terms of non-protected characteristics should receive similar predictions, and (2) differences in predictions across groups of people can only be as large as justified by their non-protected characteristics.

The first condition is related to direct discrimination or disparate treatment [18], and the second condition ensures that there is no indirect discrimination or no disparate impact [18]. In addition, [20] pointed out, for example, that

> (The first condition) can be illustrated by so called twin test: if gender is the protected attribute and we have two identical twins that share all characteristics, but gender, they should receive identical predictions.

The statement "two identical twins that share *all* characteristics, but gender" is a counterfactual situation implying that statistical discrimination measures based on observed data cannot evaluate direct or indirect discrimination; the idea from causal inference is necessary to develop discrimination-aware data mining algorithms. Nevertheless, most of AI researchers and practitioners have proposed various discrimination measures from statistical viewpoints [13]. Such discrimination measures have drawbacks. First, most of the existing discrimination measures are not designed to have a value on the range $[0,1]$ without any assumptions: they cannot be interpreted as the 'proportion' in the mathematical meaning and thus may not provide the comparable evaluation of the discrimination level, because they are formulated based on the different target populations. Thus, it is required to interpret them based on not only the values taken by these measures but also the characteristics of the relevant parameters used in the discrimination measures, such as the values, the signs, and the interactions. Second, it is not obvious how they reflect the dependences between sensitive and non-sensitive features. Third, it is too strict and unlikely to satisfy no discrimination completely in actual situations, and thus it is necessary to formulate new discrimination measures that describe how much of discrimination is based on a sensitive feature directly, indirectly or totally.

To solve these problems, we propose causal discrimination criteria based on natural direct and indirect effects [12]. Causal discrimination criteria are useful to clarify causal aspects of the discrimination mechanism and highlight the importance of causal inference in the field of discrimination-aware data mining. In addition, under the assumption that there is no confounder, we clarify the difference between the proposed discrimination criteria and the statistical discrimination criteria. Furthermore, to evaluate the discrimination level, referring to [14,19], we formulate three novel proportion measures of discrimination based on natural direct and indirect effects: the proportion of the total variation explained by direct discrimination ($\text{PTV}^{\text{direct}}$), the proportion of total discrimination explained by direct discrimination ($\text{PTD}^{\text{direct}}$), and the proportion of the total variation explained by total discrimination ($\text{PTV}^{\text{total}}$). Finally, the effectiveness of the proposed discrimination measures is confirmed on Adult Census Data [2]. Unlike the existing discrimination-aware data mining based on causal inference, e.g., [8,9], the results of this paper contribute to the reliable evaluation of how much of discrimination is based on a sensitive feature directly, indirectly, or totally, and thus are also applicable to evaluating the degree of a sensitive feature appeared in discrimination when we wish to judge whether or not discrimination-unaware data mining is appropriate to analyze observed data.

2   Structural Causal Model

In this paper, we assume that the readers are familiar with the language of counterfactuals from the semantics of structural causal models, as given in [12].

Given a set $\boldsymbol{V} = \{V_1, V_2, \ldots, V_m\}$ of random variables, let v_i and $\boldsymbol{v}$ represent the values taken by V_i and $\boldsymbol{V}$, respectively. In addition, for $V_i, V_j \in \boldsymbol{V}$, let $P(V_i = v_i) = P(v_i)$, $P(\boldsymbol{V} = \boldsymbol{v}) = P(\boldsymbol{v})$, and $P(V_i = v_i \mid V_j = v_j) = P(v_i | v_j)$ denote the (marginal) probability of $V_i = v_i$, the joint probability of $\boldsymbol{V} = \boldsymbol{v}$, and the conditional probability of $V_i = v_i$ given $V_j = v_j$, respectively. Furthermore, for $V_i, V_j \in \boldsymbol{V}$, let $E_{V_i}[V_i]$ and $E_{V_i|v_j}[V_i \mid V_j = v_j] = E_{V_i|v_j}[V_i|v_j]$ denote the (marginal) expectation of V_i and the conditional expectation of V_i given $V_j = v_j$, respectively. Similar notation is used for other probabilities and expectations. Here, it is noted that the discussion of this paper is mainly based on survival functions such as $P(V_i > v_i)$, because we have

$$E_{V_i}[V_i] = \sum_{v_i=0}^{\infty} \{P(V_i > v_i) - P(V_i \leq -v_i)\} \tag{1}$$

for $E_{V_i}[|V_i|] < \infty$, where the symbol $|V_i|$ indicates the absolute value of V_i. In addition, summation symbols are replaced by integrals whenever the summed variables are continuous unless noted otherwise.

The structural causal model is then defined as follows:

Definition 1 (Structural Causal Model). *A structural causal model is the four-tuple $\langle \boldsymbol{U}, \boldsymbol{V}, \boldsymbol{F}, P(\boldsymbol{u}) \rangle$, where*

(1) $\boldsymbol{U}$ is a set of exogenous variables, determined by factors outside the model;
(2) $\boldsymbol{V}$ is a set of endogenous variables, determined by variables in $\boldsymbol{U} \cup \boldsymbol{V}$;
(3) $\boldsymbol{F}$ is a set of functions, where each $f_{v_i} \in \boldsymbol{F}$ is a data generating process that assigns a value

$$v_i = f_{v_i}(\boldsymbol{v} \setminus \{v_i\}, \boldsymbol{u}) \tag{2}$$

to $V_i \in \boldsymbol{V}$ in response to the values $\boldsymbol{u} \cup \boldsymbol{v} \setminus \{v_i\}$ taken by $\boldsymbol{U} \cup \boldsymbol{V} \setminus \{V_i\}$; and
(4) $P(\boldsymbol{u})$ is the probability of $\boldsymbol{U} = \boldsymbol{u}$.

Through this paper, we assume that the structural causal model does not have feedback, i.e., the output of one variable in the data generating process (2) is not returned as the input of the same variable.

In the framework of structural causal models, the external intervention of $X = x$, denoted by $\mathrm{do}(X = x)$, represents a model manipulation that X is set to some fixed value x, regardless of how the value is ordinarily determined by the function f_x. The probability of $Y > y$ when the external intervention $\mathrm{do}(X = x)$ is conducted, denoted by $P(Y > y \mid \mathrm{do}(X = x))$, is called a causal risk of $X = x$ on $Y > y$ in this paper. When causal quantities such as $P(Y > y \mid \mathrm{do}(X = x))$ are given as a non-trivial function of both x and y, it is said that X has an effect on $Y > y$, or Y is affected by X.

Using the semantics of structural causal models, $P(Y > y \mid \mathrm{do}(X = x))$ can be translated into the probability of the potential outcome variable Y_x. Here, the potential outcome $Y_x = y$ represents the counterfactual sentence "Y would have the value y, had X been x." Similar notation is applied to other potential outcome variables. Then, we have

$$P(Y > y \mid \mathrm{do}(X = x)) = P(Y_x > y).$$

When a randomized experiment is conducted, X is independent of Y_x for any x, denoted as

$$Y_x \perp\!\!\!\perp X$$

for any x. This condition is often called exogeneity. Under the assumption of exogeneity, by the consistency property [12], i.e.,

$$X = x \Longrightarrow Y_x = Y, \tag{3}$$

$P(Y_x > y)$ is identifiable and is given by

$$P(Y_x > y) = P(Y > y \mid x).$$

Here, 'identifiable' means that causal quantities such as $P(Y_x > y)$ can be estimated consistently from a joint probability of observed variables. When a randomized experiment is difficult to conduct, $P(Y_x > y)$ can still be identified in accordance with conditional ignorability, or graphically, the back door criterion [12]: when there exists a set $\boldsymbol{Z}$ of variables such that X is conditionally independent of Y_x given $\boldsymbol{Z}$ for any x, denoted as

$$Y_x \perp\!\!\!\perp X \mid \boldsymbol{Z}$$

for any x, and $P(x \mid \boldsymbol{z}) > 0$ for any x and $\boldsymbol{z}$, we say that $\boldsymbol{Z}$ satisfies the conditional ignorability condition relative to (X, Y). Then, by the consistency property, $P(Y_x > y)$ can be estimated using a set $\boldsymbol{Z}$ of observed variables as follows:

$$P(Y_x > y) = E_{\boldsymbol{z}}\left[P(Y > y \mid x, \boldsymbol{Z})\right],$$

where $E_{\boldsymbol{z}}\left[P(Y > y \mid x, \boldsymbol{Z})\right]$ stands for the expectation of $P(Y > y \mid x, \boldsymbol{Z})$ regarding $\boldsymbol{Z}$. The other identification conditions of causal risks are given in [12].

3 Effect Measures

To propose new discrimination measures, we let X be a sensitive feature to be protected and Y be an outcome variable. A variable that is not sensitive is called a non-sensitive feature. In addition, we let S stand for an intermediate variable that would be affected by X and could have an effect on Y. Note that our discussion is based on a single intermediate variable, but the extension of our results to multiple intermediate variables is straightforward. Let $P(X = x) = P(x)$ and $P(Y > y \mid X = x) = P(Y > y \mid x)$ denote the (marginal) probability of $X = x$ and the conditional probability of $Y > y$ given $X = x$, respectively. Similar notation is used for other probabilities.

One of representative discrimination measures is the total variation (TV) [14, 20]:

Definition 2 (Total Variation). *The total variation (TV) on $Y > y$ comparing $X = x_1$ to $X = x_0$, denoted by $\mathrm{TV}_y(x_1, x_0)$, is defined as*

$$\mathrm{TV}_y(x_1, x_0) = P(Y > y \mid x_1) - P(Y > y \mid x_0)$$

for the risk difference scale, and

$$\mathrm{TV}_y(x_1, x_0) = P(Y > y \mid x_1)/P(Y > y \mid x_0)$$

for the risk ratio scale assuming $P(Y > y \mid x_0) \neq 0$.

The TV is nothing more than a statistical measure, since it is the comparison between the probabilities of Y in the passively observed groups of $X = x_1$ and $X = x_0$. Thus, the TV is identifiable without causal knowledge if observed probabilities $P(y \mid x)$ are available.

Let Y_x be the potential outcome variable that represents the counterfactual sentence "Y would have the value y, had X been x." Similar notation is applied to other potential outcome variables. Then, another representative discrimination measure is the total effect (TE):

Definition 3 (Total Effect). *The total effect (TE) on $Y > y$ comparing $X = x_1$ to $X = x_0$, denoted by $\mathrm{TE}_y(x_1, x_0)$, is defined as*

$$\mathrm{TE}_y(x_1, x_0) = P(Y_{x_1} > y) - P(Y_{x_0} > y)$$

for the risk difference scale, and

$$\mathrm{TE}_y(x_1, x_0) = P(Y_{x_1} > y)/P(Y_{x_0} > y)$$

for the risk ratio scale assuming $P(Y_{x_0} > y) \neq 0$.

Differently from the TV, the TE measures the change on the probability of Y when X changes from $X = x_0$ to $X = x_1$ by the external intervention while S is allowed to track the change in X.

Referring to [12], we define the NDE and NIE as follows:

Definition 4 (Natural Direct and Indirect Effects). *The natural direct effect (NDE) on $Y > y$ comparing $X = x_1$ to $X = x_0$ when S is set to S_{x_1}, denoted by $\mathrm{NDE}_y(x_1, x_0 \mid S)$, is defined as*

$$\mathrm{NDE}_y(x_1, x_0 \mid S) = P(Y_{x_1, S_{x_1}} > y) - P(Y_{x_0, S_{x_1}} > y)$$

for the risk difference scale, and

$$\mathrm{NDE}_y(x_1, x_0 \mid S) = P(Y_{x_1, S_{x_1}} > y)/P(Y_{x_0, S_{x_1}} > y)$$

for the risk ratio scale assuming $P(Y_{x_0,S_{x_1}} > y) \neq 0$. The natural indirect effect (NIE) on $Y > y$ comparing S_{x_1} to S_{x_0} when X is set to x_0, denoted by $\text{NIE}_y(x_1, x_0|S)$, is defined as

$$\text{NIE}_y(x_1, x_0|S) = P(Y_{x_0,S_{x_1}} > y) - P(Y_{x_0,S_{x_0}} > y)$$

for the risk difference scale, and

$$\text{NIE}_y(x_1, x_0|S) = P(Y_{x_0,S_{x_1}} > y)/P(Y_{x_0,S_{x_0}} > y)$$

for the risk ratio scale assuming $P(Y_{x_0,S_{x_0}} > y) \neq 0$.

The NDE measures the change on the probability of Y as X changes from $X = x_0$ to $X = x_1$ by the external intervention while setting S to whatever value it would have obtained under $X = x_1$. Contrary, the NIE measures the change on the probability of Y when the X is held constant at $X = x_0$, and S changes to whatever value it would have attained under $X = x_1$. Here, when we focus on direct and indirect discrimination based on S, note that the idea of the 'twin test' stated in [20] is reflected in the NDE and NIE. In fact, in an employment-discrimination case known as Carson versus Bethlehem Steel Corp. (70 FEP Cases 921, 7th Cir. (1996)), which was introduced by [11] to discuss the NDE and NIE, the court wrote

> The central question in any employment-discrimination case is whether the employer would have taken the *same action* had the employee been of a *different race* (age, sex, religion, national origin, etc.) and *everything else had been the same*,

which implies that the twin test is required to evaluate the discrimination level, and such a court ruling is taken into account as the unit-level NDE and NIE.

If we have

$$P(Y_x > y) = P(Y > y \mid x) \qquad (4)$$

for all x and y, then it is said that X and Y are not confounded [12]. Based on this situation, we define the spurious effect (SE) as follows:

Definition 5 (Spurious Effect). *The spurious effect (SE) on $Y > y$ of comparing $X = x_1$ to $X = x_0$, denoted by $\text{SE}_y(x_1, x_0)$, is defined as*

$$\text{SE}_y(x_1, x_0) = \text{TV}_y(x_1, x_0) - \text{TE}_y(x_1, x_0)$$

for the risk difference scale, and

$$\text{SE}_y(x_1, x_0) = \text{TV}_y(x_1, x_0)/\text{TE}_y(x_1, x_0)$$

for the risk ratio scale assuming $\text{TE}_y(x_1, x_0) \neq 0$.

Intuitively, Eq. (4) states that X and Y are not confounded whenever the observationally witnessed association between them is the same as the association that would be measured in a randomized experiment. Note that the SE does not depend on the selection of intermediate variables.

Referring to [14,19], the following theorem is straightforward:

Theorem 1. *The TV, NDE, NIE, and SE obey the following relationships:*

$$\text{TV}_y(x_1, x_0) = \text{NDE}_y(x_1, x_0|S) + \text{NIE}_y(x_1, x_0|S) + \text{SE}_y(x_1, x_0) \tag{5}$$

for the risk difference scale, and

$$\text{TV}_y(x_1, x_0) = \text{NDE}_y(x_1, x_0|S) \times \text{NIE}_y(x_1, x_0|S) \times \text{SE}_y(x_1, x_0) \tag{6}$$

for both the risk ratio.

Hereafter, we focus on the risk difference scale because the risk ratio scale can be written as the risk difference scale through the logarithm transformation of the TV, i.e.,

$$\log \text{TV}_y(x_1, x_0) = \log \text{NDE}_y(x_1, x_0|S) + \log \text{NIE}_y(x_1, x_0|S) + \log \text{SE}_y(x_1, x_0) \tag{7}$$

from Eq. (6). From Eq. (5), if the NDE, NIE, and SE (or the TE and SE) are zero simultaneously then TV is also zero. This fact would be useful to detect the possibility of discrimination from observed data because $\text{TV} \neq 0$ implies that at least one of NDE, NIE, and SE is non-zero. On the contrary, $\text{TV} = 0$ does not imply that the NDE, NIE, and SE are zero simultaneously, because of the parametric cancellation. In addition, if a set $\{X, Y, S\} \cup \mathbf{Z}$ of observed variables satisfies the sequential ignorability condition (e.g., [6]), i.e.,

$$\{Y_{x,s}, S_{x'}\} \perp\!\!\!\perp X \mid \mathbf{Z}, \quad Y_{x,s} \perp\!\!\!\perp S_{x'} \mid \mathbf{Z}, \quad Y_{x,s} \perp\!\!\!\perp S \mid \{X\} \cup \mathbf{Z},$$

$P(x \mid \mathbf{z}) > 0$, and $P(s \mid x, \mathbf{z}) > 0$ for $x, x' \in \{x_1, x_0\}$, then TE, NDE, and NIE are identifiable from the consistency property and are given by

$$\text{TE}_y(x_1, x_0) = \sum_{\mathbf{z}} \left(E[Y|x_1, \mathbf{z}] - E[Y|x_0, \mathbf{z}] \right) P(\mathbf{z}), \tag{8}$$

$$\text{NDE}_y(x_1, x_0|S) = \sum_{\mathbf{z}, s} \left(E[Y|x_1, s, \mathbf{z}] - E[Y|x_0, s, \mathbf{z}] \right) P(s|x_0, \mathbf{z}) P(\mathbf{z}), \tag{9}$$

$$\text{NIE}_y(x_1, x_0|S) = \sum_{\mathbf{z}, s} E[Y|x_1, s, \mathbf{z}] \left(P(s|x_1, \mathbf{z}) - P(s|x_0, \mathbf{z}) \right) P(\mathbf{z}), \tag{10}$$

respectively. Here, it is noted that summation symbols are replaced by integrals whenever the summed variables are continuous unless noted otherwise.

4 Discrimination Criteria

In this section, we propose the discrimination criteria based on the semantics of structural causal models as follows:

Definition 6 (Causal Discrimination Criteria).
  Letting X, S, and Y be a sensitive feature, an intermediate variable, and an outcome variable, respectively, we say that

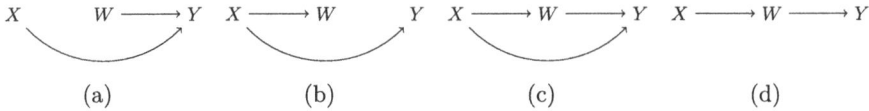

Fig. 1. Graphical representations of statistical discrimination criteria (I) and (II): (a) no indirect discrimination; (b) no indirect discrimination by statistical discrimination criteria (I), but both direct and indirect discrimination by statistical discrimination criteria (II); (c) both direct and indirect discrimination; and (d) no direct discrimination.

(1) there is no causal direct discrimination via S if $\text{NDE}_y(x_1, x_0|S) = 0$ holds for any y,
(2) there is no causal indirect discrimination via S if $\text{NIE}_y(x_1, x_0|S) = 0$ holds for any y,
(3) there is no spurious discrimination if $\text{SE}_y(x_1, x_0) = 0$ holds for any y.

Here, if the assumption of (1) does not hold, we say that there is causal direct discrimination not via S. Similarly, if the assumption of (2) does not hold, we say that there is causal indirect discrimination via S. In addition, if the assumption of (3) does not hold, there is spurious discrimination. Especially, there is no causal total discrimination if $\text{TE}_y(x_1, x_0) = 0$ holds; otherwise, we say that there is causal total discrimination.

The sequential ignorability condition plays an important role in clarifying the difference between causal discrimination criteria and the existing statistical discrimination criteria. To see this, assuming that $\boldsymbol{Z}$ is empty in the sequential ignorability condition, note that $\text{NDE}_y(x_1, x_0|S)$ is zero if $X \perp\!\!\!\perp Y \mid S$ holds and $\text{NIE}_y(x_1, x_0|S)$ is zero if $X \perp\!\!\!\perp S$ or $S \perp\!\!\!\perp Y \mid X$ holds. Based on the consideration, we propose statistical discrimination criteria as follows:

Definition 7 (Statistical Discrimination Criteria (I)). *Letting X, $\boldsymbol{W}$ and Y be a sensitive feature, a set of non-sensitive features, and an outcome variable, respectively, we say that*

(1) there is no statistical direct discrimination given $\boldsymbol{W}$ if $X \perp\!\!\!\perp Y \mid \boldsymbol{W}$, $X \not\perp\!\!\!\perp \boldsymbol{W}$, and $\boldsymbol{W} \not\perp\!\!\!\perp Y \mid X$ hold,
(2) there is no statistical indirect discrimination given $\boldsymbol{W}$ if $X \perp\!\!\!\perp \boldsymbol{W}$ or $\boldsymbol{W} \perp\!\!\!\perp Y \mid X$ holds.

Statistical discrimination criteria (I) is similar to Wang and Taylor's criteria for validating surrogate endpoints [17] in the context of randomized clinical trials (RCTs). Especially, [7] introduced the condition of $\boldsymbol{W} \not\perp\!\!\!\perp Y \mid X$ as the statistical concept of "direct prejudice" into discrimination-aware data mining.

Statistical discrimination criteria (I) can be considered to reflect statistical aspect of causal discrimination criteria through the sequential ignorability condition. Contrary, to derive statistical discrimination criteria which refer to the existing discrimination criteria, according to [7] and [20], consider the following conditions:

(1) The probabilities of the outcome variable are equal for all possible values taken by the sensitive feature. Statistically, this is interpreted as $X \perp\!\!\!\perp Y$.
(2) The probabilities of the outcome variable are equal for all possible values taken by the sensitive feature given a specific value of a non-sensitive feature. Statistically, this is interpreted as $X \perp\!\!\!\perp Y \mid \boldsymbol{W}$.

The condition of $X \perp\!\!\!\perp Y$ is called the independence criterion in the sense that the information on X is not necessary to predict Y. Meanwhile, the condition of $X \perp\!\!\!\perp Y \mid \boldsymbol{W}$ is called the sufficiency criterion in the sense that we do not need to see X when we know $\boldsymbol{W}$ to predict Y. Here, note that these discrimination criteria do not consider the statistical dependence between X and $\boldsymbol{W}$. To solve the problem, consider statistical conditions of $X \not\perp\!\!\!\perp \boldsymbol{W}$ and $Y \not\perp\!\!\!\perp \boldsymbol{W}$ which are introduced as the concepts of "latent prejudice" and "indirect prejudice", respectively, by [7]. Then, we re-define the existing statistical discrimination criteria as follows:

Definition 8 (Statistical Discrimination Criteria (II)). *Letting X, $\boldsymbol{W}$, and Y be a sensitive feature, a set of non-sensitive features, and an outcome variable, respectively, we say that*

(1) there is no statistical direct discrimination given $\boldsymbol{W}$, if $X \perp\!\!\!\perp Y \mid \boldsymbol{W}$, $X \not\perp\!\!\!\perp \boldsymbol{W}$, and $\boldsymbol{W} \not\perp\!\!\!\perp Y$ hold,
(2) there is no statistical indirect discrimination given $\boldsymbol{W}$, if $X \perp\!\!\!\perp \boldsymbol{W}$ or $\boldsymbol{W} \perp\!\!\!\perp Y$ holds.

Statistical discrimination criteria (II) is similar to Prentice's criteria for validating surrogate endpoints [15] in RCTs.

To test statistical discrimination criteria (II) in a simplified manner, we can use the quasi-TE, NDE and NIE, which are derived by replacing $\boldsymbol{Z}$ and S in Eqs. (8), (9) and (10) with an empty set and $\boldsymbol{W}$, respectively:

$$\text{qTE}_y(x_1, x_0 \mid \boldsymbol{W}) = \sum_{\boldsymbol{w}} \left(E[Y \mid x_1, \boldsymbol{w}] - E[Y \mid x_0, \boldsymbol{w}] \right) P(\boldsymbol{w}), \quad (11)$$

$$\text{qNDE}_y(x_1, x_0 \mid \boldsymbol{W}) = \sum_{\boldsymbol{w}} \left(E[Y \mid x_1, \boldsymbol{w}] - E[Y \mid x_0, \boldsymbol{w}] \right) P(\boldsymbol{w} \mid x_0), \quad (12)$$

$$\text{qNIE}_y(x_1, x_0 \mid \boldsymbol{W}) = \sum_{\boldsymbol{w}} E[Y \mid x_1, \boldsymbol{w}] \left(P(\boldsymbol{w} \mid x_1) - P(\boldsymbol{w} \mid x_0) \right), \quad (13)$$

where "q" is an abbreviation "quasi-". We also define the quasi-SE as

$$\text{qSE}_y(x_1, x_0 \mid \boldsymbol{W}) = \text{TV}_y(x_1, x_0) - \text{qTE}_y(x_1, x_0 \mid \boldsymbol{W}). \quad (14)$$

The distinction between sensitive and non-sensitive features is important to evaluate Eqs. (11)–(14), however, it is not necessary to identify causal quantities. In other words, neither statistical discrimination criteria (I) nor (II) may reflect causal aspects of the discrimination mechanism, but are helpful to judge the discrimination class from observed data. When qNDE, qNIE, and qSE are simultaneously zero, it is said that there is statistically no discrimination or simply no discrimination in this paper.

When W is an intermediate variable W, the difference between statistical discrimination criteria (I) and (II) can be clarified in a causal diagram [12] that illustrates the direct and indirect effects via different paths between X and Y (see Fig. 1). In this setting, using either statistical discrimination criteria, Fig. 1(a) and (d) show the situations judged as no indirect and no direct discrimination, respectively. In contrast, Fig. 1(b) shows the situation judged as no indirect discrimination from statistical discrimination criteria (I), but as both direct and indirect discrimination from statistical discrimination criteria (II), because $W \not\perp\!\!\!\perp Y$, $X \not\perp\!\!\!\perp Y \mid W$, and $W \perp\!\!\!\perp Y \mid X$ hold. This consideration implies that causal discrimination criteria and the existing statistical discrimination criteria are derived from the different motivations for a sensitive feature, even if no unmeasured confounders exist.

Contrary, regarding the equivalence between statistical discrimination criteria (I) and (II), the following theorem is derived:

Theorem 2. *When the probabilities of X, Y, and W are strictly positive and $X \not\perp\!\!\!\perp Y$ and $X \perp\!\!\!\perp Y \mid W$ hold, there is no statistical direct discrimination given W in the sense of both statistical discrimination criteria (I) and (II).*

Proof. Assuming that the probabilities of X, Y, and W are strictly positive and $X \perp\!\!\!\perp Y \mid W$ holds, then $W \perp\!\!\!\perp Y \mid X$ and $X \perp\!\!\!\perp Y \mid W$ imply that $\{X, W\} \perp\!\!\!\perp Y$ by the intersection property [10], which induces both $X \perp\!\!\!\perp Y$ and $W \perp\!\!\!\perp Y$ from the decomposition property [10]. However, as both $X \not\perp\!\!\!\perp Y$ and $X \perp\!\!\!\perp Y \mid W$ are assumed, $W \not\perp\!\!\!\perp Y \mid X$ holds by contraposition. Similarly, $W \perp\!\!\!\perp Y$ and $X \perp\!\!\!\perp Y \mid W$ imply $\{X, W\} \perp\!\!\!\perp Y$ by the contraction property [10]. As both $X \not\perp\!\!\!\perp Y$ and $X \perp\!\!\!\perp Y \mid W$ are assumed, $W \not\perp\!\!\!\perp Y$ also holds by contraposition. Finally, if both $X \perp\!\!\!\perp Y \mid W$ and $X \perp\!\!\!\perp W$ imply $X \perp\!\!\!\perp \{Y, W\}$ by the contraction property [10]. However, from the assumption of both $X \not\perp\!\!\!\perp Y$ and $X \perp\!\!\!\perp Y \mid W$, $X \not\perp\!\!\!\perp W$ holds by contraposition.

Under the assumption of $X \not\perp\!\!\!\perp Y$, the relationships between statistical dependencies and statistical discrimination criteria (I) and (II) are shown in Table 1(a) and 1(b), respectively. Table 1 demonstrates that the statistical judgment of direct, indirect, or both direct and indirect discrimination depends on which criteria are used to detect or explain how discrimination occurs.

Theorem 2 does not mean that there are both direct and indirect discrimination in the sense of statistical discrimination criteria (I) if and only if there are both direct and indirect discrimination in the sense of statistical discrimination criteria (II). Based on the consideration, by introducing the separation criterion $X \perp\!\!\!\perp W \mid Y$ [18], the equivalence condition between statistical discrimination criteria (I) and (II) is derived as follows:

Theorem 3. *When the probabilities of X, Y, and W are strictly positive and $X \perp\!\!\!\perp W \mid Y$ holds under the assumption $X \not\perp\!\!\!\perp Y$, statistical discrimination criteria (I) and (II) are equivalent.*

Table 1. Comparison between statistical discrimination criteria (I) and (II). "ND", "NI", "ID", and "—" mean no direct discrimination, no indirect discrimination, both direct and indirect discrimination, and contradiction against $X \not\!\perp\!\!\!\perp Y$, respectively.

(a) Statistical discrimination criteria (I).

	$X \perp\!\!\!\perp Y \mid W$		$X \not\!\perp\!\!\!\perp Y \mid W$	
	$W \perp\!\!\!\perp Y \mid X$	$W \not\!\perp\!\!\!\perp Y \mid X$	$W \perp\!\!\!\perp Y \mid X$	$W \not\!\perp\!\!\!\perp Y \mid X$
$X \perp\!\!\!\perp W$	—	—	NI	NI
$X \not\!\perp\!\!\!\perp W$	—	ND	NI	ID

(b) Statistical discrimination criteria (II).

	$X \perp\!\!\!\perp Y \mid W$		$X \not\!\perp\!\!\!\perp Y \mid W$	
	$W \perp\!\!\!\perp Y$	$W \not\!\perp\!\!\!\perp Y$	$W \perp\!\!\!\perp Y$	$W \not\!\perp\!\!\!\perp Y$
$X \perp\!\!\!\perp W$	—	—	NI	NI
$X \not\!\perp\!\!\!\perp W$	—	ND	NI	ID

Proof. The combination of $W \perp\!\!\!\perp Y \mid X$ and $X \perp\!\!\!\perp W \mid Y$ induces $W \perp\!\!\!\perp \{X, Y\}$ by the intersection property [10]. Similarly, the combination of $W \perp\!\!\!\perp Y$ and $X \perp\!\!\!\perp W \mid Y$ induces $W \perp\!\!\!\perp \{X, Y\}$ by the contraction property [10]. In addition, the combination of $X \perp\!\!\!\perp Y \mid W$ and $X \perp\!\!\!\perp W \mid Y$ induces $X \perp\!\!\!\perp \{W, Y\}$ by the intersection property [10]. However, as $X \not\!\perp\!\!\!\perp Y$ holds, $X \not\!\perp\!\!\!\perp Y \mid W$ can be derived by the contradiction. This implies that there is direct discrimination. Thus, under the assumption that both $X \not\!\perp\!\!\!\perp Y$ and $X \perp\!\!\!\perp W \mid Y$ hold, $X \perp\!\!\!\perp W$ implies that there is no indirect discrimination in the sense of both statistical discrimination criteria (I) and (II), and $X \not\!\perp\!\!\!\perp W$ implies that there are both direct and indirect discrimination in the sense of both statistical discrimination criteria (I) and (II).

5 Proportion Measures of Discrimination

In actual situations, it would be rare to strictly satisfy causal discrimination criteria so that it is reasonable to evaluate the discrimination level representing how much of discrimination is based on the sensitive feature directly, indirectly or totally. However, most of the existing discrimination measures cannot be interpreted as the "proportion" in the mathematical meaning and thus may not provide the comparable evaluation of the discrimination level, because they are formulated based on the different target populations. To solve the problem, we propose three types of novel proportion measures of discrimination: the proportion of the TV explained by direct discrimination ($\text{PTV}^{\text{direct}}$), the proportion of total discrimination explained by direct discrimination ($\text{PTD}^{\text{direct}}$), and the proportion of the TV explained by total discrimination ($\text{PTV}^{\text{total}}$). These measures are defined as follows:

$$\text{PTV}_y^{\text{direct}}(x_1, x_0|S) = \frac{\text{NDE}_y(x_1, x_0|S)^2}{\text{SE}_y(x_1, x_0)^2 + \text{NIE}_y(x_1, x_0|S)^2 + \text{NDE}_y(x_1, x_0|S)^2}, \tag{15}$$

$$\text{PTD}_y^{\text{direct}}(x_1, x_0|S) = \frac{\text{NDE}_y(x_1, x_0|S)^2}{\text{NIE}_y(x_1, x_0|S)^2 + \text{NDE}_y(x_1, x_0|S)^2}, \tag{16}$$

$$\text{PTV}_y^{\text{total}}(x_1, x_0) = \frac{\text{TE}_y(x_1, x_0)^2}{\text{SE}_y(x_1, x_0)^2 + \text{TE}_y(x_1, x_0)^2}, \tag{17}$$

where $0/0$ is defined as 0 in this paper. As seen from Eqs. (15), (16), and (17), the proposed discrimination measures are defined based on a single target population, and always fall inside the range $[0, 1]$ without any assumptions.

The higher values of the proposed discrimination measures show a more severe situation in the sense that most part of discrimination is attributed to the sensitive feature alone and thus may not be removable by adjusting the non-sensitive features (in this paper, "severe" does not always mean the degree of social seriousness based on the sensitive feature). In this sense, the proposed discrimination measures help us to classify the severity of discrimination, as shown in the last part of this section. In addition, the proposed discrimination measures are applicable to evaluating how much of discrimination is explained by causal direct or total discrimination, in order to judge whether or not the discrimination-unaware data mining algorithm is appropriate to analyze observed data.

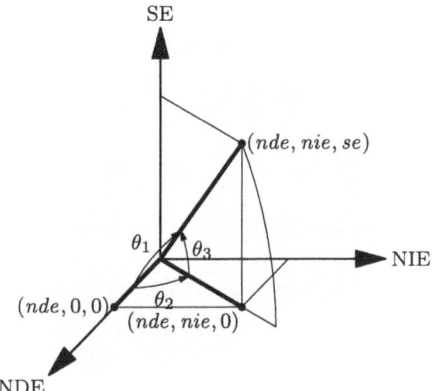

Fig. 2. Motivating idea of the proposed discrimination measures.

Here, it would be worthwhile to state that the motivating idea behind the proposed discrimination measures comes from the similarity measure used in cluster analysis, i.e., the cosine similarities between the TV and the NDE and between the TE and the NDE, as shown in Fig. 2. Letting TV, TE, and NDE correspond to the vectors (nde, nie, se), $(nde, nie, 0)$, and $(nde, 0, 0)$, respectively, the cosine similarities between the TV and the NDE and between the TE and the NDE correspond to $\text{PTV}_y^{\text{direct}}(x_1, x_0|S)$ and $\text{PTD}_y^{\text{direct}}(x_1, x_0|S)$, respectively. Indeed, denoting the angle between the TV and the NDE as θ_1, which is the angle between the vectors (nde, nie, se) and $(nde, 0, 0)$ and the angle between the TE and the NDE as θ_2, which is the angle between the vectors $(nde, nie, 0)$ and $(nde, 0, 0)$, we have

$$\text{PTV}_y^{\text{direct}}(x_1, x_0|S) = \cos^2 \theta_1,$$
$$\text{PTD}_y^{\text{direct}}(x_1, x_0|S) = \cos^2 \theta_2.$$

Then, letting θ_3 be the angle between (nde, nie, se) and $(nde, nie, 0.0)$, which is interpreted as the angle between the TV and the TE, we have $\cos^2 \theta_1 = \cos^2 \theta_2 \cos^2 \theta_3$, i.e.,

$$\text{PTV}_y^{\text{direct}}(x_1, x_0|S) = \text{PTD}_y^{\text{direct}}(x_1, x_0|S) \cos^2 \theta_3$$

from "Theorem of Three Perpendiculars". This shows that these discrimination measures are not sufficient for determining each other, but are not entirely independent. In addition, $\cos^2 \theta_3$ can be considered as another discrimination measure representing the proportion of the TV explained by total discrimination.

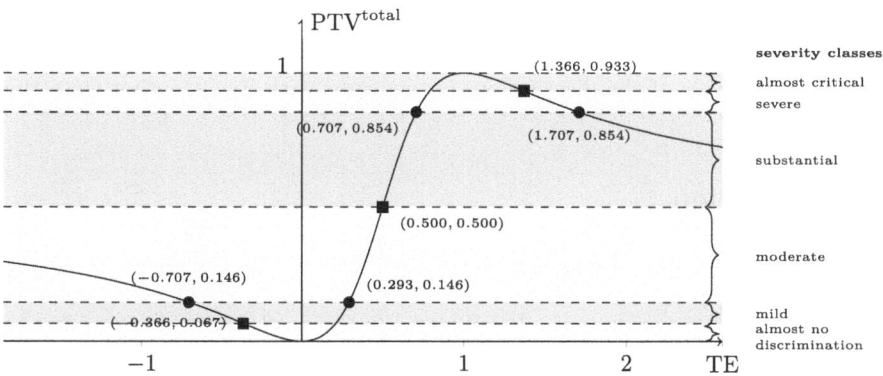

Fig. 3. $\text{PTV}^{\text{total}}$ versus TE graph for TV = 1.000. ■ and • indicate inflection and jerk points, respectively.

From a mathematical viewpoint, unlike the existing discrimination measures, the proposed discrimination measures can provide cut-off values for judging the severity class of discrimination based on the derivatives of $\text{PTD}^{\text{direct}}$ and $\text{PTV}^{\text{total}}$. To see this, we consider that $\text{PTV}^{\text{total}}$ is a function of the $\text{TE}_y(x_1, x_0)(= \text{TE})$ for a given value of the $\text{TV}_y(x_1, x_0)(= \text{TV})$. Then, the

Table 2. Summary statistics of the discrimination measures from Adult Census Dataset.

	TV	NMD	slift	elift	PTD^{direct}	PTV^{direct}	PTD^{total}
estimate	−0.196	0.545	0.358	1.270	0.935	0.934	1.000
s.e	0.004	0.011	0.010	0.006	0.038	0.042	0.029
lcl	−0.197	0.545	0.358	1.269	0.922	0.914	0.977
mean	−0.196	0.545	0.358	1.270	0.924	0.916	0.978
ucl	−0.196	0.546	0.359	1.270	0.926	0.917	0.980

inflection points of the function are obtained when $\text{TE} = \frac{1}{2}\text{TV}, \frac{1 \pm \sqrt{3}}{2}\text{TV}$ by taking the second derivative of PTV^{total} with respect to TE. The jerk points of the function are obtained when $\text{TE} = \frac{1}{\sqrt{2}}\text{TV}, \frac{2 \pm \sqrt{2}}{2}\text{TV}$ by taking the third derivative of PTV^{total} with respect to TE. With this consideration, we suggest several severity classes of discrimination shown in Fig. 3. As the ranges of "substantial" and "moderate" may be considered practically too wide, noting that the fourth derivative of the PTV^{total} does not provide the value in the range $[0.500, 0.854)$ or $[0.146, 0.500)$, we divide the range $[0.500, 0.854)$ into two parts, $[0.500, 0.634)$ and $[0.634, 0.854)$ based on the fifth derivative of the PTV^{total}. Similarly, we divide the range $[0.146, 0.500)$ into two parts, $[0.146, 0.366)$ and $[0.366, 0.500)$. Considering the sampling variability, it would be better to judge the severity class of discrimination by comparing the upper/lower limits of the confidence intervals of the discrimination measures with Fig. 3, rather than the point estimates.

6 Application

We apply the discrimination measures to the Adult Census Dataset, available from the UCI Repository of Machine Learning Databases [2]. The data consists of 1994 census information in the US. The Adult Census Dataset contains 48,842 instances with 6 numerical and 8 categorical features. A predictive model developed on this data was expected to determine whether a person's income makes over 50K a year.

The individual's income Y is a dichotomous outcome variable indicating whether a person's income makes over 50,000 a year, i.e., his/her income above 50,000 ($Y = 1$) or below 50,000 ($Y = 0$). Following [4], "sex" is considered as a sensitive feature X. Here, $X = 1$ indicates the disadvantaged group (female for "sex") and $X = 0$ indicates the advantaged group (male for "sex"). For details, refer to [4].

In order to evaluate the discrimination measures, under the sequential ignorability condition, we assume the logistic regression model of Y on X and $\boldsymbol{W}$ as a predictive model:

$$P(Y=1\mid X=x, \boldsymbol{W}=\boldsymbol{w}) = \frac{1}{1+\exp\{-(\theta_0+\theta_x x+\boldsymbol{\theta}_w^\top \boldsymbol{w})\}},$$

where $(\theta_0, \theta_x, \boldsymbol{\theta}_w^\top)^\top$ is a coefficients vector of the logistic regression model and $\boldsymbol{W}$ includes 13 features (excluding the sensitive feature X).

Let x_n, y_n, and $\boldsymbol{w}_n$ be the values taken by X, Y, and $\boldsymbol{W}$ for individual n, respectively, and let N be the sample size. We also use a logistic regression model to obtain a consistent estimate $\hat{x}(\boldsymbol{w}_n)$ of the propensity score $E[X|\boldsymbol{w}] = P(X=1|\boldsymbol{w})$ for individual n. From Eqs. (11) and (12), we have

$$\mathrm{qTE}_y(1,0|\boldsymbol{W}) = E\left[\frac{YX}{P(X=1|\boldsymbol{W})} - \frac{Y(1-X)}{1-P(X=1|\boldsymbol{W})}\right],$$

$$\mathrm{qNDE}_y(1,0|\boldsymbol{W}) = E\left[\left(\frac{YX}{P(X=1|\boldsymbol{W})} - \frac{Y(1-X)}{1-P(X=1|\boldsymbol{W})}\right)\frac{P(X=0|\boldsymbol{W})}{P(X=0)}\right]$$

$$= E\left[\frac{YX(1-P(X=1|\boldsymbol{W}))}{P(X=1|\boldsymbol{W})(1-P(X=1))} - \frac{Y(1-X)}{1-P(X=1)}\right].$$

Thus, the consistent estimates of qTE, qNDE, and qNIE are given by

$$\widehat{\mathrm{qTE}}_y(1,0|\boldsymbol{W}) = \left(\sum_{n=1}^N \frac{y_n x_n}{\hat{x}(\boldsymbol{w}_n)}\right) \Big/ \left(\sum_{n=1}^N \frac{x_n}{\hat{x}(\boldsymbol{w}_n)}\right)$$

$$- \left(\sum_{n=1}^N \frac{y_n(1-x_n)}{1-\hat{x}(\boldsymbol{w}_n)}\right) \Big/ \left(\sum_{n=1}^N \frac{1-x_n}{1-\hat{x}(\boldsymbol{w}_n)}\right),$$

$$\widehat{\mathrm{qNDE}}_y(1,0|\boldsymbol{W}) = \left(\sum_{n=1}^N \frac{y_n x_n(1-\hat{x}(\boldsymbol{w}_n))}{\hat{x}(\boldsymbol{w}_n)}\right) \Big/ \left(\sum_{n=1}^N \frac{x_n(1-\hat{x}(\boldsymbol{w}_n))}{\hat{x}(\boldsymbol{w}_n)}\right)$$

$$- \left(\sum_{n=1}^N y_n(1-x_n)\right) \Big/ \left(\sum_{n=1}^N (1-x_n)\right),$$

$$\widehat{\mathrm{qNIE}}_y(1,0|\boldsymbol{W}) = \widehat{\mathrm{qTE}}_y(1,0|\boldsymbol{W}) - \widehat{\mathrm{qNDE}}_y(1,0|\boldsymbol{W}),$$

respectively [5]. On the basis of these estimates, we calculate the proportion measures of discrimination $\mathrm{PTD}^{\mathrm{direct}}$, $\mathrm{PTV}^{\mathrm{direct}}$, and $\mathrm{PTV}^{\mathrm{total}}$.

In this scenario, the performances of the existing and proposed discrimination measures are listed in Table 2. Here, the normalized mean difference (NMD) is a representative discrimination measure, whereas "slift" and "elift" are known as the impact (risk) ratio and the ratio of additive interactions, respectively [20]. In addition, Table 2 shows the sample estimates from the original data (denoted by "estimate"), the standard errors (denoted by "s.e.") and the 95% bootstrap confidence intervals (CIs) (denoted by "lcl" for the lower confidence limits, and "ucl" for the upper confidence limits) evaluated by 2,000 bootstrap replications.

From Table 2, the 95% CIs of the TV and NMD do not include zero, and the 95% CIs of the slift and elift do not include one. It seems that the association between the sensitive feature and the outcome is statistically significant, and thus we can consider that $X \not\!\perp\!\!\!\perp Y$ holds. However, from the existing discrimination measures, it is uncertain how much of discrimination is based on the sensitive feature not via non-sensitive features. On the contrary, the estimates of the $\text{PTD}^{\text{direct}}$, $\text{PTV}^{\text{direct}}$, and $\text{PTV}^{\text{total}}$ show that most of total discrimination and total variation are based on the sensitive feature directly and totally. In addition, the 95% lower confidence limits of $\text{PTD}^{\text{direct}}$, $\text{PTV}^{\text{direct}}$, and $\text{PTV}^{\text{total}}$ are above 0.854. This shows that "sex" may be judged as "severe", "almost critical" or "critical" sensitive feature from the viewpoint of the proposed measures.

7 Conclusion

Most of the existing measures proposed for discrimina-tion-aware data mining have the deficiencies stated in Sect. 1. To overcome these deficiencies, we proposed causal discrimination measures based on the natural direct and indirect effects. The proposed discrimination measures are not estimable from observed data without causal knowledge, but the bounding formulas for the causal quantities, e.g., [1,3] would play an important role in evaluating the discrimination level. In addition, although [19] introduced the idea of the effect decomposition based on the "effect of treatment on the treated" into discrimination-aware data mining, the application of our results to their framework is straightforward.

Acknowledgments. We would like to thank the two anonymous reviewers for their helpful comments.

Disclosure of Interests. This research was funded by JFE Engineering Corporation and Japan Society for the Promotion of Science (JSPS), Grant Number 19K11856 and 21H03504.

References

1. Balke, A., Pearl, J.: Bounds on treatment effects from studies with imperfect compliance. J. Amer. Statist. Assoc. **92**(439), 1171–1176 (1997)
2. Becker, B., Kohavi, R.: Adult. UCI Machine Learning Repository (1996). https://doi.org/10.24432/C5XW20
3. Cai, Z., Kuroki, M., Pearl, J., Tian, J.: Bounds on direct effects in the presence of confounded intermediate variables. Biometrics **64**(3), 695–701 (2008)
4. Hamilton, E.: Benchmarking four approaches to fairness-aware machine learning. Ph.D. thesis, Haverford College. Department of Computer Science (2017). https://scholarship.tricolib.brynmawr.edu/handle/10066/19295
5. Huber, M.: Identifying causal mechanisms (primarily) based on inverse probability weighting. J. Appl. Economet. **29**(6), 920–943 (2014)
6. Imai, K., Keele, L., Tingley, D., Yamamoto, T.: Unpacking the black box of causality: learning about causal mechanisms from experimental and observational studies. Am. Polit. Sci. Rev. **105**(4), 765–789 (2011)
7. Kamishima, T., Akaho, S., Asoh, H., Sakuma, J.: Fairness-aware classifier with prejudice remover regularizer. In: Flach, P.A., De Bie, T., Cristianini, N. (eds.) ECML PKDD 2012. LNCS (LNAI), vol. 7524, pp. 35–50. Springer, Heidelberg (2012). https://doi.org/10.1007/978-3-642-33486-3_3

8. Kilbertus, N., Rojas-Carulla, M., Parascandolo, G., Hardt, M., Janzing, D., Schölkopf, B.: Avoiding discrimination through causal reasoning. In: Proceedings of the 31st International Conference on Neural Information Processing Systems, NIPS 2017, pp. 656–666. Curran Associates Inc., Red Hook (2017)
9. Kusner, M.J., Loftus, J., Russell, C., Silva, R.: Counterfactual fairness. In: Guyon, I., et al. (eds.) Advances in Neural Information Processing Systems, NIPS 2017, vol. 30, pp. 4066–4076. Curran Associates, Inc. (2017)
10. Pearl, J.: Probabilistic Reasoning in Intelligent Systems: Networks of Plausible Inference. Morgan Kaufmann, Burlington (1988)
11. Pearl, J.: Direct and indirect effects. In: Proceedings of the Seventeenth Conference on Uncertainty in Artificial Intelligence, pp. 411–420. Morgan Kaufmann Publishers Inc., San Francisco (2001)
12. Pearl, J.: Causality: Models, Reasoning and Inference, 2nd edn. Cambridge University Press, Cambridge (2009)
13. Pessach, D., Shmueli, E.: A review on fairness in machine learning. ACM Comput. Surv. **55**(3), 1–44 (2022)
14. Plecko, D., Bareinboim, E.: Causal fairness analysis. Technical report, R-90, Causal Artificial Intelligence Lab, Columbia University (2022)
15. Prentice, R.L.: Surrogate endpoints in clinical trials: definition and operational criteria. Stat. Med. **8**(4), 431–440 (1989)
16. Toreini, E., Aitken, M., Coopamootoo, K., Elliott, K., Zelaya, C.G., van Moorsel, A.: The relationship between trust in AI and trustworthy machine learning technologies. In: Proceedings of the 2020 Conference on Fairness, Accountability, and Transparency, FAT* 2020, pp. 272–283. Association for Computing Machinery, New York (2020)
17. Wang, Y., Taylor, J.M.G.: A measure of the proportion of treatment effect explained by a surrogate marker. Biometrics **58**(4), 803–812 (2002)
18. Zafar, M.B., Valera, I., Gomez Rodriguez, M., Gummadi, K.P.: Fairness beyond disparate treatment & disparate impact: learning classification without disparate mistreatment. In: Proceedings of the 26th International Conference on World Wide Web, WWW 2017, pp. 1171–1180. International World Wide Web Conferences Steering Committee, Republic and Canton of Geneva, CHE (2017)
19. Zhang, J., Bareinboim, E.: Fairness in decision-making - the causal explanation formula. In: Proceedings of the AAAI Conference on Artificial Intelligence, vol. 32, no. 1 (2018)
20. Žliobaitė, I.: Measuring discrimination in algorithmic decision making. Data Min. Knowl. Disc. **31**, 1060–1089 (2017)

Addressing Discretization Artifacts in Tomography by Accessing and Balancing Pixel Coverage of Projections

Csaba Olasz[✉]

University of Szeged, Szeged, Hungary
olaszcs@inf.u-szeged.hu

Abstract. In computed tomography, the reconstructed object and the measured detector values are represented in a discrete domain. As the real world is not discrete, this restriction of the model usually creates artifacts (i.e., *interpolation errors*) on the reconstructed images. To formulate the projection model, we use so-called interpolation methods which enumerate the interactions between beams and the image as rays pass through the pixels of the image. The type of interpolation method we select strongly influences the artifacts in the reconstructed image. In this paper, we show a connection between pixel coverage to get a better understanding of interpolation errors, and we also propose an effective correction method to reduce the effects of interpolation errors in reconstructed images. We tested our proposed method in a comprehensive experiment, where we found that our proposed correction method can significantly improve the quality of the image.

Keywords: Computed tomography · Projection matrix · FBP · SIRT · Pixel coverage

1 Introduction and Related Work

Computed Tomography (CT) is an imaging technique in which an object can be reconstructed slice by slice in a non-destructive way [11]. To do so, we expose the object of study to some kind of penetrating wave or radiation (e.g., X-ray, seismic waves, ultrasound, neutron beams, radio waves, etc.) and use mathematical tools to derive the structure of the object from the changes of the waves passing through the object or reflected from it [8,11].

In the case of transmission X-ray tomography, measurements are taken with penetrating X-ray radiation. As the radiation passes through the object, it attenuates proportionally to the amount and density of the material it crosses. There

---

This research was supported by project TKP2021-NVA-09. Project no. TKP2021-NVA-09 has been implemented with the support provided by the Ministry of Innovation and Technology of Hungary from the National Research, Development and Innovation Fund, financed under the TKP2021-NVA funding scheme.

© The Author(s), under exclusive license to Springer Nature Switzerland AG 2024
R. P. Barneva et al. (Eds.): IWCIA 2024/ISAIM 2024, LNCS 14494, pp. 182–205, 2024.
https://doi.org/10.1007/978-3-031-63735-3_11

are multiple sensors placed behind the object, where the total attenuation of an X-ray beam is detected at a given view angle. One can reconstruct the attenuation coefficient at any given point of an object if a large enough number of measurements are available at different view angles.

The object is reconstructed as a digital image, and the measurements are also modeled as discrete values. The reconstruction algorithm must solve the following question: "*Which ray passes through which pixels?*" Due to the discrete representations, there will be many cases where we can only give an approximate formulation, and consequently one has to use the so-called *"interpolation methods"* to estimate the projection geometry. Interpolation methods are not perfect, and hence different kinds of interpolation result in different structured artifacts in the reconstructed image.

There are many interpolation methods in the literature, and researchers have done extensive research to find the best one. For example, the authors of [5] compared three linear interpolation approaches. In [3] the projection matrix was analyzed on real data created by the Siddon [20] and Joseph [10] methods. Researchers of binary tomography are developed algorithms that work with hexagonal [16] or triangular [14] pixels. In non-binary tomography, alternatives to the square pixels representation of the image space were introduced in [13], while the contribution of the authors of [23] was the hexagonal-2x-oversampled scheme of the discrete grid. In [6] the connection between pixels and rays was defined with the help of local basis functions. The authors of [15] invented their own distance-driven method. As one can see, there are already numerous approaches available. Researchers not only investigated the different interpolation methods (i.e. possible connections between pixels and rays), but also experimented with different ray or pixel representations. In the end, every method has advantages and disadvantages, depending on the task and subjective preferences.

In the field of Cross-Hole Tomography [19], the geometric configuration of the measurements is very different from that of transmission X-ray tomography. Due to its different nature, the spatial coverage of pixels by projection lines is widely discussed [9] in this field. During our investigation of the interpolation methods, we found a connection between the pixel coverage and the structured artifacts of the given interpolation method. To capitalize on that, we propose an artifact removal approach based on balancing the pixel coverage within projections. With correction, the pixel coverage becomes uniform, which reduces structural artifacts in the reconstructed results. Note that the authors of [9] use an algorithm to divide the reconstructed image - consisting of equal pixel size - into distinct pixel sizes with uniform pixel coverage. This is similar to our correction presented here, but we find it easier to leave the pixel with the same size and rather correct the weights of the pixels in the projection matrix using the calculated pixel coverage.

This paper is organized as follows: we formulate the reconstruction task in Sect. 2 using the algebraic representation of tomography. Here, we also provide the definition of the projection matrix and how to calculate its weights with respect to the interpolation methods. The calculation of the pixel coverage and details of the proposed correction of the projection matrix are also found in Sect. 2. Next, we describe our experimental setup in Sect. 3. After that, we present our results in Sect. 4. Finally, we draw our conclusions in Sect. 5.

2 Mathematical Background and Projection Matrix

Our experiments were carried out using a parallel beam geometry. Furthermore, we utilized the algebraic representation of the reconstruction problem, which can be formulated as follows. Let us say that we have an $N \times N$ image f on a square grid indexed in a row-major order, as shown in Fig. 1. N represents the number of pixels in the rows and columns. Furthermore, let the sinogram be a grid with M_p projections each containing M_d detector values indexed in a detector-major order. In this way, let p_i ($i \in M$, $M = M_p \cdot M_d$) be the sum of the rays of the i-th ray shown in Fig. 1. The following linear equation can be stated:

$$\sum_{j=1}^{N^2} W_{ij} f_j = p_i, \quad i = 1, 2, ..., M , \qquad (1)$$

where W is the projection matrix (or system matrix), which contains the connection between all rays and pixels ordered by view angles.

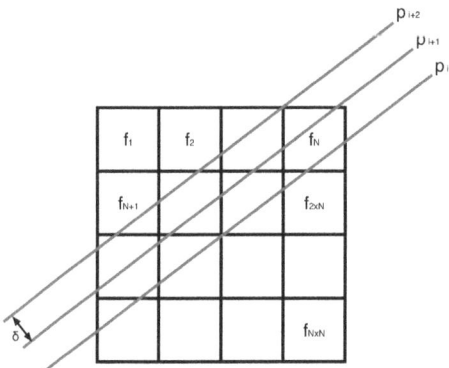

Fig. 1. The reconstruction problem as a system of linear equations. f_N is the pixels of an image on a square grid. p_i is a ray at a given angle. δ represents the distance between neighboring rays.

2.1 Interpolation During the Projection Matrix Calculation

Various interpolation techniques exist in the literature, with different characteristics and performance. In this paper, we decided to analyze some of the best performing and most common techniques to describe interpolation errors, and propose a way for their corrections. In [18] the authors concluded that the

Linear and the Gaussian Distance Interpolation (*GDI*) resulted in the most accurate reconstructions. Others reasoned that images with *Strip* interpolation can be more appealing to human eyes [6] and we argue that it better represents the physical phenomenon during data acquisition. Therefore, we examine these methods and also include the *Line* interpolation because of its popularity. Figure 2 illustrates the calculation of the weights of the projection matrix with the different interpolations.

In the case of *Line* interpolation, projection beams are modeled by straight lines [2]. The weight W_{ij} is the length of the section of the p_i ray that travels through the f_j pixel.

When calculating W_{ij} using *Linear* interpolation, we first need to take into account the slopes of the beams [5]. Let x be the horizontal axis and y be the vertical axis on an Euclidean coordinate system holding our image f. Let us also assume that the projection beam is a line defined by the equation $y = m \cdot x + c$ with some m slope and c constant. In this way, we have two cases. When $|m| > 1$ (i.e., the projection beam is closer to vertical), we go through the rows of the image, and in each row the W_{ij} coefficients are calculated to perform a linear interpolation of the two pixels closest to the intersection of the beam and the row. In $|m| \leq 1$ we do the same, but by stepping through columns of the image. Therefore, the W_{ij} weight of the f_j-th pixel will be d_2 looking at Fig. 2. The advantage of this interpolation technique is that it is relatively easy to implement and is computationally efficient compared to some other techniques.

The *Strip* interpolation behaves as if the p_i ray has width [2]. The weight W_{ij} of f_j-th is the area in which the p_i and the f_j-th pixels overlap. This technique has a more complex and computationally demanding implementation, but if the space is fully covered by beams, then it behaves very well with respect to the interpolation error.

The interpolation technique we call *GDI* uses a parameterised Gauss function [6] to provide the weight W_{ij}. As illustrated in Fig. 2 the W_{ij} weight of f_j is defined by the Gaussian of the distance between the projection line p_i and the center of the f_j-th pixel. In our implementation, the Gaussian σ parameter of *GDI* was set to 0.7 and we only calculated the coefficient for pixels within 2 units of distance from the projection lines (for other pixels, the weights were set to zero).

In this paper, we used the ASTRA Toolbox [1] for the calculation of the projection matrix in Matlab software [17], except for *GDI* interpolation. For *GDI* we used our own implementation because this option is not available in ASTRA.

2.2 Pixel Coverage and Our Proposed Correction Step

We found that the interpolation artifacts are diverse properties of the interpolation methods but are closely related to the pixel coverage. Let us define two versions of the pixel coverage using the notations introduced in (1). Let the global coverage of the j-th pixel of f be defined as

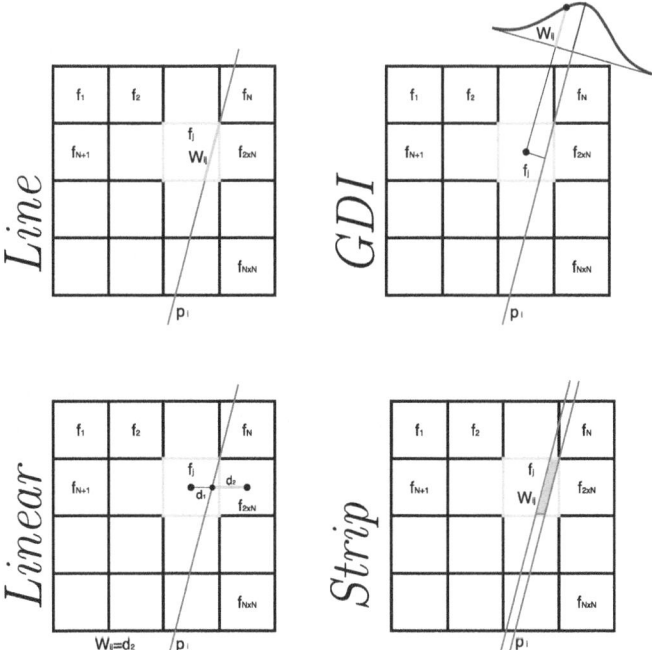

Fig. 2. The four interpolation models tested in our experiments.

$$\mathcal{C}_j = \sum_{i=1}^{M} W_{ij} \ , \qquad (2)$$

giving the sum of coefficients belonging to all the pixels in all the projections. Moreover, let Θ_k be the set of projection line indices i belonging to the k-th projection angle, and let the $\mathcal{C}_{j,\Theta_k}$ directed coverage of the j-th pixel be calculated as

$$\mathcal{C}_{j,\Theta_k} = \sum_{i \in \Theta_k} W_{ij} \ . \qquad (3)$$

For each pixel this gives us the sum of projection coefficients within a projection. We can also say that the global and directed pixel coverage are related as

$$\mathcal{C}_j = \sum_{k=1}^{M_p} \mathcal{C}_{j,\Theta_k} \ . \qquad (4)$$

The directed and global pixel coverage determines how much a pixel contributes to a projection or the sum of projections. It also reflects how much intensity is propagated into the specific pixel when reconstructed. We argue that imbalances in the local and global pixel coverage show how uneven the representation of pixels in the geometry is and thus can cause artifacts.

To show this phenomenon, we calculated projection matrices with the same geometry but different interpolations. We then calculated the global coverage of each f_j pixel for each interpolation. Note that the global coverage map is what we refer to simply as pixel coverage in the following.

Our proposal is to even out the global coverage map by dividing the weights of the projection matrix in a given view angle by the directed coverage, i,e,

$$W'_{ij} = W_{ij}/\mathcal{C}_{j,\Theta_k} \quad | \quad i \in \Theta_k \ . \tag{5}$$

Basically, we normalize the weights by the sum of the weights within the angles. After normalization, the global coverage map of W'_{ij} contains only ones. As the pixel coverage map contains only ones in the case of the *Strip* interpolation originally, we expect no significant change after correction in that case.

There are multiple images in Fig. 3. The pixel coverage maps are shown in the first column. In the second and third columns, there is an example to demonstrate the effect of the correction made on the projection matrix. The second column denoted by *Raw - GT* contains the uncorrected reconstructed images minus ground truth, namely stop_sign in Fig. 4. The result after the proposed correction is shown in the third column, where the *Corrected* minus *GT* images are presented. Note that the patterns in the first and second columns of Fig. 3 correlate with each other.

3 Experimental Setup

In our research we examined the merits of the interpolation methods and their correction (detailed in Sect. 2) in various circumstances by changing multiple parameters. Table 1 summarizes all parameters that we tested in all combinations. The parameters are explained in detail in the following subsections.

Table 1. The variable parameters we used in our tests. (*FBP*: Filtered backprojection, *SIRT*: Simultaneous Iterative Reconstruction Technique, *GDI*: Gaussian Distance Interpolation, G: Gaussian, P: Poisson, **MAE**: Mean Absolute Error, **SSIM**: Structural Similarity, **HHom**: Haralick Homogeneity)

Parameter name	Tested settings
Algorithm	*FBP, SIRT*
Interpolation	*Line, Linear, Strip, GDI*
Geometry	*Basic, View++, Ray++*
Noise	*G 36 dB, G 42 dB, G 48 dB, Noiseless, P 48 dB, P 42 dB, P 36 dB*
Normalization	*Raw, Corrected*
Metric	**MAE, SSIM, HHom**

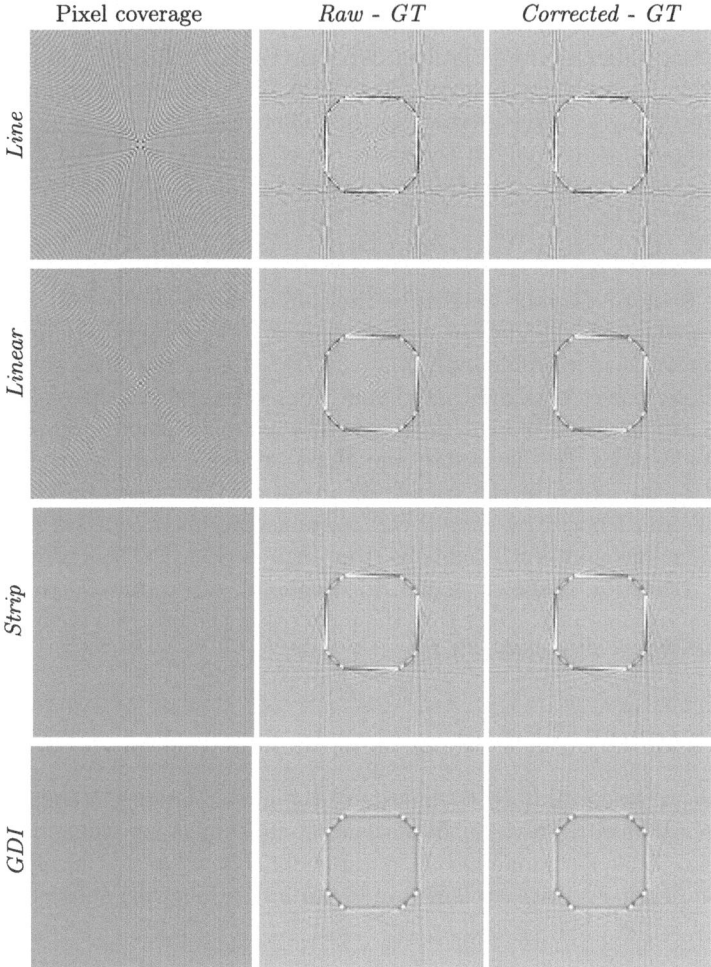

Fig. 3. The first column shows the pixel coverage according the indicated interpolation method. Pixel coverage images are normalized between the interval [0, 1]. The standard deviations of the pixel coverage maps: *Line*: 0.0682, *Linear*: 0.0423, *Strip*: 0.00002, *GDI*: 0.00037. The second and third columns contain the corresponding difference images of the stop_sign phantom (see Fig. 4), where *Raw* is the uncorrected reconstruction, *Corrected* is the reconstruction with correction and *GT* is the ground truth, stop_sign in Fig. 4. The display window of the difference images is [−0.15, 0.15].

3.1 Reconstruction Algorithms

In our paper, we perform reconstructions with the Filtered Backprojection (*FBP*) algorithm [11] or with the Simultaneous Iterative Reconstruction Technique (*SIRT*) [21].

FBP is an analytic reconstruction algorithm that is the most popular choice in commercial CT scanners. *FBP* can produce a good quality reconstruction in a short time if a large number of projections are available. In this paper we used the following implementation:

$$f = W^T \cdot h \cdot p ,\qquad(6)$$

where W^T is the transposed projection matrix, f is the reconstructed image, p is the projections, and h is the Ram-Lak filter. We performed the filtering in Fourier space.

The other algorithm we used was the Simultaneous Iterative Reconstruction Technique (*SIRT*), which is a popular algebraic method for the reconstruction of CT images. It usually takes more time to get a result with *SIRT* as it works in an iterative manner, but it proved to be superior compared to *FBP* [25] because of its potential to improve image quality with repeated steps as long as necessary. We let the algorithm run for a fixed number of 100 iterations. We used the following implementation:

$$f_{k+1} = f_k + CW^T R(p - W f_k) ,\qquad(7)$$

where k is iteration index, and f_k is the intermediate reconstructed image – or in our case f_{100} is the result –, p is the projections, and W^T is the transposed projection matrix. Furthermore, C and R are diagonal matrices giving additional weighting to ensure the convergence of the algorithm. The C and R matrices are constructed as

$$R \in \mathbb{R}^{M \times M} ,\quad r_{ii} = 1/\sum_{j=1}^{N^2} W_{ij} ,\qquad(8)$$

$$C \in \mathbb{R}^{N^2 \times N^2} ,\quad c_{jj} = 1/\sum_{i=1}^{M} W_{ij} ,\qquad(9)$$

where $N \times N$ is the number of pixels and M is the number of rays.

3.2 Geometry

If the reconstructed image consists of square pixels with sides of unit length, then in the case of an image $N \times N$ the diagonal length is $\sqrt{2}N$. Based on the sampling criteria, we need $2\sqrt{2}N$ projection lines to cover the entire length of the diagonal (i.e. $M_d = 2\sqrt{2}N$). We sampled the projections positioned equiangularly in the half-circle. According to the sampling criteria [11] we need at least

$$M_p = \frac{N}{\sqrt{2}}\pi ,\qquad(10)$$

projections for an accurate reconstruction. We used three different settings considering the geometry, all of which had at least the required number of projection lines and the minimal number of projection angles. In the case that we will call

"*Basic*" geometry, we set the number of lines to 366 and the number of views to 285. When we used the geometry called "*Ray++*", we increased the number of lines to 732, while the number of viewing angles remained the same. The so-called "*View++*" geometry was implemented with 570 view angles and 366 lines. Taking into account the distance between lines (that is, δ in Fig. 1), we have set the value 0.5 for *Ray++* and 1.0 for the rest. Table 2 summarizes the exact values of the geometries.

Table 2. Parameter values of the tested geometries.

	Number of lines	Number of views	Distance among lines
Basic	366	285	1
View++	366	570	1
Ray++	732	285	0.5

3.3 Test Images

Our test database consisted of 12 images with a size of 128 × 128 pixels. The intensities of the images were selected from the [0, 1] interval. All images are shown in Fig. 4. We built four groups of phantoms to construct a data set for structured experiments. First, we placed the same circle on the images, but in slightly different places (see the first row of Fig. 4). Second, we altered the various shapes in some images to ensure that the objects had roughly the same area (second row of Fig. 4). Third, we used binary and non-binary phantoms including the Shepp-Logen [12] and Forbild [24] head phantoms. The projections of the images were produced by an analytical computer simulation that calculates the line intersection of the geometric shapes with the projection lines; therefore, the projection values were not calculated from rasterized phantoms. In this task, we utilized the codes of [24].

3.4 Random Noise on Projections

In practice, tomographic imaging is always exposed to some type and level of noise. Commonly accepted solution in simulation studies [4] is to model shot noise with Poisson distribution, while the detector electrical noise is usually modeled by Gaussian distribution. We examine both types of distributions in our experiments, one at a time. Altogether, we have three noise levels with each distribution and the noiseless case. We set the noise levels based on the signal-to-noise (SNR) ratio of the simulated projections. The noise levels are 36 dB, 42 dB, and 48 dB.

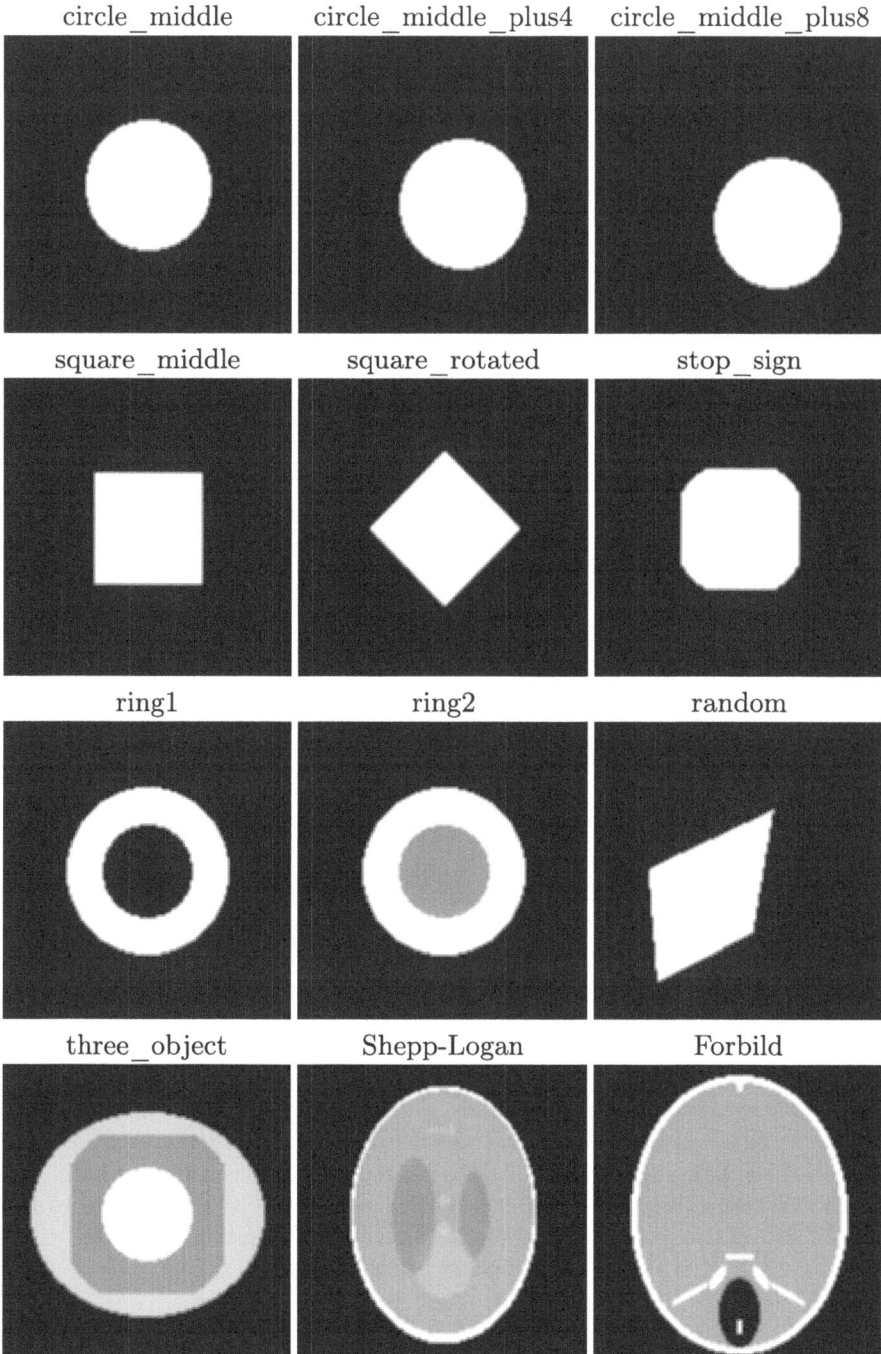

Fig. 4. Image dataset.

3.5 Metrics

We calculated three quality metrics (namely, Mean Absolute Error /MAE/; Structural Similarity /SSIM/ and Haralick Homogenity /HHom/) to objectively evaluate the performance of the interpolation methods in the different setups. In this paper, we focus on artifacts that are caused by the interpolation error. These artifacts can be detected on homogeneous regions of objects. Therefore, we calculated all metrics only inside the biggest object of the image. To do that, we applied per-image masks, which were initiated by the area of the biggest object. Then, we morphologically eroded the object to exclude the pixels of the borders. The border pixels of the object are naturally highly variable, which we do not consider as a heterogeneous region. Although the border reconstructing capability of the interpolation methods can be an interesting question, this is not in the scope of our present paper. Figure 5 shows an example of the ground truth (GT) image and its mask. Only the blue pixels will be included in the calculation of the metrics.

Fig. 5. Example for an image and its mask. (Colors: black - background; white - object pixels on original phantom; blue - mask) We included only the pixels with the color of blue at the calculation of the metrics.) (Color figure online)

The Mean Absolute Error (MAE) is a pixel-wise metric for compering the reconstructed result to a GT image. If the reconstructed image matches perfectly to its GT, the MAE gives a value of zero. There is no upper limit with MAE.

The Structural Similarity (SSIM) index also requires a GT image, but the original aim of this metric was to model the visual perception of people by the values of SSIM [22]. The formula of SSIM has three terms:

$$\text{SSIM}(\mathbf{f}, \hat{\mathbf{f}}) = [l(\mathbf{f}, \hat{\mathbf{f}})]^\alpha [c(\mathbf{f}, \hat{\mathbf{f}})]^\beta [s(\mathbf{f}, \hat{\mathbf{f}})]^\gamma \quad, \tag{11}$$

namely the $l(\mathbf{f}, \hat{\mathbf{f}})$ luminance, the $c(\mathbf{f}, \hat{\mathbf{f}})$ contrast and the $s(\mathbf{f}, \hat{\mathbf{f}})$ structural terms. The pictures to be compared are denoted with $\mathbf{f}$ and $\hat{\mathbf{f}}$. The contribution of the three terms can be regulated by α, β, and γ. As we wanted to focus here on the contrast and structure related terms, we set α to zero while β and γ get the value of one. Originally, the range of the measure is $\text{SSIM}(\mathbf{f}, \hat{\mathbf{f}}) \in (-1, 1]$. If $\text{SSIM}(\mathbf{f}, \hat{\mathbf{f}}) = 1$, then the two compared images are identical. If $\text{SSIM}(\mathbf{f}, \hat{\mathbf{f}})$ is negative, then there is similarity, but in a complementer manner. If $\text{SSIM}(\mathbf{f}, \hat{\mathbf{f}}) = 0$, then no similarity was found between the two images. Since reconstruction techniques are not intended to produce complementer reconstructions, negative values should be regarded as errors, so negative values were replaced by zero. The final value of SSIM is the average of the locally calculated similarity values of pixels inside the object with the most outside borders. Here, we note that we calculated SSIM with the Matlab build-in function, where we set the'Radius' parameters to 0.5 and'DynamicRange' to 1.1. The 1.1 value for dynamic range is necessary because the intensity values of the reconstructed images can be above the maximum value of the GT images.

The Haralick texture features evaluate one single image without a GT [7]. We must calculate the co-occurrence matrix of the image to define the Haralick features. This matrix contains the frequency of two pixel intensities appearing together next to each other at a given offset. We always used the offset $[0, 1]$, therefore, the two pixels to be evaluated are the pixel of interest and its right neighbor. Regarding the other parameters of the co-occurrance matrix calculation we used 255 gray levels and the $[0, 1.1]$ interval as gray limits according to the Matlab inside function. We will present the results of the Haralick Homogeneity feature (HHom), which is between the $[0, 1]$ interval. For a constant image $\text{HHom}(\mathbf{f}) = 1$.

4 Results and Discussion

In this section, we will present our results and share our finding. Note that given the high number of parameters in our experiment, we got too much data to present every detail individually. Therefore, we will present only slices from the results, while we try our best not to lose the whole picture.

Table 3 contains the results for all phantoms individually in a setup, where we use *FBP* reconstruction, *Basic* projection geometry, Noiseless projection data, and evaluate results with HHom. Statistics are provided for both the *Raw* version of the interpolation methods, and the global coverage map balanced variances obtained by (5). Moreover, Table 3 shows the average (AVG) and standard deviation (STD) of the corresponding columns. To better visualize the magnitude of STD compared to the difference in interpolation methods, we plotted the AVG-s and STD-s together in Fig. 6. We noticed that the STD is relatively high compared to the difference of interpolation methods, but we argue that this is due to the variability of the phantoms in our dataset (see Table 3). To support this conclusion, we calculated the Pearson correlation among the interpolation methods, which were calculated including the entire result set. Table 4 shows the pairwise correlations between the methods. We understand that the high value of the pairwise correlation (always above 0.9474) means that the interpolation methods follow the same trend through the test phantoms. Therefore, in the following sections we will discuss only the *AVG* values.

Table 3. Results of reconstructing all phantoms with the settings: *FBP* reconstruction, Noiseless data, *Basic* projection geometry, and HHom measure. Average (*AVG*) and standard deviation (*STD*) values are also presented. In the case of *Raw*, we used the uncorrected projection matrix. Phantoms marked with * have roughly the same area.

FBP, *Basic*, Noiseless, HHom	*Line*		*Linear*		*Strip*		*GDI*	
	Raw	*Corrected*	*Raw*	*Corrected*	*Raw*	*Corrected*	*Raw*	*Corrected*
AVG	**0.4291**	**0.6261**	**0.5590**	**0.7004**	**0.7463**	**0.7463**	**0.8940**	**0.8821**
STD	**0.0799**	**0.1820**	**0.0961**	**0.1610**	**0.1467**	**0.1467**	**0.1326**	**0.1162**
circle_middle*	0.4961	0.9438	0.6750	0.9470	0.9514	0.9514	0.9756	0.9756
circle_plus4*	0.4870	0.8210	0.6147	0.8629	0.8831	0.8831	0.9978	0.9607
circle_plus8*	0.5063	0.7839	0.6166	0.8195	0.8350	0.8350	0.9883	0.9169
square_middle*	0.3796	0.4925	0.5425	0.6367	0.7707	0.7707	0.9830	0.9878
square_rotated	0.3151	0.4867	0.4702	0.6327	0.7462	0.7462	0.9675	0.9688
stop_sign*	0.4301	0.6035	0.5530	0.6488	0.6751	0.6751	0.9892	0.9338
ring1	0.4639	0.6607	0.6387	0.8087	0.8533	0.8533	0.9601	0.9655
ring2	0.4863	0.7584	0.6387	0.8164	0.8386	0.8386	0.8560	0.8564
random	0.3595	0.4521	0.4598	0.5291	0.6012	0.6012	0.9273	0.8619
three_object	0.4722	0.6145	0.5806	0.6821	0.7050	0.7048	0.7148	0.7837
Shepp-Logen	0.4865	0.6059	0.5824	0.6603	0.6914	0.6914	0.7778	0.7793
Forbild	0.2662	0.2896	0.3356	0.3601	0.4045	0.4045	0.5904	0.5945

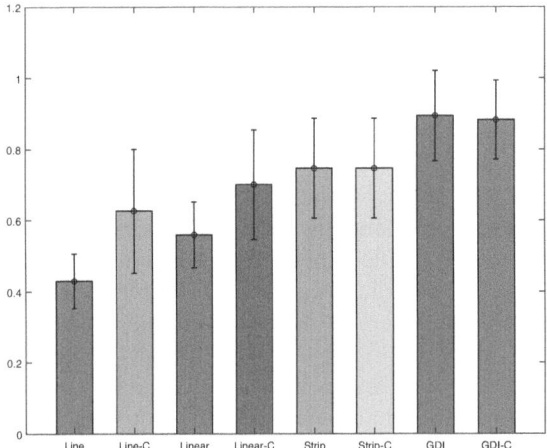

Fig. 6. The average values from Table 3 (settings: *FBP*, Noiseless, *Basic*, and HHom). Standard deviations are shown as error bars.

Table 4. The Pearson correlation between the interpolation methods.

Full dataset		Line		Linear		Strip		GDI	
Pearson correlation		Raw	Corrected	Raw	Corrected	Raw	Corrected	Raw	Corrected
Line	Raw	1.0000	0.9982	0.9963	0.9961	0.9843	0.9843	0.9612	0.9613
	Corrected	0.9982	1.0000	0.9945	0.9963	0.9886	0.9886	0.9605	0.9605
Linear	Raw	0.9963	0.9945	1.0000	0.9994	0.9815	0.9815	0.9776	0.9777
	Corrected	0.9961	0.9963	0.9994	1.0000	0.9853	0.9853	0.9772	0.9773
Strip	Raw	0.9843	0.9886	0.9815	0.9853	1.0000	1.0000	0.9474	0.9475
	Corrected	0.9843	0.9886	0.9815	0.9853	1.0000	1.0000	0.9474	0.9475
GDI	Raw	0.9612	0.9605	0.9776	0.9772	0.9474	0.9474	1.0000	1.0000
	Corrected	0.9613	0.9605	0.9777	0.9773	0.9475	0.9475	1.0000	1.0000

4.1 Reconstruction Algorithm and Noise

Let's look at the results considering only the *Basic* geometry from the geometries we detailed in Sect. 3.2. Figure 7, 8 and 9 shows 2 grouped bar charts each with the average values of the corresponding metric. Regarding the reconstruction algorithm, one can see that the *SIRT* has higher SSIM and HHom, except for a few cases with zero or low level Gaussian noise using the *GDI*. This better performance of the *SIRT* is expected according to the literature [25]. With MAE the images produced by *SIRT* have less error even at the highest level of Gaussian noise, but with increasing the strength of the Poisson noise the *FBP* has smaller values of average except for *Line* interpolation. A rather interesting phenomenon also occurs when looking at the noiseless and right-hand side (Poisson noise) part of the bar chart with MAE. Compared to the noiseless case, one can see a drop in the MAE values when we added a low level of noise (48 dB). In our opinion, when we randomly added the low-level noise to the image, it could change the values of individual pixels with the same volume but in the opposite way as the interpolation artifacts did. Or it can be just a simple smoothing effect caused by the small inconsistencies in the projection matrix due to the random noise. With that exception stated, generally we can say that adding noise to the projections has a negative effect on the image quality according to all metrics. Moreover, the quality of the reconstructed image dropped at a higher rate when Poisson noise was added. Examples of reconstructed images from noisy and noiseless data can be found in Fig. 10.

Switching to the difference between the *Raw* and the *Corrected* versions (third column in Fig. 10) we can see that applying the correction step efficiently removed the structural patters from the images, indicating the removal of interpolation artifacts. In the *Raw* minus *Corrected* image, the effect of correction is clearly visible despite the type of noise. Still, only looking at the reconstructed images in the Poisson noise affected case, we can say that strong noise can dominate the interpolation errors, making them invisible on the results. On one note, this might have been the reason why interpolation errors have not been extensively researched so far. When the random noise in the measured data is strong,

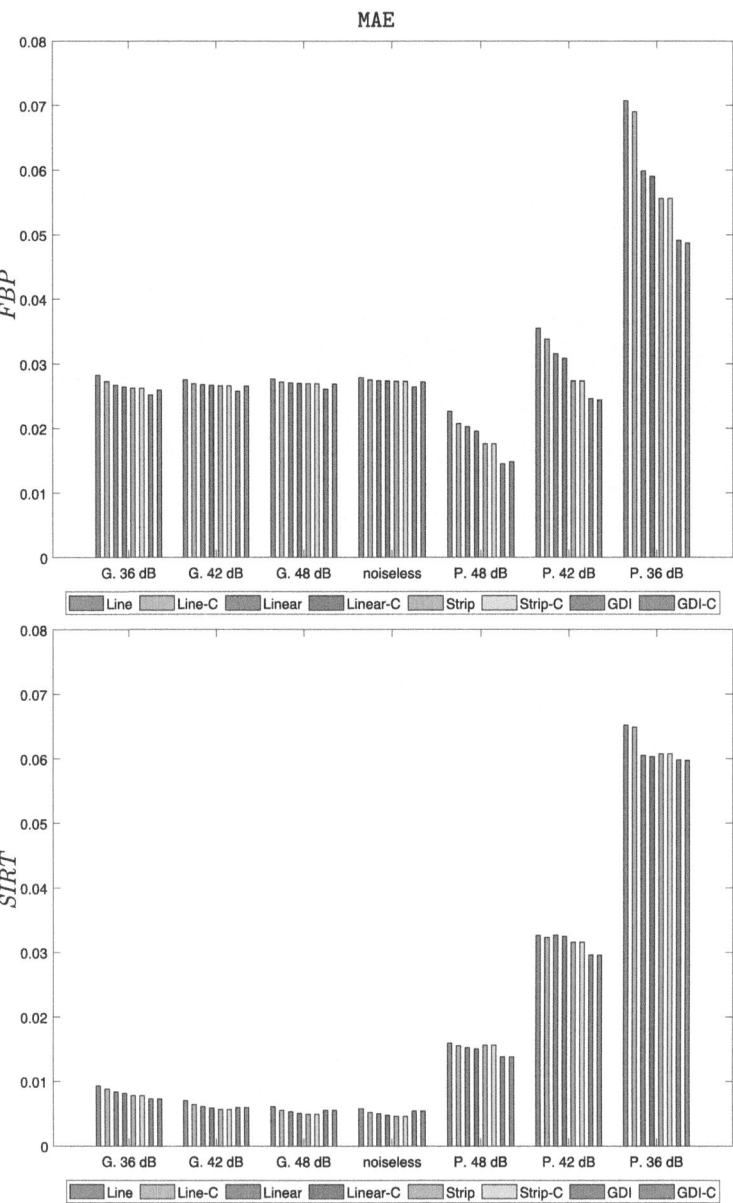

Fig. 7. There are two grouped bar charts on this figure belonging to the given reconstruction algorithm with *Basic* geometry. Values are MAE averages of the entire dataset. Each chart contains seven groups representing the seven noise levels we tested. In the middle, one can see the noiseless case. To the right the Poisson noise and to the left the Gaussian noise are raising. In each group, the four interpolation methods are represented with different colors and their *Corrected* versions with brighter shade of the same color. In the legend, the *Corrected* versions are marked with the "−C" suffix.

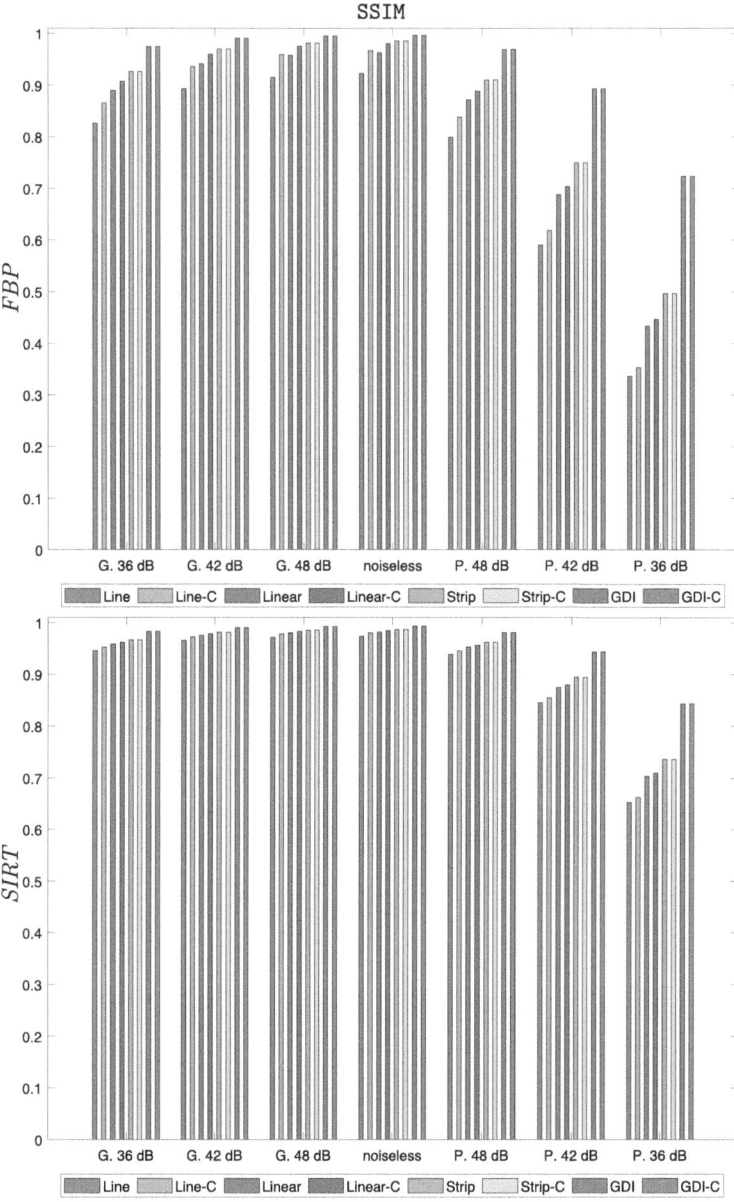

Fig. 8. There are two grouped bar charts on this figure belonging to the given reconstruction algorithm with *Basic* geometry. Values are SSIM averages of the entire dataset. Each chart contains seven groups representing the seven noise levels we tested. In the middle, one can see the noiseless case. To the right the Poisson noise and to the left the Gaussian noise are raising. In each group, the four interpolation methods are represented with different colors and their *Corrected* versions with brighter shade of the same color. In the legend, the *Corrected* versions are marked with the "−C" suffix.

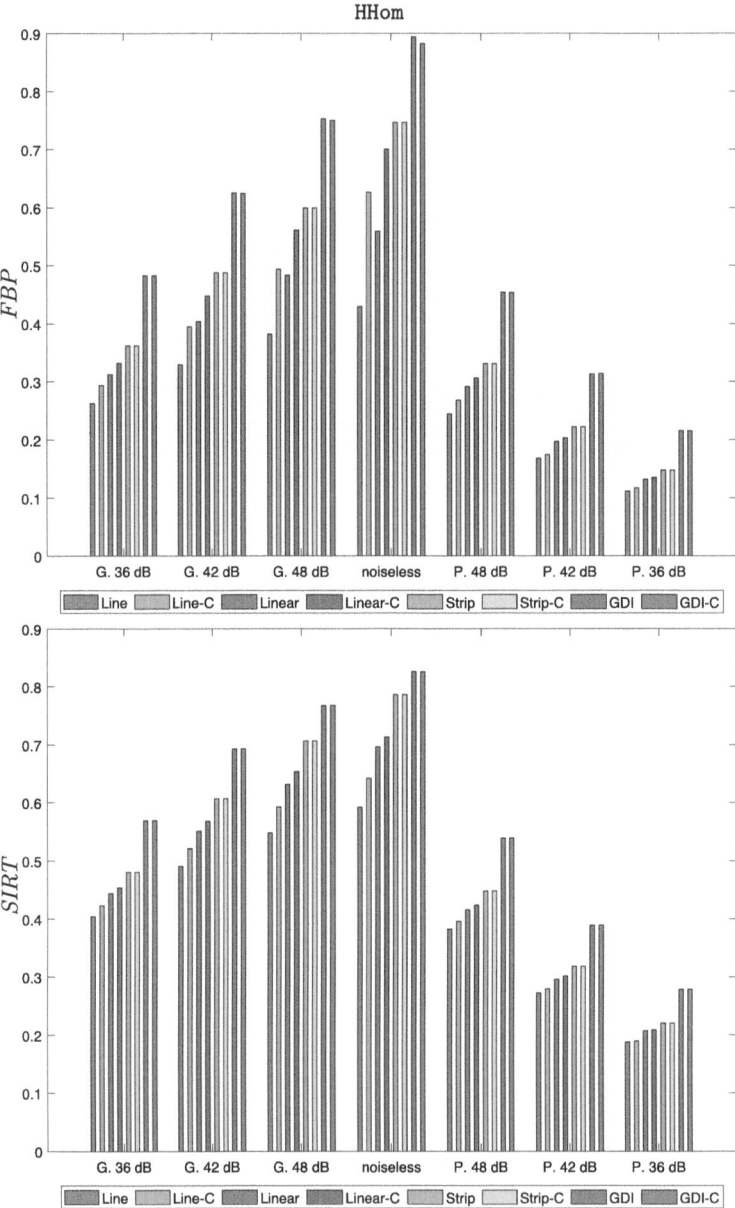

Fig. 9. There are two grouped bar charts on this figure belonging to the given reconstruction algorithm with *Basic* geometry. Values are HHom averages of the entire dataset. Each chart contains seven groups representing the seven noise levels we tested. In the middle, one can see the noiseless case. To the right the Poisson noise and to the left the Gaussian noise are raising. In each group, the four interpolation methods are represented with different colors and their *Corrected* versions with brighter shade of the same color. In the legend, the *Corrected* versions are marked with the "−C" suffix.

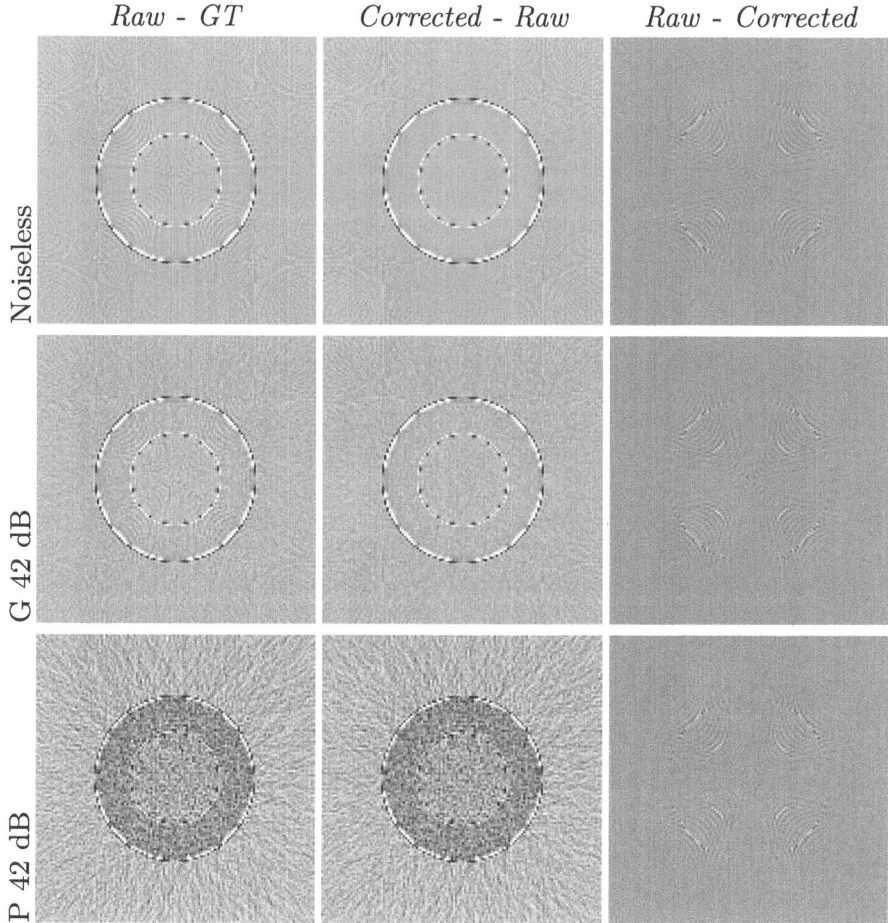

Fig. 10. An example from our dataset. The settings: *Basic* geometry, *FBP* and *Line* interpolation. The figure contains the noiseless case, and the result with middle-level Gaussian and Poisson noise. The *Raw* and *Corrected* versions are also shown in the first and second columns, respectively, while the third column is the difference between the two. Display window [−0.06, 0.06] for the differences and [0.47, 1.06] for the others.

the interpolation errors are not significant compared to other distortions. With improving data acquisition techniques, on the other hand, good quality imaging techniques in the future may make it more vital to account for interpolation errors in tomographic reconstruction.

For a further detailed analysis of the proposed correction technique, Fig. 11 shows a phantom reconstructed by all interpolation methods with and without correction. We also provide the MAE maps in the second and fourth rows (that is, essentially the pixelwise absolute difference between the phantom and the reconstructed image). The *Line* and the *Linear* interpolation methods are

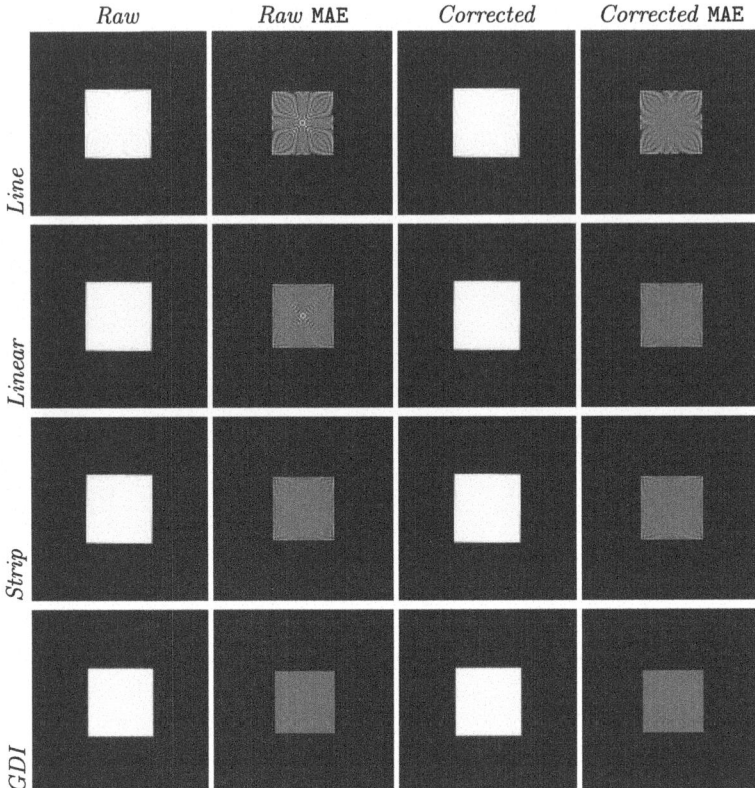

Fig. 11. An example from our dataset, namely the square_middle from Fig. 4. The settings: *Basic* geometry, *FBP* and Noiseless. All interpolation methods are presented with and without correction with the corresponding MAE map. Display window [0, 0.09] for the MAE maps and [0.47, 1.06] for the reconstructed images.

affected the most by the interpolation error. The structures of the interpolation errors are clearly visible on those images. This supports the explained connection between the interpolation error and the pixel coverage in Sect. 2. One can recognize the same patterns as we presented in Fig. 2. As the *Strip* interpolation has an uniform pixel coverage originally, the effect of the correction is not visible. Regarding the balance in the pixel coverage, the *Strip* interpolation is a perfect interpolation technique. Note that despite the *Strip* model being perfectly balanced, the resulted image with *GDI* is still likely to be more appealing. We argue that this is because the *GDI* has an inbuilt smoothing capability that can suppress other artifacts not caused by coverage imbalance but coming from the filter of the Filtered Back-Projection algorithm. Therefore, it is not directly connected to the pixel coverage.

4.2 Geometries

We were interested in how the increased number of view angles or projection rays above the necessary number may affect the quality of the reconstructed image. We found that the pixel coverage improves when there are more rays or view angles. Still we can say that the improvement is more significant if we increased the number of rays. Figure 12 shows the coverage of the pixels with *Line* interpolation according to the three geometries. The main trends remain the same when switching from *Basic* to *View++* geometry, but the *STD* decreased from 0.0682 to 0.0658 due to the doubled view angels. One can see that some of the noise outside the center of the image disappeared from the images of the *View++* geometry. In the case of the *Ray++* geometry, a more significant improvement in *STD* and a change in the pattern of the pixel coverage were recognized. Note that the number of rays and view angles cannot be raised arbitrarily in real-world applications. These numbers are physically limited by the equipment that is used for the acquisition of the data. We also point out that the necessary number of rays and view angles are depending on the size of the reconstructed image. Therefore, the physical limits of the equipment has a greater effect if an image with higher resolution is reconstructed. For example, the required number of ray and view angles according to Sect. 3.2 is 1138 and 1449 respectively, in the case of a 512×512 pixels sized image.

Figure 13 and Fig. 14 show the values MAE and HHom using the *View++* and *Ray++* geometries. The results of SSIM are omitted, because their conclusions coincide with the results of HHom. As pixel coverage is predicted, the *Ray++* geometry results in values and trends similar to the *Basic* geometry. Looking at the MAE with the *Ray++* geometry, the effect of the correction is not significant. The *Corrected* and uncorrected of the interpolation method are moving together regardless of the noise or the algorithm. On the contrary, when examining the HHom values, the *Corrected* versions of the *Line* and *Linear* methods are showing improvement in the noiseless and the Gaussian noise affected cases.

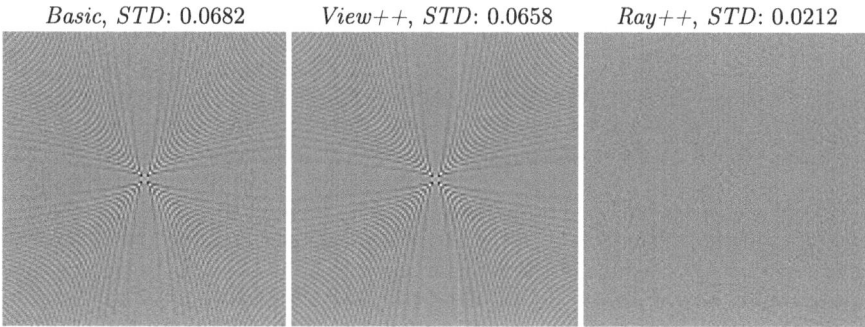

Fig. 12. The pixel coverage of the *Line* interpolation with the three geometries. Images were normalized between the [0, 1] interval.

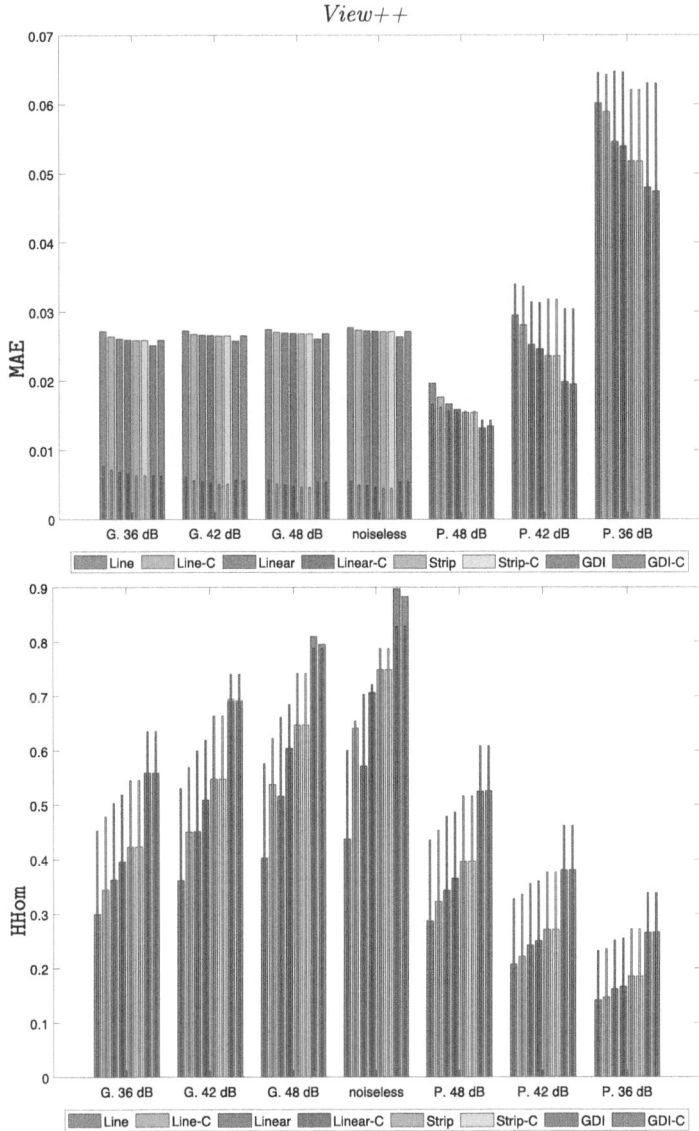

Fig. 13. There are two grouped bar charts on this figure belonging to the given reconstruction algorithm with *View++* geometry. Values are calculated by the given error metric and they are averages of the entire dataset. Each chart contains seven groups representing the seven noise levels we tested. In the middle, one can see the noiseless case. To the right the Poisson noise and to the left the Gaussian noise are raising. In each group, the four interpolation methods are represented with different colors, and their *Corrected* versions are represented with brighter shade of the same color. In the legend, the *Corrected* versions are marked with the "−C" suffix. The two bars in every position indicate the average with *FBP* (wider) and *SIRT* (thinner).

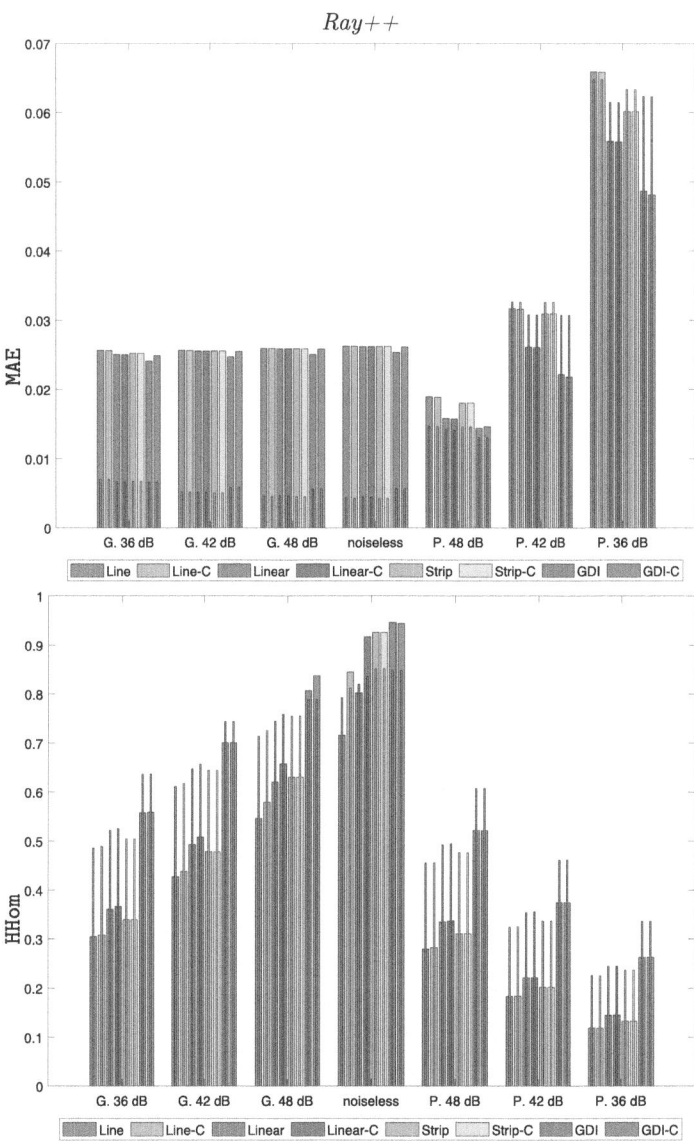

Fig. 14. There are two grouped bar charts on this figure belonging to the given reconstruction algorithm with *View++* geometry. Values are calculated by the given error metric and they are averages of the entire dataset. Each chart contains seven groups representing the seven noise levels we tested. In the middle, one can see the noiseless case. To the right the Poisson noise and to the left the Gaussian noise are raising. In each group, the four interpolation methods are represented with different colors, and their *Corrected* versions are represented with brighter shade of the same color. In the legend, the *Corrected* versions are marked with the "−C" suffix. The two bars in every position indicate the average with *FBP* (wider) and *SIRT* (thinner).

5 Conclusion

In this study, we examined the effects of using discretized approximate models in transmission X-ray tomography. We proposed applying the pixel coverage to get a better understanding of the interpolation errors and also to visualize the different artifact structures caused by the different interpolation methods. We proposed an effective correction formula for the projection matrix to eliminate interpolation errors from the reconstructed images. We tested our proposed method in a comprehensive experiment.

We found that our proposed correction method significantly improved the quality of the image if the applied interpolation was the *Line* or the *Linear*. Although the increased noise levels can suppress the artifacts caused by the discretization, the calculated metrics indicated improvements due to our proposed correction at higher level of noise. *SIRT* was shown to be more robust against interpolation errors. Nevertheless, we also concluded that some discretized projection interpolation models, by design, behave better in terms of interpolation errors.

The experiments we elaborate in this paper were made in a context of transmission X-ray tomography. But we think that our results could be useful in the broad landscape of tomography.

Acknowledgements. I thank my colleagues Antal Nagy and László G. Varga for their support and suggestions.

References

1. van Aarle, W., et al.: Fast and flexible X-ray tomography using the astra toolbox. Opt. Express **24**(22), 25129–25147 (2016)
2. Bleichrodt, F., van Leeuwen, T., Palenstijn, W.J., van Aarle, W., Sijbers, J., Batenburg, J.: Easy implementation of advanced tomography algorithms using the astra toolbox with spot operators. Numer. Algorithms **71**(3), 673–697 (2016)
3. Flores, L., Vidal, V., Verdu, G.: System matrix analysis for computed tomography imaging. PLoS ONE **10**, 1252–1255 (2015)
4. Foi, A., Trimeche, M., Katkovnik, V., Egiazarian, K.: Practical poissonian-gaussian noise modeling and fitting for single-image raw-data. IEEE Trans. Image Process. **17**(10), 1737–54 (2008)
5. Hahn, K., Schondube, H., Stierstorfer, K., Hornegger, J., Noo, F.: A comparison of linear interpolation models for iterative CT reconstruction. Med. Phys. **43**(12), 6455–6473 (2016)
6. Hanson, K.M., Wecksung, G.W.: Local basis-function approach to computed tomography. Appl. Opt. **24**, 4028 (1985)
7. Haralick, R., Shanmugam, K., Dinstein, I.: Textural features for image classification. IEEE Trans. Syst. Man Cybern. SMC **3**(6), 610–621 (1973)
8. Herman, G.: Fundamentals of Computerized Tomography: Image Reconstruction from Projections, 2nd edn. Springer, London (2009). https://doi.org/10.1007/978-1-84628-723-7

9. Imhof, A.L., Calvo, C.A., Santamarina, J.C.: Seismic data inversion by cross-hole tomography using geometrically uniform spatial coverage. Revista Brasileira de Geofísica **28**(1), 79-88 (2010). https://doi.org/10.1590/S0102-261X2010000100006
10. Joseph, P.: An improved algorithm for reprojecting rays through pixel images. IEEE Trans. Med. Imag. **1**(3), 192–6 (1982)
11. Kak, A., Slaney, M.: Principles of Computerized Tomographic Imaging. IEEE Press, New York (1999)
12. Shepp, L.A., Logan, B.: The fourier reconstruction of a head section. IEEE Trans. Nuclear Sci. **21**(3), 21–43 (1974)
13. Lewitt, R.: Alternatives to voxels for image representation in iterative reconstruction algorithms. Phys. Med. Biol. **37**(3), 705–16 (1992)
14. Lukić, T., Nagy, B.: Deterministic discrete tomography reconstruction by energy minimization method on the triangular grid. Pattern Recognit. Lett. **49**, 11–16 (2014). https://www.sciencedirect.com/science/article/pii/S016786551400169X
15. Man, B.D., Basu, S.: Distance-driven projection and backprojection in three dimensions. Phys. Med. Biol. **49**, 2463–2475 (2004)
16. Matej, S., Vardi, A., Herman, G.T., Vardi, E.: Binary Tomography Using Gibbs Priors, pp. 191–212. Birkhäuser Boston, Boston (1999). https://doi.org/10.1007/978-1-4612-1568-4_8
17. MATLAB: Matlab r2022a. The MathWorks, Inc., Natick, Massachusetts, United States
18. Olasz, C., Varga, L.G., Nagy, A.: Evaluation of the interpolation errors of tomographic projection models. In: Bebis, G., et al. (eds.) ISVC 2019. LNCS, vol. 11845, pp. 394–406. Springer, Cham (2019). https://doi.org/10.1007/978-3-030-33723-0_32
19. Sheriff, R., Geldart, L.: Exploration Seismology. Cambridge University Press, Cambridge (1995). https://books.google.hu/books?id=wRYgAwAAQBAJ
20. Siddon, R.: Fast calculation of the exact radiological path for a three-dimensional CT array. Med. Phys. **12**, 252–255 (1985)
21. van der Sluis, A., van der Vorst, H.: SIRT- and CG-type methods for the iterative solution of sparse linear least-squares problems. Linear Algebra Appl. **130**, 257–303 (1990)
22. Wang, Z., Bovik, A., Sheikh, H., Simoncelli, E.: Image quality assessment: from error visibility to structural similarity. IEEE Trans. Image Process. **13**(4), 600–612 (2004)
23. Xu, F., Mueller, K.: A comparative study of popular interpolation and integration methods for use in computed tomography. In: 3rd IEEE International Symposium on Biomedical Imaging: Nano to Macro 4, pp. 1252–1255 (2006)
24. Yu, Z., Noo, F., Dennerlein, F., A. Wunderlich, G.L., Hornegger, J.: Simulation tools for two-dimensional experiments in X-ray computed tomography using the FORBILD head phantom. Phys. Med. Biol. **57**(13), 237–252 (2012)
25. Zarour, L., Malykhina, G.F.: Comparison of analytical BP-FBP and algebraic SART-SIRT image reconstruction methods in computed tomography for the oil measurement system. In: Arseniev, D.G., Overmeyer, L., Kälviäinen, H., Katalinić, B. (eds.) CPS&C 2019. LNNS, vol. 95, pp. 335–343. Springer, Cham (2020). https://doi.org/10.1007/978-3-030-34983-7_32

Finding the Straight Skeleton for 3D Orthogonal Polyhedrons: A Combinatorial Approach

Anukul Maity[1], Mousumi Dutt[2(✉)], and Arindam Biswas[3]

[1] Narula Institute of Technology, Kolkata, India
[2] St. Thomas' College of Engineering and Technology, Kolkata, India
duttmousumi@gmail.com
[3] Indian Institute of Engineering Science and Technology, Shibpur, Howrah, India

Abstract. A combinatorial algorithm is presented here to determine the straight skeleton of a given 3D orthogonal polyhedron in $O(n \log n)$ time, where n is the number of vertices in the 3D orthogonal polyhedron. The 3D orthogonal polyhedron is traversed by considering a cuboid at each step. One of the combinatorial rule is applied and a part of the 3D orthogonal polyhedron is discarded as per the reduction rule. When the whole 3D polyhedron is traversed, the straight skeleton is obtained. The straight skeleton has many applications in shape analysis, shape matching, and shape retrieval.

Keywords: Straight skeleton · Orthogonal polyhedron · Combinatorial rule · Shape analysis · Shape retrieval · Shape matching

1 Introduction

The straight skeleton is introduced in 1995 by Aichholzer et al. [1,2]. It preserves the underlying geometric and topological structure of the shape in terms of connected line segments. This linear structure is the most powerful feature of the straight skeleton compared to the curve skeleton [19] which is applicable in solving many industrial and automated applications in computer science [20]. The straight skeleton of an orthogonal polygon is identical to medial axes [9] which represent the internal structure of shapes and have many applications including designing the roof of building [2,12,28], generate 3D building models [17,18,21,32], origami constructions [23], city Voronoi diagram [3,22], 3D navigation network [14], 3D indoor topology network [27], motorcycle graph [34], motion planning [24,30], computer-aided manufacturing [15,16,31], graph drawing [5], surface folding [4], polygon decomposition [33], and geographic information systems for shape reconstruction [1].

The straight skeleton for 3D objects are stated in [7,25,26]. Barequet et al. [6] propose the algorithms for constructing the 3D straight skeleton of voxel-based polyhedra whose running time is $O(min(n^2 \log n, k \log^{O(1)} n))$. Blanding et

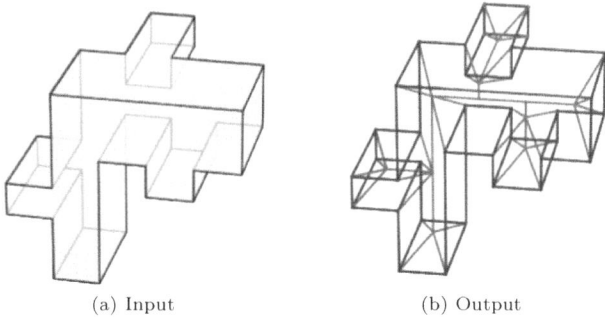

Fig. 1. (a) Orthogonal Polyhedron, P, (b) Straight Skeleton of P, where polyhedron edges are marked in blue color and skeleton edges are marked in red color. (Color figure online)

al. [8] state the skeleton based 3D geometric morphing in the field of engineering and industrial application for generating 3D solid models. Demaine et al. [11] describe the fundamental properties of 3D straight skeletons. The 3D Voronoi diagram is a well-known concept in computational geometry. Since extraction of an exact 3D skeleton of shapes is difficult in terms of robustness, accuracy and computational efficiency, some of approximate 3D skeleton or 3D medial axis representations can be the alternative to design an efficient algorithm [10,13,29]. Here, an efficient combinatorial algorithm is proposed which takes orthogonal polyhedron as input (Fig. 1(a)) and generates the straight skeleton as output in $O(n \log n)$ time, where n is the number of vertices of the polyhedron as shown in Fig. 1(b).

The paper is organised as follows. Section 2 presents definitions and preliminaries to understand the paper. A set of combinatorial rules is proposed in Sect. 3. The algorithm to determine the straight skeleton is stated in Sect. 4. The experimental results and its analysis are proposed in Sect. 5. The concluding remarks are stated in Sect. 6.

2 Definitions and Preliminaries

Definition 1 (cubic grid and voxel). *In 3D, the grid point set is $\mathbb{Z}^3$. A grid vertex is shifted by $(0.5, 0.5, 0.5)$ w.r.t. a grid point in 3D. A pair of adjacent grid vertices having Euclidean distance one from each other defines a grid edge. A grid square is defined by four grid edges that form a square, and a grid cube or voxel in 3D is defined by six grid squares that form a cube. The cubic grid is either a voxel or a set of voxels whose union is a cuboid in 3D Euclidean Space.*

Definition 2 (orthogonal polyhedron). *An orthogonal polyhedron is one all of whose faces meet at right angles, and all of whose edges are parallel to axes of a Cartesian coordinate system.*

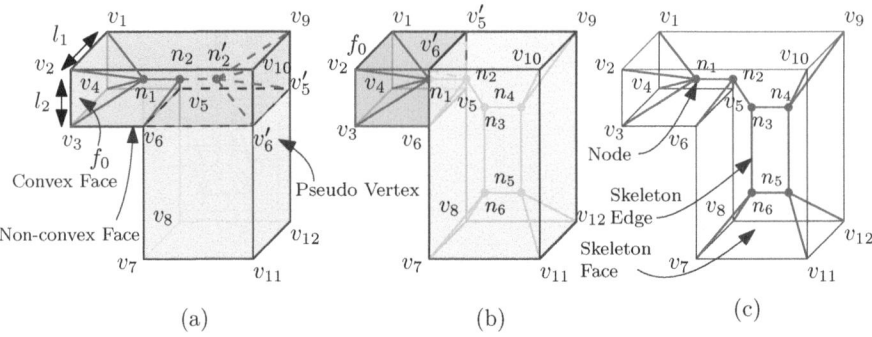

Fig. 2. (a) The top-left convex face of orthogonal polyhedron is f_0 is detected and corresponding cuboid is formed (green color). The straight skeleton is shown in red color. (b) The portion of the cuboid in gray color is deleted and next convex face is detected. The cuboid is detected (yellow color) and the resulting straight skeleton is formed. (c) The resulting straight skeleton is merged with the previously obtained result. The nodes, edges, and faces of the straight skeleton is marked. (Color figure online)

Definition 3 (simple orthogonal polyhedron). *A simple orthogonal polyhedron has axis-parallel faces that are simple polygons and three perpendicular edges at each vertex. It is not the polyhedron which has the topology of a sphere but its surface (its boundary).*

The orthogonal polyhedron has two types of vertices.

Definition 4 (convex vertex and reflex vertex). *Each polyhedron vertex is associated with three planes. If the dihedral angle of the planes w.r.t. a vertex is 90^0 (270^0), then the vertex is termed as convex vertex (reflex vertex).*

Definition 5 (convex face). *The face which is rectangle and four of its dihedral angles are $90°$, is called convex face.*

The faces which are not convex are termed as non-convex faces. In Fig. 2(a), the convex and non-convex faces are shown.

Definition 6 (straight skeleton for orthogonal polyhedron). *The straight skeleton of an orthogonal polyhedron P is defined by a continuous shrinking process such that all faces are parallelly translated at the same speed inward, with each vertex following the line which is equidistant from the three edge associated with the corresponding vertex. The reflex vertices also travel along the line which is equidistant from the three edge associated with the corresponding vertex, which implies that the incident edge grows in length at that endpoint. The shrinking continues until one of three events occurs.*

- A face shrinks to zero area.
- An edge shrinks to zero length.

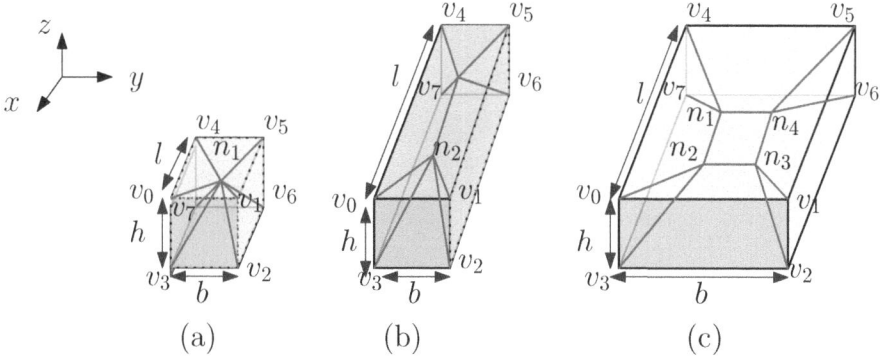

Fig. 3. The basic rules to determine straight skeleton of cuboids (a) $l = h = b$, (b) $l \neq h = b$, and (c) $l \neq h \neq b$.

– *A reflex vertex collides with an edge. At this point, the original polyhedron is "pinched off", creating two new polyhedrons. The shrinking process then continues on the two polyhedrons independently.*

It is to be noted here that the points in the straight skeleton which are not the polyhedron vertices are termed as nodes. The straight skeleton contains vertices, nodes, edges, and faces. The doubly connected edge list (DCEL) is used to store the straight skeleton of 3D polyhedron.

In this paper, a combinatorial technique is applied to obtain the straight skeleton of a given orthogonal polyhedron. At each step a cuboid is considered and its straight skeleton is determined (see Fig. 2(a)-(b)). While the cuboid is formed some vertices are temporarily considered which are termed as pseudo vertices (v'_5 and v'_6 in Fig. 2(a)). While determining the straight skeleton some nodes are created which do not belong to the final result of the straight skeleton (n'_2 in Fig. 2(a)). These nodes are created for the part of faces of the cuboid which does not belong to the orthogonal polyhedron. Some part is deleted as per the reduction rule and the next face of traversal is determined. Again, the cuboid is formed and by applying the basic rule straight skeleton for the cuboid is obtained which is merged with the previously obtained straight skeleton as shown in Fig. 2(b). In Fig. 2(c), the resulting straight skeleton is shown in red color.

3 Combinatorial Rules to Derive Straight Skeleton

Given an orthogonal polyhedron, P, its top-left-front convex face is identified and the convex face is swept until an obstacle is reached. The faces of the orthogonal polyhedron, P is sorted lexicographically w.r.t. each plane. Let the convex face, f, be in YZ-plane. f is swept along YZ-plane until it hits another face, say, f'. A cuboid, C, is considered from f to f'. The basic rules are applied on C and

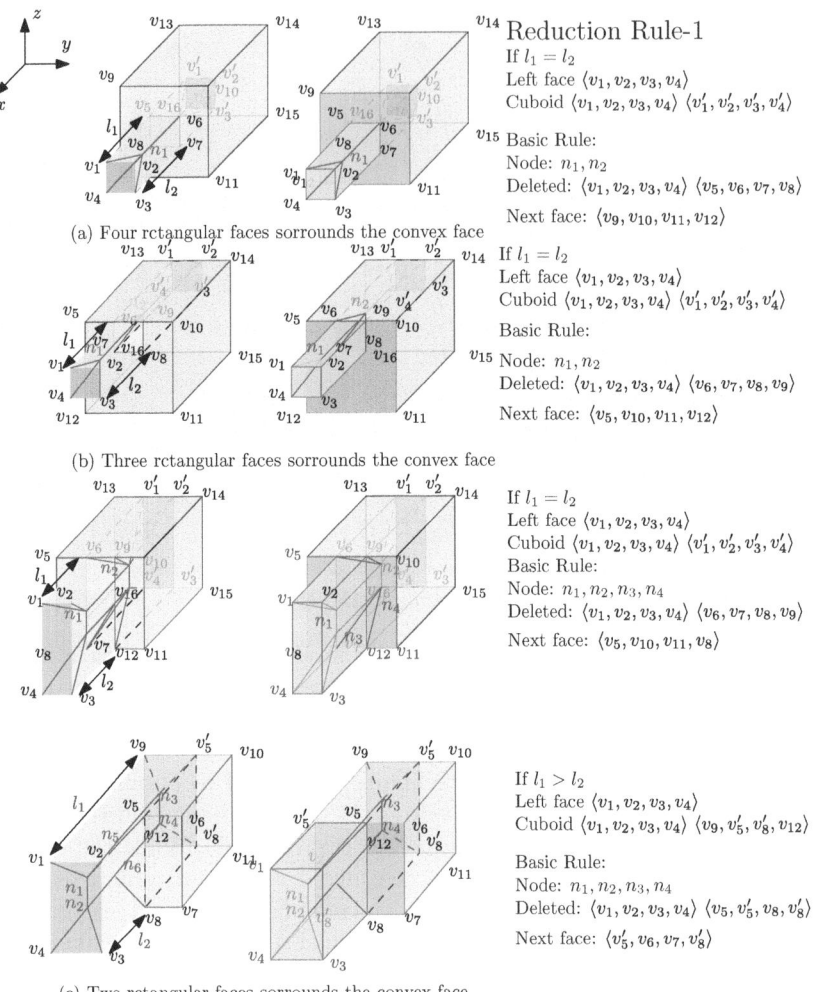

Fig. 4. Reduction Rule 1 for different cases.

the straight skeleton is obtained. The reduction rule is applied on it and some part of it is deleted. The next face of traversal is determined and the algorithm proceeds. The resulting straight skeleton is merged step by step.

The basic rule states the straight skeleton for a cuboid. The straight skeleton for the cuboid having same length, breadth, and height is shown in Fig. 3(a) where all the edges of the straight skeleton meet at a node. The straight skeleton for the cuboid for which any two of height, breadth, and length are equal is shown in Fig. 3(b) where all the edges of the straight skeleton meet at two nodes. The straight skeleton for the cuboid for which height, breadth, and length all are unequal is shown in Fig. 3(c) where all the edges of the straight skeleton meet at four nodes.

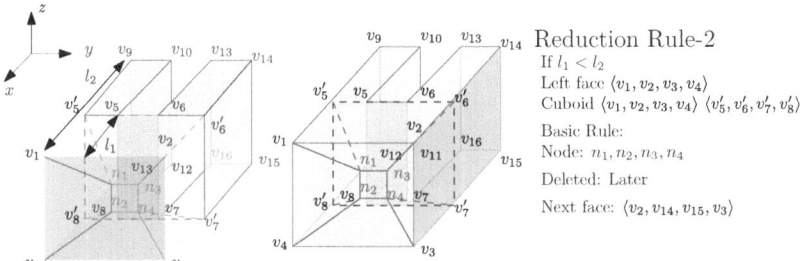

Fig. 5. Reduction Rule 2.

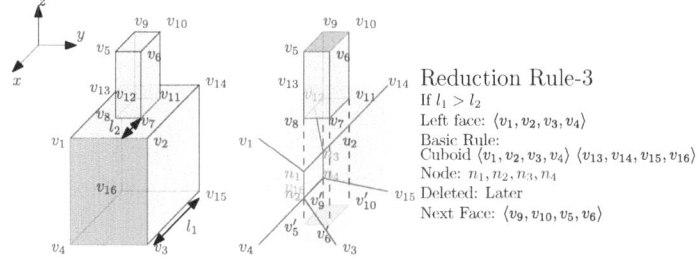

Fig. 6. Reduction Rule 3.

The reduction rules are depicted in Fig. 4, Fig. 5, and Fig. 6 which are discussed as follows.

Rule 1. This rule is applied when the area of f is less than that of f' and the deletion of the considered cuboid does not disconnect the polyhedron as shown in Fig. 4. The straight skeleton is determined for the cuboid by applying any of the suitable basic rules as shown in Fig. 3. The convex face, f, is colored orange and f' is colored green in the left column of Fig. 4. Some of the different orientation of the convex faces are shown. When the face, f is swept along the plane (say YZ-plane), a face, f'', may appear before f'. The cuboid from convex face f to f'' is deleted as shown in gray color in Fig. 4. The next convex face of traversal is f'' (orange color in the right column of Fig. 4).

Rule 2. When the area of f is greater than that of f' and the deletion of the considered cuboid disconnects the polyhedron, then Rule 2 is applied. The straight skeleton is determined for the cuboid by applying any of the suitable basic rules. No part is deleted as of now. The next face of traversal is the right side face (considering that f is in YZ-plane) as shown in orange color in the right side of Fig. 5.

Rule 3. When one of the face (except f and f') of the cuboid, C, has hole, then Rule 3 is applied (Fig. 6). The straight skeleton for C is obtained and no part is deleted. The part of the polyhedron through the hole is resolved first and the next face of traversal is the convex face through the hole as shown in orange color in the right side of Fig. 6.

Algorithm 1: STSKL-3DPOLY

   Input: P
   Output: S

1 $S \leftarrow \emptyset$
2 $F_{xy}, F_{yz}, F_{zx} \leftarrow$ SORTFACE(P)
3 $f \leftarrow$ FIND-TOPLEFTFRONTFACE(F_{xy}, F_{yz}, F_{zx})
4 $P' \leftarrow P$
5 **while** $P' \neq \emptyset$ **do**
6     $f' \leftarrow$ FIND-ENDFACE($f, F_{xy}, F_{yz}, F_{zx}$)
7     $C \leftarrow$ FIND-CUBOID(f, f', P')
8     $S \leftarrow S \cup$ APPLYBASICRULE(C, P')
9     $x \leftarrow$ CHECK-FACES(C, P')
10   **if** $x = 0$ **then**
11       $f'', P' \leftarrow$ APPLY-R3(C, P', S)
12   **else if** AREA(f)<AREA(f') **then**
13       $f'', P' \leftarrow$ APPLY-R1(C, P', S)
14   **else**
15       $f'', P' \leftarrow$ APPLY-R2(C, P', S)
16    $f \leftarrow f'$
17 **return** S

4 Algorithm for Determining Straight Skeleton

The Algorithm STSKL-3DPOLY generates the straight skeleton, S, on 3D orthogonal polyhedron, P. The doubly connected edge list (DCEL) of P is considered as input. S stores the straight skeleton as DCEL and initially it is set to $\emptyset$ (step 1). The faces are sorted w.r.t. each plane and the lists F_{xy}, F_{yz}, and F_{zx} are generated (step 2). The face f is the top-left-front face of P which is detected by applying the procedure FIND-TOPLEFTFACE (step 3). P is assigned to P' in step 4. The loop is executed until $P' = \emptyset$ (steps 5–16). In step 6, the procedure FIND-ENDFACE finds the opposite face of f and assigned to f'. The cuboid is determined as stated in Sect. 3 by using the procedure FIND-CUBOID (step 7). The basic rule is applied on C and the result is merged with S (step 8). In step 9, the procedure CHECK-FACES checks whether any of the faces except f or f' of C has hole. The procedure returns a value which is assigned in x. If $x = 0$, the Reduction Rule 3 is applied by using the procedure APPLY-R3 (steps 10–11). The straight skeleton for the cuboid is obtained by applying the basic rules and merged with the previous result (assigned to S). After deletion of some part as per the reduction rule, the updated polyhedron is assigned to P' and the next convex face of traversal is assigned to f'. If $x \neq 0$ and the area of f is less than that of f', then the Reduction Rule 1 is applied by using the procedure APPLY-R1 (steps 12–13). Otherwise, the Reduction Rule 2 is applied by using the procedure APPLY-R2 (steps 14–15). In step 16, f' is assigned to f. The resulting straight skeleton S is returned (step 17). It is to be noted here that

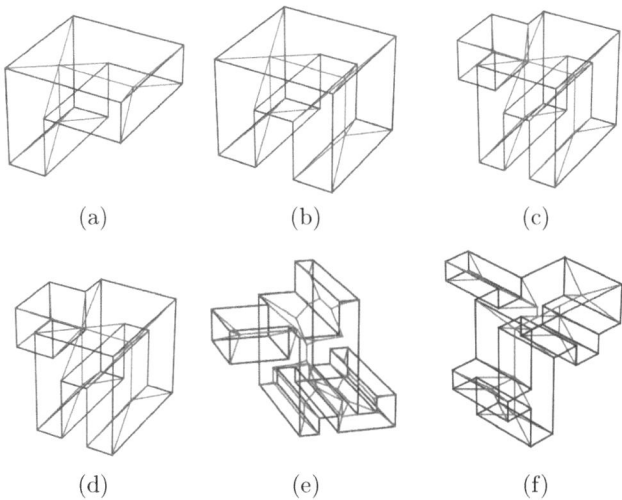

Fig. 7. Experimental results of 3D straight skeleton for a set of orthogonal polyhedrons.

the resulting straight skeleton is rotation invariant, where the rotation angles are $k \cdot \pi$ ($k = \pm 1, \pm 2, \ldots$). If the starting face is different one, then also the same straight skeleton is obtained.

Time Complexity. Let the polyhedron P contains n number of vertices. To sort the list of faces and to find top-left-front face $O(n \log n)$ time is needed. While the faces of cuboid is checked for application of rules, the DCEL is traversed. The basic rules can be applied in constant time and reduction rule can be applied in linear time. Thus the overall time complexity is $O(n \log n)$.

5 Experimental Results

The experimental results are generated on a computer system with Intel core i5 processor and OS Ubuntu 16.04. Some of the results of the straight skeleton on different 3D orthogonal polyhedrons are given in Fig. 7 and Fig. 8. The straight skeleton of 3D orthogonal polyhedron is useful for shape analysis. Several parameters of the straight skeleton can be used as shape descriptors for 3D orthogonal polyhedron. The number of vertices, nodes, edges, faces are useful parameters for shape descriptors. In Table 1, all the above mentioned parameters are shown for the straight skeleton shown in Fig. 7 and Fig. 8. The CPU time is also calculated. For the complex polyhedrons, the complexity in the shape of the polyhedron is proportional to number of edges and nodes in the straight skeleton. The length of edges and the distance among nodes are also important features. Given two straight skeletons of the two orthogonal polyhedrons, the features of the straight skeletons can be analysed. A skeleton matching algorithm can be developed in

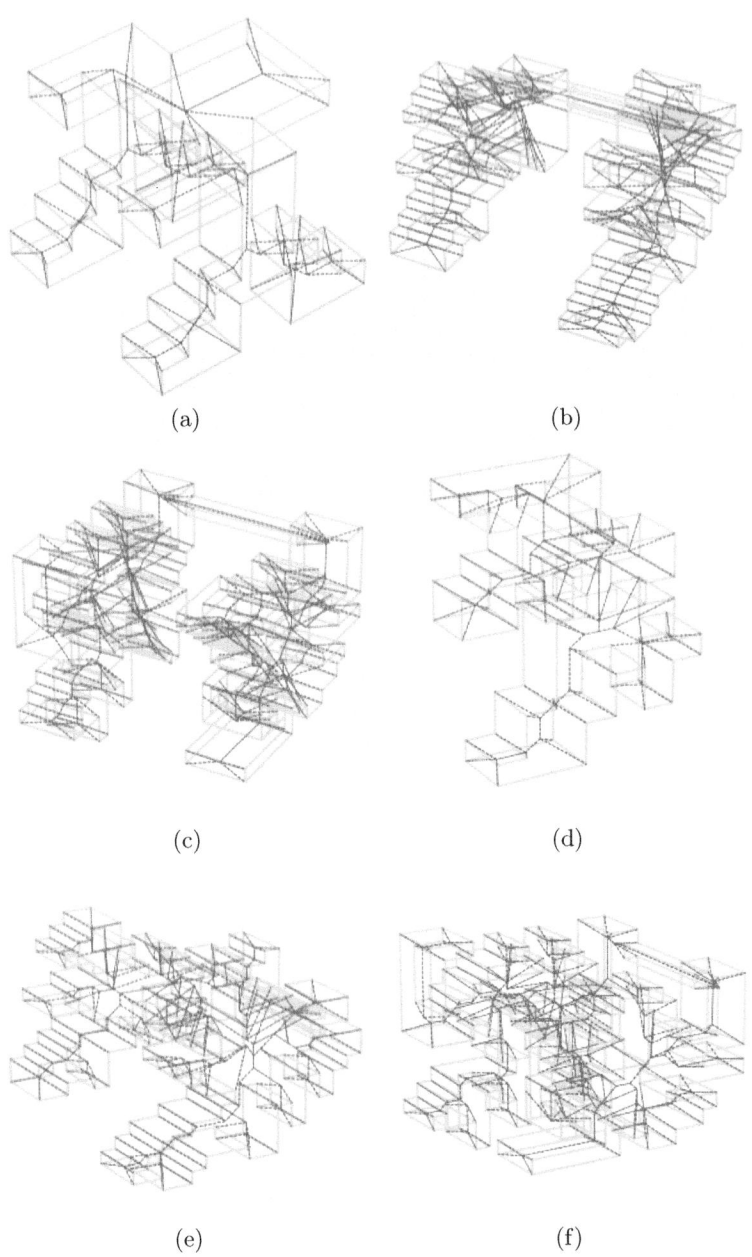

Fig. 8. Experimental results of 3D straight skeleton for another set of large complex orthogonal polyhedrons.

Table 1. The data of experimental results shown in Fig. 7 and Fig. 8.

Object	Polyhedron			Straight skeleton				CPU Time(ms)
	Vertices	Edges	Faces	Vertices	Edges	Nodes	Planes	
Figure 7(a)	12	18	7	12	14	3	16	0.949
Figure 7(b)	16	24	10	16	18	5	20	1.426
Figure 7(c)	16	24	10	16	19	4	26	1.624
Figure 7(d)	24	34	14	24	28	5	31	1.833
Figure 7(e)	36	50	18	36	38	12	40	2.209
Figure 7(f)	36	54	18	36	40	11	44	2.294
Figure 8(a)	94	128	54	94	127	36	139	3.892
Figure 8(b)	280	416	138	280	356	76	366	7.802
Figure 8(c)	288	432	142	288	384	96	392	8.987
Figure 8(d)	104	156	54	104	137	36	143	4.157
Figure 8(e)	276	412	134	276	342	66	352	7.468
Figure 8(f)	284	428	144	284	366	82	378	8.556

future to measure the similarity between two straight skeletons. Thus several features of the straight skeleton can be used in shape analysis, shape matching, etc.

6 Conclusion

This paper presents a combinatorial algorithm that uniquely determines the straight skeleton of 3D orthogonal polyhedron in $O(n \log n)$ time, where n is the number of vertices of the polyhedron. The resulting straight skeleton is starting face invariant and rotation invariant. At each step a cuboid is formed and part by part straight skeleton is obtained by application of combinatorial rules. At each step, the resulting straight skeleton is merged with the previous result. The reconstruction of orthogonal polyhedron is also possible from the straight skeleton. The straight skeleton can be used for shape descriptor of 3D orthogonal polyhedron. Several features of the straight skeleton can be used for shape descriptor. The straight skeleton of 3D orthogonal polyhedron can also be represented by a 3D graph which contains cycle since the straight skeleton contains faces. The straight skeleton can be used for shape matching in future.

References

1. Aichholzer, O., Aurenhammer, F.: Straight skeletons for general polygonal figures in the plane. In: Cai, J.-Y., Wong, C.K. (eds.) COCOON 1996. LNCS, vol. 1090, pp. 117–126. Springer, Heidelberg (1996). https://doi.org/10.1007/3-540-61332-3_144
2. Aichholzer, O., Aurenhammer, F., Alberts, D., Gärtner, B.: A novel type of skeleton for polygons. J. Univ. Comput. Sci. **1**(12), 752–761 (1995)

3. Aichholzer, O., Aurenhammer, F., Palop, B.: Quickest paths, straight skeletons, and the city voronoi diagram. In: Proceedings of the Eighteenth Annual Symposium on Computational Geometry, SCG 2002, pp. 151–159. Association for Computing Machinery, New York (2002)
4. Asao, Y., et al.: Folding and punching paper. J. Inf. Process. **25**, 590–600 (2017)
5. Bagheri, A., Razzazi, M.: Drawing free trees inside simple polygons using polygon skeleton. Comput. Inf. **23**, 239–254 (2004)
6. Barequet, G., Eppstein, D., Goodrich, M.T., Vaxman, A.: Straight skeletons of three-dimensional polyhedra. In: Halperin, D., Mehlhorn, K. (eds.) ESA 2008. LNCS, vol. 5193, pp. 148–160. Springer, Heidelberg (2008). https://doi.org/10.1007/978-3-540-87744-8_13
7. Berg, M.D., Cheong, O., Kreveld, M.V., Overmars, M.: Computational Geometry: Algorithms and Applications, 3rd edn. Springer-Verlag, Santa Clara (2008). https://doi.org/10.1007/978-3-540-77974-2
8. Blanding, R.L., Turkiyyah, G.M., Storti, D.W., Ganter, M.A.: Skeleton-based three-dimensional geometric morphing. Comput. Geom. **15**(1–3), 129–148 (2000)
9. Blum, H.: A transformation for extracting new descriptors of shape. In: Wathen-Dunn, W. (ed.) Models for the Perception of Speech and Visual Form, pp. 362–380. MIT Press, Cambridge (1967)
10. Culver, T., Keyser, J., Manocha, D.: Exact computation of the medial axis of a polyhedron. Comput. Aided Geom. Des. **21**(1), 65–98 (2004)
11. Demaine, E.D., Demaine, M.L., Lindy, J.F., Souvaine, D.L.: Hinged dissection of polypolyhedra. In: Dehne, F., López-Ortiz, A., Sack, J.-R. (eds.) WADS 2005. LNCS, vol. 3608, pp. 205–217. Springer, Heidelberg (2005). https://doi.org/10.1007/11534273_19
12. Eppstein, D., Erickson, J.: Raising roofs, crashing cycles, and playing pool: applications of a data structure for finding pairwise interactions. Discret. Comput. Geom. **22**, 569–592 (1999)
13. Foskey, M., Lin, M.C., Manocha, D.: Efficient computation of a simplified medial axis. J. Comput. Inf. Sci. Eng. **3**(4), 274–284 (2003)
14. Fu, M., Liu, R., Qi, B., Issa, R.R.: Generating straight skeleton-based navigation networks with industry foundation classes for indoor way-finding. Autom. Constr. **112**, 103057 (2020)
15. Held, M., Lukács, G., Andor, L.: Pocket machining based on contour-parallel tool paths generated by means of proximity maps. Comput. Aided Des. **26**(3), 189–203 (1994)
16. Held, M.: Voronoi diagrams and offset curves of curvilinear polygons. Comput. Aided Des. **30**(4), 287–300 (1998)
17. Held, M., Palfrader, P.: Straight skeletons with additive and multiplicative weights and their application to the algorithmic generation of roofs and terrains. Comput. Aided Des. **92**, 33–41 (2017)
18. Held, M., Palfrader, P.: Skeletal structures for modeling generalized chamfers and fillets in the presence of complex miters. Comput.-Aided Des. Appl. **16**(4), 620–627 (2019)
19. Karmakar, N., Mondal, S., Biswas, A.: Determination of 3d curve skeleton of a digital object. Inf. Sci. **499**, 84–101 (2019)
20. Kirkpatrick, D.: Efficient computation of continuous skeletons. In: Proceedings of the 20th Annual IEEE Symposium FOCS, pp. 18–27. IEEE, San Juan (1979)
21. Koźniewski, E., Banaszak, K.: Roof geometry in building design. Open Eng. **10**, 839–845 (2020)

22. Koźniewski, E., Orłowski, M.: Pre-determination of prediction of yield-line pattern of slabs using voronoi diagrams. Open Eng. **12**, 647–661 (2022)
23. Lang, R.J.: A computational algorithm for origami design. In: Proceedings of the Twelfth Annual Symposium on Computational Geometry, SCG 1996, pp. 98–105. Association for Computing Machinery, New York (1996)
24. Ó'Dúnlaing, C., Yap, C.: A retraction method for planning the motion of a disc. J. Algor. **6**(1), 104–111 (1985)
25. O'Rourke, J.: Computational Geometry in C. Cambridge University Press, Cambridge (1998)
26. Preparata, F., Shamos, M.: Computational Geometry. Springer, Berlin (1985). https://doi.org/10.1007/978-1-4612-1098-6
27. Rahman, S.A.F.S.A., Maulud, K.N.A., Pradhan, B., Mustorpha, S.N.A.S.: Manifestation of lattice topology data model for indoor navigation path based on the 3D building environment. J. Comput. Des. Eng. **8**(6), 1533–1547 (2021)
28. Recuero, A., Cutiérrez, J.P.: Sloped roofs for architectural cad systems. Comput.-Aided Civil Infrastruct. Eng. **8**(2), 147–159 (1993)
29. Sherbrooke, E.C., Patrikalakis, N.M., Brisson, E.: An algorithm for the medial axis transform of 3D polyhedral solids. IEEE Trans. Visual Comput. Graphics **2**(1), 44–61 (1996)
30. Storer, J.A., Reif, J.H.: Shortest paths in the plane with polygonal obstacles. J. ACM **41**(5), 982–1012 (1994)
31. Sudhalkar, A., Gürsöz, L., Prinz, F.: Box-skeletons of discrete solids. Comput. Aided Des. **28**(6–7), 507–517 (1996)
32. Sugihara, K.: Automatic generation of 3-D building models by straight skeleton. In: SIGGRAPH Asia 2011 Sketches, SA 2011. Association for Computing Machinery, New York (2011)
33. TĂnase, M., Veltkamp, R.C.: Polygon decomposition based on the straight line skeleton. In: Asano, T., Klette, R., Ronse, C. (eds.) Geometry, Morphology, and Computational Imaging. LNCS, vol. 2616, pp. 247–268. Springer, Heidelberg (2003). https://doi.org/10.1007/3-540-36586-9_16
34. Vigneron, A., Yan, L.: A faster algorithm for computing motorcycle graphs. In: Proceedings of the Twenty-Ninth Annual Symposium on Computational Geometry, SoCG 2013, pp. 17–26. Association for Computing Machinery, New York (2013)

Towards a Unifying View on Monotone Constructive Definitions

Linde Vanbesien[1]( ), Samuele Pollaci[2], Bart Bogaerts[2], and Marc Denecker[1]

[1] Department of Computer Science, KU Leuven, Leuven, Belgium
{linde.vanbesien,marc.denecker}@kuleuven.be
[2] Department of Computer Science, Vrije Universiteit Brussel, Ixelles, Brussel, Belgium
{samuele.pollaci,bart.bogaerts}@vub.be

Abstract. Constructive definitions, including inductive and recursive definitions, are ubiquitous in mathematical texts and occur in a wide variety of computer science fields and Knowledge Representation applications. While in different areas there is a high level of familiarity with certain types of constructive definitions, fairly little interaction between different areas seems to exist, resulting in a lack of deep understanding of principles and their applications. This paper aims to fill this void by laying the foundations for a single unifying framework, bringing together a wide variety of definitions. First, we recall the principle of (monotone) inductive definition and its formalization in fixpoint theory. We discuss the constructive and the non-constructive interpretation of inductive definitions and the induction process. We then analyze examples, including but not limited to (co)inductive and (co)recursive definitions, found in a wide range of areas through the lens of our proposed framework.

Keywords: Constructive definitions · Construction space · Induction process

1 Introduction

In mathematics, there are perhaps few concepts so enigmatic as that of inductive or recursive definitions.[1] Students become familiar with them through examples such as the transitive closure of a binary relation, the Fibonacci function or the satisfaction relation of propositional or predicate logic.

Common to such definitions is that they define a concept by describing how to construct it through iterated application of rules, starting from the empty set. This construction process is often called the *induction process*. For definitions of sets, the defined set is often explained non-constructively as the least set satisfying the rules; the constructive and non-constructive ways are known to be equivalent. While usually, it is not

---

This work was supported by Fonds Wetenschappelijk Onderzoek – Vlaanderen (project G0B2221N), by the Flanders AI Impulse programme (project DTAI-EWI-IMPULS-AI) and the KU Leuven C1 grant C14/19/082.

[1] Is there a difference between *inductive* and *recursive* definition? According to some there is, according to others not. In this paper, we propose a way to distinguish *inductive* and *recursive* definitions that is sensible and seems to match with intuitions of some.

formally explained what inductive definitions mean, students apparently learn to understand them and reason with them. Brouwer [6], the famous constructionist, observed that many fundamental objects in mathematics were defined by describing how to *construct* them and that in understanding these constructions, we rely on our basic cognitive skills for *temporal reasoning*.[2] It was later argued that also our skills for *causal reasoning* play a role here [10, 13, 15]: the hypothesis is that our understanding and reasoning capabilities for inductive definitions stem from our understanding of the induction process as a causal process, idealized and generalized to an (often infinite) universe of mathematical objects. This suggests a strong, but not well-known link between mathematics and common sense knowledge. While recently a lot of effort in the domain of large language models has resulted in surprisingly good commonsense reasoners, it is well-known that these are not reliable enough for sensitive applications where exactness and correctness are crucial. For these applications a logic-based approach that includes different constructive definitions is desired. Therefore the study of the principle of inductive definition is a worthy topic in Knowledge Representation (KR).

It is clear that the concept of inductive definition plays an important role in mathematics and foundations of computer science. We claim it also plays an important role in KR, at the meta-level (e.g., in the many inductive definitions used to define syntax and semantics for logics in KR), and at the object-level, since definitions constitute an important, common and precise form of human knowledge. In an important class of applications, the definition is inductive, in which case it is often not expressible in first-order logic (FO), yielding a second reason for studying inductive definitions [10]. A third reason is the intuition of some researchers that inductive and recursive definitions form the declarative understanding of two well-known declarative programming paradigms, logic and functional programming [11, 19]. Finally, due to the close connection between inductive definitions and causal information, studying inductive definitions is useful for expressing common sense causal knowledge [10, 13].

There exists extensive research on inductive definitions [2, 18, 23, 24, 33]. Also (co-)recursive definitions have received a lot of attention in relation to functional programming languages [26, 27, 32], as well as in *domain theory* where the functions, and the domain they are defined on are defined simultaneously [1, 29]. While there is a high level of familiarity with certain types of constructive definitions, in the current state of the art, fairly little interaction between research on different types of definitions seems to exist, resulting in a lack of deep understanding of common principles and applications. Many researchers seem aware that their theories only cover part of the topic. Already a long time ago, Moschovakis [24] explained how Kleene [21] in early papers had consciously studied constructive definitions[3] but *explicitly had drawn back from studying all of them*. Another complicating factor has been that inductive/recursive definitions have often been studied from a recursion-theoretic point of view, as programs to compute truth or function values, rather than as plain definitions of a concept.

---

[2] While we follow Brouwer in his views on the nature and importance of constructive definitions, we use standard mathematics and set theory (also in this paper) whenever suitable.
[3] In his work, Kleene used the term *inductive definitions* to denote the overarching class which we call constructive definitions.

This paper contributes to the study of monotone constructive definitions by introducing some key concepts as the foundations for a unifying set-theoretical framework. This offers the key insight that all these different types are instances of the same basic constructive principles. This has important practical implications. Firstly, it entails that research on a particular type of definition might transcend its class and actually be applicable to all constructive definitions. E.g., *non-monotone* inductive definitions have been researched algebraically [14], but non-monotone recursive definitions remain uncharted territory. Our correspondence suggests a way to generalise the study of non-monotonic definitions to other types of constructive definitions. Secondly, we believe this framework will be instrumental to integrate different types of definitions in a single knowledge representation language. The main contribution of this paper is to show how a whole range of examples from different areas can be reduced to instantiations of the same fundamental principles, using standard set-theoretic constructions. First, we recall the principle of (monotone) inductive definition and its formalization in fixpoint theory, which will involve a *semantic operator* on a so-called *construction space*, which is often richer than the *exact space*, in which the *defined object* naturally lives. We then analyze examples found in a wide range of areas. In each example, we describe the exact space, the construction space, the monotone semantic operator and the defined entity. We will see how the construction space can be used as the key factor to distinguish between classes of definitions from different research areas. We focus mostly on (co)inductive definitions of sets and (co)recursive definitions of functions, but also briefly discuss some more complex types of constructive definitions.

2 Algebraic Formalisation

We now introduce the algebraic formalism needed for an in-depth presentation of various types of constructive definitions and illustrate them with a first detailed example.

A *partially ordered set (poset)* $\langle C, \leq \rangle$ is a set C equipped with a partial order $\leq$, i.e., a reflexive, antisymmetric, transitive relation. When $\leq$ is clear from the context, we sometimes just write C to refer to $\langle C, \leq \rangle$. As usual, we write $x < y$ for $x \leq y \wedge x \neq y$. If S is a subset of C, then x is an *upper bound* of S if $s \leq x$ for each $s \in S$; it is a *least upper bound* ($\text{lub}(S)$) of S if moreover it is smaller than every other upper bound. We call a poset $\langle C, \leq \rangle$ a *chain-complete partial order (cpo)* if every chain of C (i.e., every subset of C for which $\leq$ is total) has a least upper bound. Each cpo has a least element $\bot$, which is the least upper bound of $\emptyset$.

A function $f \colon C_1 \to C_2$ between cpo's is *monotone* if for all $x, y \in C_1$ such that $x \leq_1 y$, it holds that $f(x) \leq_2 f(y)$. We refer to functions $O \colon C \to C$ with domain equal to the codomain as *operators*. An element $x \in C$ is a *prefixpoint*, resp. a *fixpoint* of O if $O(x) \leq x$, resp. $O(x) = x$ [31]. By Tarski's least fixpoint theorem [34], every monotone operator O on a cpo has a least fixpoint, that we denote $\text{lfp}(O)$. It is also the least prefixpoint of O and it can be *constructed* as the limit of the possibly transfinite sequence $(O^i)_{i \geq 0}$, where $O^{i+1} = O(O^i)$ and $O^\lambda = \text{lub}(\{O^j \mid j < \lambda\})$ for limit ordinals λ (in particular, this means $O^0 = \bot$). This allows for a first, algebraic formalization of constructive definitions [2]. A constructive definition for a concept $\mathcal{D}$ is (formalized as) an operator $O : \mathcal{C} \to \mathcal{C}$ on a cpo $\mathcal{C}$. It defines the object D representing $\mathcal{D}$ by describing how to construct it. The construction, normally called the *induction process*, is the sequence $(O^i)_{i\geq 0}$. The defined object D is the limit of this sequence. This limit can be obtained by construction but it can also be characterized non-constructively, as the least (pre)fixpoint of O, yielding the duality between the constructive and non-constructive view on inductive definitions.

Let us illustrate this abstract formalization of constructive definitions on a prototypical example. To streamline the presentation of various examples, we initially present a constructive definition as a set $\mathcal{R}$ of rules[4] which resemble the style used in logic programming, as well as in functional programming. We believe this will lead to an improved understanding of our examples. Moreover, it gives an idea of how constructive definitions in natural language can be formalised, which is essential when developing knowledge representation languages that include them.

Example 1 (Transitive closure). Let $\mathcal{G} = (V, E)$ be a directed graph. The set F of edges of the *transitive closure* $\mathcal{T} = (V, F)$ of $\mathcal{G}$ is defined inductively:

- $(x, y) \in F$ if $(x, y) \in E$;
- $(x, y) \in F$ if there exists a vertex z such that $(x, z) \in F$ and $(z, y) \in F$.

The set of rules $\mathcal{R}_F$ defining $\mathcal{T} = (V, F)$ is as follows.

$$\left\lfloor \begin{array}{l} \forall x \forall y : F(x, y) \leftarrow E(x, y). \\ \forall x \forall y : F(x, y) \leftarrow \exists z : F(x, z) \wedge F(z, y). \end{array} \right\rfloor$$

We expect this definition to construct a set $F \subseteq V^2$ of edges. Hence, we consider the cpo $\mathcal{C}_F = \langle 2^{V^2}, \subseteq \rangle$, with the power set of V^2 as underlying set with subsetorder. Moreover, the rules in $\mathcal{R}_F$ suggest an operator $O_F : \mathcal{C}_F \to \mathcal{C}_F$ showcasing rule application, by mapping a set $S \in \mathcal{C}_F$ to

$$O_F(S) = E \cup \{(x, y) \mid (x, z), (z, y) \in S \text{ for some } z \in V\}.$$

It is not hard to prove that O_F is monotone and that its least fixpoint is the set F of edges of the transitive closure of $\mathcal{G}$.

Proposition 1. *O_F is a monotone operator.*

Proof. Let $S_1 \subseteq S_2$ be two subsets of V^2. We have to show that $O_F(S_1) \subseteq O_F(S_2)$. By definition of O_F, we have that $O_F(S_1) = E \cup \{(x, y) \mid \exists z : (x, z) \in S_1 \wedge (z, y) \in S_1\}$. Let $(x, y) \in O_F(S_1)$. If $(x, y) \in E$, then $(x, y) \in O_F(S_2)$. If $(x, y) \notin E$, then there exists $z \in V$ such that $(x, z), (z, y) \in S_1$. Since $S_1 \subseteq S_2$, $(x, y) \in O_F(S_2)$, as desired.

In other words, O_F formalizes the constructive definition of F. The defined set F can be characterised non-constructively as the least fixpoint of O_F, or constructively as the limit of the induction process, i.e., the sequence built by iterative application of O_F starting from the empty set.

Denecker, et al. [15] remarked that given the informal rules of Example 1, we most likely picture the induction process as a sequence of applications of rule instances,

---

[4] We use different brackets to indicate the kind of definition: inductive and recursive definitions will be enclosed in floor-brackets $\lfloor \mathcal{R} \rfloor$, coinductive and corecursive definitions in ceil-brackets $\lceil \mathcal{R} \rceil$, and any other kind of constructive definition in curly brackets $\{\mathcal{R}\}$.

rather than iterations of O_F. In this view of the induction process, the elementary step is the application of a rule instance (or perhaps more generally, the application of a set of rule instances). This natural view of the induction process raises two issues. First, it identifies the *rule* as the modular unit of the definition and its induction process. This modularity is abstracted away when formalizing the definition as an operator O. Second, it leads to a highly non-deterministic notion of induction process, since rules can be applied in different orders. This non-determinism is of great pragmatical use when reasoning on the definition, since it allows us to *steer* the induction process towards a particular goal, e.g., towards computing whether a specific pair (a, b) belongs to F. On the other hand, the non-determinism raises the question whether all these different induction processes are *confluent* (i.e., have the same limit). This should be the case, otherwise the definition would be ambiguous!

In Fig. 1, the start of two such induction processes for Example 1 are visualized. In the top sequence $(F_0, \ldots, F_3)$, all applicable rules are applied at every step of the construction, making it the fastest process. This corresponds to $(O_F^0, \ldots, O_F^3)$, the first four iterations of the operator O_F. In the bottom sequence, a slower induction process $(F'_0, \ldots, F'_3)$ is shown, one that first applies all instances of the base rule, then a single instance of the transitivity rule per iteration.

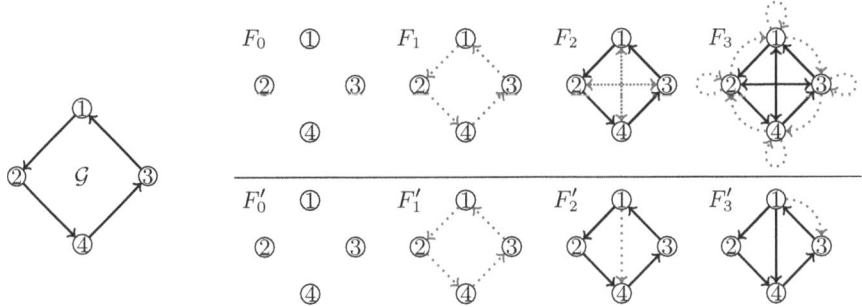

Fig. 1. A graph $\mathcal{G}$ (left) and the start of two monotone inductions of the definition of its transitive closure (right). Dotted red arrows indicate newly derived edges at each state. (Color figure online)

This more natural approach is formalized as a *monotone induction* of O: an increasing sequence $(x_i)_{i \leq \beta}$ satisfying

- $x_i \leq x_{i+1} \leq O(x_i)$, for successor ordinals $i + 1 \leq \beta$,
- $x_\lambda = \text{lub}(\{x_i \mid i < \lambda\})$, for limit ordinals $\lambda \leq \beta$ (in particular, $O^0 = \bot$).

Here, $x_i \leq x_{i+1} \leq O(x_i)$ formalizes the idea of applying *some* rule instances in x_i, but not necessarily all. We say a monotone induction of O is *terminal* if there does not exist a strictly greater refinement of its limit, i.e., if $x_\beta \not< O(x_\beta)$. It is straightforward to prove that all terminal monotone inductions are confluent (see, e.g., [5, Corollary 3.7]).

In the next section, we present several examples of constructive definitions. While they originate from very different fields, they can be presented in a uniform way using the following mathematical objects:

- A mathematical object D corresponding to the concept defined by the constructive definition. We call D the *defined object*.
- A set $\mathcal{E}$ where D lives. This should be naturally identified by the specifications in the constructive definition. We call $\mathcal{E}$ the *exact space*.
- A cpo $\mathcal{C} = \langle C, \leq \rangle$ with an injection $\theta : \mathcal{E} \hookrightarrow 2^C \setminus \{\emptyset\}$ such that for all $e_1, e_2 \in \mathcal{E}$ with $e_1 \neq e_2$, $\theta(e_1) \cap \theta(e_2) = \emptyset$, i.e., different elements of the exact space are mapped to disjoint subsets of C. The elements of $\theta(e)$ are (potentially different) representations of $e \in \mathcal{E}$ in $\mathcal{C}$. We call $\mathcal{C}$ the *construction space*.
- An operator $O \colon \mathcal{C} \to \mathcal{C}$ on the construction space, of which the least fixpoint coincides with the defined object D: $\mathrm{lfp}(O) \in \theta(D)$. We call O the *semantic operator*.

The elements of the construction space are meant to approximate the elements of the exact space. Some, or all, elements $c \in \mathcal{C}$ are representations of exact elements $e \in \mathcal{E}$, namely those for which $c \in \theta(e)$. Inversely, θ determines a surjective partial function $\pi \colon C \to \mathcal{E}$ such that for $c \in \mathcal{C}$, $\pi(c)$ exists and is equal to e iff $c \in \theta(e)$. In practice, we will use π to project away any additional information that was needed for construction and to derive the associated value in the exact space from the least fixpoint of the operator, i.e., $D = \pi(\mathrm{lfp}(O))$. Often but not always, the exact and the construction space are the same and π is the identity function. E.g., in Example 1, the defined object is the set F of edges of the transitive closure, the exact space $\mathcal{E}_F$ is the set 2^{V^2} of all sets of possible edges; the construction space is $\mathcal{C}_F = \langle 2^{V^2}, \subseteq \rangle$; the semantic operator is O_F.

3 Different Flavours of Constructive Definitions

In this section, we instantiate the earlier introduced framework for a range of constructive definitions coming from different areas. We bring them together to show that indeed, in different fields, the design of the exact and construction space is the key point. Once this choice is made explicit, typically the definition of the operator follows straightforwardly, and the defined object is constructed by the fixpoint theory. The final step may be to project the fixpoint from the construction space to an element of the exact space, i.e., the defined object, using π. In the majority of the proposed examples, this is not needed, since the injection θ just sends an exact element $e \in \mathcal{E}$ to the singleton $\{e\} \in 2^C$. However, Example 7 illustrates where the projection π plays a role.

3.1 (Co)inductive Definitions of Sets

Inductive definitions are ubiquitous in mathematical texts. Concepts such as the transitive closure, the natural numbers, ordinals, and formulas in logic, are usually defined inductively [2,3]. On the other hand, many common infinite datatypes such as infinite streams, infinite trees and coterms, are typically defined coinductively [22]. In general, these definitions define sets of elements of a certain type $\mathcal{T}$, given by the context. Naturally, the exact space then consists of all sets of elements of $\mathcal{T}$, i.e., it is $2^{\mathcal{T}}$. Intuitively, the construction process associated with inductive definitions gradually grows the defined set, starting from the empty set. In contrast, the construction process for coinductive definitions puts stronger restrictions on the defined set in every step, resulting in a gradually shrinking set. In both cases the power set contains all elements necessary for the construction, since we are only adding or removing elements from a subset of $\mathcal{T}$. By endowing the exact space with the subset order $\subseteq$ and the superset order $\supseteq$, we capture the respective behaviours of growing and shrinking associated with induction and coinduction. For inductive definitions we obtain the power set lattice $\langle 2^{\mathcal{T}}, \subseteq \rangle$ as a construction space. This is a complete lattice and thus a cpo. The same holds for the construction space $\langle 2^{\mathcal{T}}, \supseteq \rangle$.

Let us consider the domain of (finite or infinite) lists of natural numbers. The set of all such lists is denoted by $<$. We use a well-known notation for lists where Nil represents the empty list and $[x \mid y]$ represents the list starting with $x \in \mathbb{N}$ (often referred to as the *head*) followed by the list y (often referred to as the *tail* of the list).

Example 2 (Prime array). The set PA of all *prime arrays* is defined inductively:

- $Nil \in PA$.
- If x is a prime number and $y \in PA$, then $[x \mid y] \in PA$.

This is a monotone inductive definition, formally represented by the set of rules

$$\left[\begin{array}{l} \forall y \in List : PA(y) \leftarrow y = Nil. \\ \forall x \in \mathbb{N}, \forall y \in List : PA([x \mid y]) \leftarrow P(x) \wedge PA(y). \end{array} \right]$$

with P the set of prime numbers. The exact space is the power set 2^{List}. As construction space we then have $\mathcal{C}_{PA} = \langle 2^{List}, \subseteq \rangle$. The semantic operator for this example is $O_{PA} : \mathcal{C}_{PA} \rightarrow \mathcal{C}_{PA}$, defined by mapping a set of lists $S \subseteq \mathcal{C}_{PA}$ to

$$O_{PA}(S) = \{l \mid l = Nil \text{ or } l = [x \mid y] \text{ for some } x \in P, y \in S\}$$

Proposition 2. *The operator O_{PA} is monotone.*

Proof. Let $S_1 \subseteq S_2$ be two subsets of $List$, and let $l \in O_{PA}(S_1)$. Either $l = Nil$, in which case $x \in O_{PA}(S_2)$, or there exist $x \in P$ and $y \in S_1$ such that $l = [x \mid y]$. Since $S_1 \subseteq S_2$, we have $y \in S_2$, which implies that $x \in O_{PA}(S_2)$ also for the latter case.

The fastest induction process corresponds to the sequence $\emptyset = PA_0 \subseteq PA_1 \subseteq \ldots \subseteq PA$, with $PA_i = \bigcup_{m<i}\{[n_0, \ldots, n_m] \mid n_0, \ldots, n_m \in P\}$, the set of lists of primes with length at most i. Thus, the defined set of prime arrays consists of all finite lists containing only prime numbers. Interestingly, the same set of rules gives rise to a sensible *coinductive* definition.

Example 3 (Prime lists). The set PL of all *prime lists* is defined coinductively:

- $Nil \in PL$.
- $[x \mid y] \in PL$, if x is a prime number and $y \in PL$.

Towards a Unifying View on Monotone Constructive Definitions   225

As suggested before, this definition corresponds to exactly the same formal set of rules as the previous example after replacing PA by PL

$$\left[\begin{array}{l} \forall y \in List : PL(y) \leftarrow y = Nil. \\ \forall x \in \mathbb{N}, \forall y \in List : PL([x \mid y]) \leftarrow PL(y) \wedge P(x). \end{array} \right]$$

Unsurprisingly, we consider the same exact space 2^{List} as in Example 2, and the construction space with inverted order, namely $\langle 2^{List}, \supseteq \rangle$. Except for its signature, the inverted order does not influence the semantic operator O_{PL}, which equals O_{PA}.

Proposition 3. *The operator O_{PL} is monotone.*

Proof. Clear by the definition of O_{PL} and Proposition 2.

Here, the fastest induction process results in the sequence $List = PL_0 \supseteq PL_1 \supseteq \ldots \supseteq PL$, with $PL_i = \bigcup_{m<i}\{[n_0,\ldots,n_m] \mid n_0,\ldots,n_m \in P\} \cup \{[n_0,\ldots,n_i,\ldots] \mid n_0,\ldots,n_i \in P\}$ where $[n_0,\ldots,n_i,\ldots]$ denotes a list with length greater than $i-1$. Intuitively, this set corresponds to all lists l of natural numbers such that no non-primes occur within the first i elements of the list. Clearly, this sequence converges to the set of all finite and infinite lists of prime numbers. A final adaptation of the list-example restricts the defined object to only the infinite lists of prime numbers.

Example 4 (Prime streams). The set PS of all *prime streams* is defined coinductively:

– $[x \mid y] \in PS$ if x is a prime number and $y \in PS$.

By excluding the case for the empty list Nil, we obtain only the infinite lists, i.e., the streams. The definition is formalised by the following coinductive rule:

$$\left[\forall x \in \mathbb{N}, \forall y \in List : PS([x \mid y]) \leftarrow PS(y) \wedge P(x). \right]$$

We keep the same exact space and construction space as in Example 3. Here, the difference lies with the semantic operator O_{PS} which maps a set of lists S to

$$O_{PS}(S) = \{[y \mid z] \mid z \in S, y \in P\}$$

Proposition 4. *The operator O_{PS} is monotone.*

Proof. Let $S_1 \subseteq S_2$ be two subsets of $List$, and let $l \in O_{PS}(S_1)$, i.e. $l = [y, z]$ for some $y \in P$ and $z \in S_1$. Since $S_1 \subseteq S_2$, we have $y \in S_2$, which implies that $x \in O_{PS}(S_2)$, as desired.

The fastest induction process for this definition starts from $List$, since by default everything belongs to the set. During the first step it will delete the empty list and all lists with a head a such that $a \notin P$. At each subsequent step i it will remove all lists for which the ith element either does not exist, or it is not a prime number, giving us the sequence The fastest induction process then gives us the sequence $List = PS_0 \supseteq PS_1 \supseteq \ldots \supseteq PS$, with $PS_i = \{[n_0,\ldots,n_i,\ldots] \mid \forall j < i, n_j \in P\}$, i.e., the set of all lists of length at least i of which the first i elements are primes. Note that interpreting this set of rules inductively rather than coinductively will not be able to derive the inclusion of a single element, i.e., the defined object would be the empty set.

Now, let us turn our attention to a different type of examples that uses an aggregate expression, known as the "company controls" problem [20].

Example 5 (Company control-relation). Given a set C of companies, each of which owns a percentage of the shares of the other companies, the control-relation is defined inductively as follows: a company x controls another company y, if the sum of the shares of y owned by x or by companies controlled by x, is strictly more than half.

In formal rule notation:

$$\left| \forall x, y \in C : Cont(x,y) \leftarrow \left(\sum_{z \in Cont^x} Sh(z,y) \right) > 0.5. \right|$$

where $Sh: C^2 \to [0,1]$ is a function that maps a pair of companies (x,y) to the fraction of shares of y owned by x and $Cont^x = \{x\} \cup \{u \mid Cont(x,u)\}$. Under the (natural) assumption that $Sh(x,y) \geq 0$, this definition is monotone. The more companies that are determined to be under control of a company x, the higher the fraction of shares controlled by x in any (other) company y. The exact space is now given by the set of binary relations over C, i.e., 2^{C^2}, as the construction space we choose $\mathcal{C}_{Cont} = \langle 2^{C^2}, \subseteq \rangle$. Once again, the semantic operator $O_{Cont}: \mathcal{C}_{Cont} \to \mathcal{C}_{Cont}$ results from rule application, i.e., it maps a binary relation R to

$$O_{Cont}(R) = \left\{ (x,y) \;\middle|\; \sum_{z \in \{x\} \cup \{u \mid (x,u) \in R\}} Sh(z,y) > 0.5 \right\}.$$

Proposition 5. *The operator O_{Cont} is monotone.*

Proof. Let $S_1 \subseteq S_2$ be two subsets of C^2, and let $(x,y) \in O_{Cont}(S_1)$, i.e.

$$\sum_{z \in \{x\} \cup \{u \mid (x,u) \in S_1\}} Sh(z,y) > 0.5.$$

Since $S_1 \subseteq S_2$, we also have the inclusion $\{x\} \cup \{u \mid (x,u) \in S_1\} \subseteq \{x\} \cup \{u \mid (x,u) \in S_2\}$. Since $Sh(z,w) \geq 0$ for all $(z,w) \in C^2$, we have

$$\sum_{z \in \{x\} \cup \{u \mid (x,u) \in S_2\}} Sh(z,y) > 0.5,$$

i.e. $(x,y) \in O_{Cont}(S_2)$.

Figure 2 visualizes the induction process for an example share-function Sh represented by a labeled directed graph. Coincidentally, the depicted induction is the only possible induction with strict increments since at every step exactly one rule is applicable.

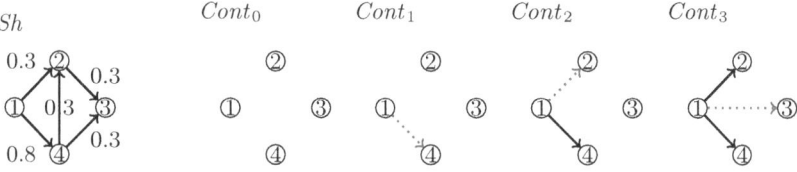

Fig. 2. An edge (a, b) in the leftmost graph indicates that $Sh(a, b) > 0$ and its label shows the exact value of $Sh(a, b)$. The other graphs show an induction process. Newly derived company pairs (represented by edges) are indicated with dotted red lines. At first only the base case is added. Later, combined ownership is derived. (Color figure online)

3.2 (Co)recursive Definitions of Functions

We now present another set of examples, this time regarding the definition of functions. Recursion and its dual corecursion are extensively used as methods to define functions in a wide variety of mathematical and computer scientific fields [26,27,32]. Some well-known mathematical functions, like the factorial or the greatest common divisor, can be defined recursively, and (co)recursive definitions of functions are supported in most functional programming languages [16,17,28].

For our formalization, the exact space of (co)recursive definitions of functions is obtained in a natural way: if we want to define a function $f: X \to Y$, then the exact space is the set of functions from X to Y, denoted by Y^X. Contrary to (co)inductive definitions, here we cannot just choose the construction space to equal the exact space. The main reason for this is that in intermediate steps of the construction process only partial functions have been constructed.

Example 6 (Fibonacci sequence). The Fibonacci sequence is viewed here as the function $Fib: \mathbb{N} \to \mathbb{N}$, where $Fib(n)$ is the nth Fibonacci number. Its recursive definition, in the formal notation, is the following:

$$\left\lfloor \begin{array}{ll} Fib(0) := 0. & Fib(1) := 1. \\ \forall n \in \mathbb{N}: Fib(n+2) := Fib(n) + Fib(n+1). \end{array} \right\rfloor$$

Clearly, the exact space is $\mathbb{N}^\mathbb{N}$. Moreover, we can get some insight into the construction process from the rules above. Note that the image of $n + 2$ under Fib depends on the image of n and $n + 1$. As long as the latter are not derived, it is impossible to determine $Fib(n+2)$. Hence, it is natural to think of Fib at an intermediate step of the construction as a partial function, defined on a subset S of $\mathbb{N}$. Equivalently, we can view such a partial function as a function from $\mathbb{N}$ to $\mathbb{N}_\bot := \mathbb{N} \cup \{\bot\}$, where $\bot$ denotes "undefined". The set $\mathbb{N}_\bot$ is naturally equipped with the definedness order $\leq_d$, given for all $n, m \in \mathbb{N}$ by $n \leq_d m$ iff $n = m$ or $n = \bot$. We take the construction space $\mathcal{C}_{Fib} = \langle (\mathbb{N}_\bot)^\mathbb{N}, \leq_d \rangle$ to be the set of functions $(\mathbb{N}_\bot)^\mathbb{N}$, equipped with the pointwise extension of $\leq_d$. This indeed forms a cpo.

Lemma 1. $\langle \mathbb{N}_\bot, \leq \rangle$ *is a cpo.*

Proof. Let $S \subseteq \mathbb{N}_\bot$ be a chain. By the definition of the order $\leq$, S has at most two elements. Hence, the $lub(S) = n$ where $\{n\} = S \cap \mathbb{N}$ if $S \cap \mathbb{N} \neq \emptyset$, or $\bot$ otherwise.

Proposition 6. $\mathcal{C}_{Fib}$ *is a cpo.*

Proof. Let $S \subseteq \mathcal{C}_{Fib}$ be a chain. Since the order $\leq_d$ is defined pointwise, for all $n \in \mathbb{N}$, $S_n := \{f(n) \mid f \in S\} \subseteq \mathbb{N}_\bot$ is a chain. By Lemma 1, for all $n \in \mathbb{N}$, there exists $lub(S_n)$. It is easy to see that the function $F : \mathbb{N} \to \mathbb{N}_\bot$ defined by $F(n) := lub(S_n)$ is the least upper bound of S.

It remains to define a monotone operator on $\mathcal{C}_{Fib}$. First, the sum $+ : \mathbb{N}^2 \to \mathbb{N}$ can be extended to $\mathbb{N}_\bot \times \mathbb{N}_\bot$ by defining for each $n \in \mathbb{N}_\bot$, $\bot + n = n + \bot = \bot$. Then, the operator $O_{Fib} : \mathcal{C}_{Fib} \to \mathcal{C}_{Fib}$ is defined to map a function $f \in \mathcal{C}_{Fib}$ to

$$O_{Fib}(f) := \begin{cases} n \mapsto n & (\text{for } n \in \{0,1\}) \\ n+2 \mapsto f(n+1) + f(n) \end{cases}$$

Proposition 7. *The operator O_{Fib} is monotone.*

Proof. Let $f, g \in \mathcal{C}_{Fib}$ such that $f \leq_d g$, i.e. for all $n \in \mathbb{N}$, $f(n) \leq g(n)$. Notice that we have

$$O_{Fib}(f)(0) = 0 = O_{Fib}(g)(0)$$
$$O_{Fib}(f)(1) = 1 = O_{Fib}(g)(1)$$
$$\forall n \in \mathbb{N} \setminus \{0,1\}, \; O_{Fib}(f)(n) = f(n-1) + f(n-2)$$
$$\leq g(n-1) + g(n-2)$$
$$= O_{Fib}(g)(n).$$

Hence, $O_{Fib}(f) \leq_d O_{Fib}(g)$, as desired.

By the definition of O_{Fib}, it is easy to see that the desired function Fib is the least fixpoint of O_{Fib}. The element $\text{lfp}(O_{Fib})$ can also be constructed as the limit of the increasing sequence of functions $f_0, f_1, \ldots$ in $\mathcal{C}_{Fib}$ obtained by iterating O_{Fib} on the bottom element of $\mathcal{C}_{Fib}$. We write here the functions in the first iterations of the process:

$$f_0(n) := \bot \qquad f_1(n) := \begin{cases} 0 & \text{if } n = 0 \\ 1 & \text{if } n = 1 \\ \bot & \text{otherwise} \end{cases} \qquad f_2(n) := \begin{cases} 0 & \text{if } n = 0 \\ 1 & \text{if } n \in \{1,2\} \\ \bot & \text{otherwise} \end{cases}$$

The enrichment of the exact space with $\bot$ allows us to deal with partially defined concepts, providing us with a suitable choice for the construction space. Nevertheless, such choice may be even more subtle, as we show next.

Example 7 (Ackermann function). $Ack : \mathbb{N}^2 \to \mathbb{N}$ is defined recursively as follows:

$$\left| \begin{array}{ll} \forall y \in \mathbb{N} : Ack(0, y) & := y + 1. \\ \forall x \in \mathbb{N} : Ack(x+1, 0) & := Ack(x, 1). \\ \forall x, y \in \mathbb{N} : Ack(x+1, y+1) & := Ack(x, Ack(x+1, y)). \end{array} \right|$$

The exact space is again the set of functions with the right signature, namely $\mathbb{N}^2 \to \mathbb{N}$. Analogously to Example 6, the function Ack is defined on every element of its domain only after infinitely many steps. Hence, it may seem natural to consider as construction space the functions from $\mathbb{N}^2$ to $\mathbb{N}_\perp$. However, this enlargement of the construction space is not sufficient: due to the third rule of the definition, during the construction, the Ackermann function might be invoked on an output of a partially constructed object, which can possibly be $\perp$. This prompts us to add $\perp$ to the *domain* of the functions in the construction space.

Just as before, we can order this expanded space $\mathbb{N}_\perp^{\mathbb{N}_\perp \times \mathbb{N}_\perp}$ by the pointwise extension of the definedness order $\leq_d$ on $\mathbb{N}_\perp$. However, the operator induced by the definition of Ack is not monotone on the full set of functions. Fortunately, it was shown that this operator is monotone on a sufficiently large subset defined next. We expand the definedness order $\leq_d$ to $\mathbb{N}_\perp \times \mathbb{N}_\perp$ as the product order of $\leq_d$ on $\mathbb{N}_\perp$, and consider the subset C_{Ack} of monotone functions of $\mathbb{N}_\perp^{\mathbb{N}_\perp \times \mathbb{N}_\perp}$. It turns out that the operator of the Ackerman definition, and the limit operation for increasing sequences of monotone functions both preserve monotoniticy of functions, making $C_{Ack} \subseteq \mathbb{N}_\perp^{\mathbb{N}_\perp \times \mathbb{N}_\perp}$ a suitable space to perform the induction process. In other words, we choose the construction space to be the cpo $\mathcal{C}_{Ack} := \langle C_{Ack}, \leq_d \rangle$. The proof of the fact that the construction space $\mathcal{C}_{Ack}$ is a cpo analogous to the proof of Proposition 6, given that $\langle \mathbb{N}_\perp \times \mathbb{N}_\perp, \leq_d \rangle$ is a cpo, which follows easily from Lemma 1.

It is important to note that for the first time, the injection θ is nontrivial, since we have enlarged the domain of the considered functions to $\mathbb{N}_\perp \times \mathbb{N}_\perp$. In particular, $\theta : \mathbb{N}^{\mathbb{N}^2} \to 2^{C_{Ack}}$ sends a function $f : \mathbb{N}^2 \to \mathbb{N}$ to the set of functions

$$\theta(f) := \{g : (\mathbb{N}_\perp)^2 \to \mathbb{N}_\perp \mid \forall (x,y) \in \mathbb{N}^2, f(x,y) = g(x,y)\}.$$

Hence, it is easy to see that the surjective partial function $\pi : C_{Ack} \to \mathbb{N}^{\mathbb{N}^2}$ associated to θ is defined only on the set of of functions whose restriction to $\mathbb{N}^2$ maps into $\mathbb{N}$ and maps each such function to its restriction to $\mathbb{N}^2$.

By the above recursive definition of Ack, the choice for the operator $O_{Ack} : \mathcal{C}_{Ack} \to \mathcal{C}_{Ack}$ is clear: for all $f \in L$,

$$O_{Ack}(f) := \begin{cases} (0, y) & \mapsto y + 1 \\ (x+1, 0) & \mapsto f(x, 1) \\ (x+1, y+1) & \mapsto f(x, f(x+1, y)) \end{cases}$$

where $+$ is extended to $\mathbb{N}_\perp \times \mathbb{N}_\perp$ as in Example 6. Note that $O_{Ack}(f)$ is indeed an element of $\mathcal{C}_{Ack}$, since the composition of monotone functions is monotone.

Proposition 8. *The operator O_{Ack} is monotone.*

Proof. Let $f, g \in C_{Ack}$ such that $f \leq_d g$. Then we have

$$\forall y \in \mathbb{N}_\perp, \ O_{Ack}(f)(0, y) = y + 1 = O_{Ack}(g)(0, y),$$
$$O_{Ack}(f)(\perp, 0) = f(\perp, 0) \leq_d g(\perp, 0) = O_{Ack}(g)(\perp, 0),$$
$$\forall x \in \mathbb{N} \setminus \{\perp, 0\}, \ O_{Ack}(f)(x, 0) = f(x-1, 0)$$
$$\leq_d g(x-1, 0) = O_{Ack}(g)(x, 0).$$

Moreover, for all $x, y \in \mathbb{N}_\perp$,

$$O_{Ack}(f)(x+1, y+1) = f(x, f(x+1, y))$$
$$\leq_d f(x, g(x+1, y))$$
$$\leq_d g(x, g(x+1, y))$$
$$= O_{Ack}(g)(x+1, y+1),$$

where the first inequality holds by the monotonicity of f. Hence, $O_{Ack}(f) \leq_d O_{Ack}(g)$, as desired.

The construction process starts from the bottom element $\perp_{Ack}$ of $\mathcal{C}_{Ack}$, namely the function $\perp_{Ack}: \mathbb{N}_\perp \times \mathbb{N}_\perp \to \mathbb{N}_\perp$ sending every pair to $\perp$. By iteratively applying the operator O_{Ack}, we obtain an increasing sequence of monotone functions $f_0, f_1, \ldots$ in $\mathcal{C}_{Ack}$, representing the partial functions of the intermediate steps of the recursion. At the first steps of the process we get the functions defined on $(x, y) \in \mathbb{N}_\perp \times \mathbb{N}_\perp$ as follows:

$$f_0(n) := \perp \qquad f_1(x, y) := \begin{cases} y+1 & \text{if } x = 0 \\ \perp & \text{otherwise} \end{cases} \qquad f_2(x, y) := \begin{cases} y+1 & \text{if } x = 0 \\ 2 & \text{if } (x, y) = (1, 0) \\ \perp & \text{otherwise} \end{cases}$$

Only after transfinitely many steps, we reach the least fixpoint of O_{Ack}. Finally, it is not hard to see that applying the projection π on the least fixpoint yields the defined object, i.e., $\pi(\text{lfp}(O_{Ack})) = \text{lfp}(O_{Ack})|_{\mathbb{N}^2} = Ack$.

We now move to examples of co-recursive definitions of functions. Once again, the choice of a suitable construction space turns out to be non-trivial.

Example 8 (Co-Fibonacci). The co-Fibonacci function $co\_Fib: \mathbb{N}^2 \to List$, which maps a pair (x, y) of natural numbers to the Fibonacci sequence starting with x, y, is defined co-recursively to send (x, y) to $[x \mid co\_Fib(y, x+y)]$.

We can present the corecursive definition of co_Fib by

$$\left[\forall x, y \in \mathbb{N} : co\_Fib(x, y) := [x \mid co\_Fib(y, x+y)]. \right]$$

In particular, $co\_Fib(0, 1)$ is the list corresponding to the Fibonacci sequence, defined recursively in Example 6. The exact space is again clear from the signature of the function we want to define: it is the set of functions from $\mathbb{N}^2$ to $List$.

As in Example 6, we need to enlarge the codomain of the considered functions in order to represent intermediate steps of the process. Thus, we define a set $List^o$, containing lists of natural numbers and finite lists of natural numbers ending with o. A list $[x_1, \ldots, x_n \mid o]$ of the latter type represents a list of natural numbers with overdefined o as tail. In other words, the list $[x_1, \ldots, x_n \mid o]$ represents the set $\{[x_1, \ldots, x_n \mid l]: l \in List\}$ of lists of natural numbers. In particular, the list o represents the overdefined list, i.e., the set of all lists of natural numbers.[5] Accordingly, on $List^o$, definedness order $\leq_d$ is defined inductively as follows:

---

[5] Note that the earlier introduced notation $[x \mid y]$ is used now to denote a list of $List^o$ with head a finite list x of natural numbers, and tail a list y of $List^o$.

- for all $t \in List^o: t \leq_d o$
- for all $x \in \mathbb{N}, t_1, t_2 \in List^o: [x \mid t_1] \leq_d [x \mid t_2]$ if $t_1 \leq_d t_2$

In this order, o is indeed "more defined" than any list. The set $List^o$ with the order $\leq_d$ is not a cpo since it has no least element, however, with the inverted order $\geq_d$ it indeed is a cpo, with "least" element o.

Lemma 2. $\langle List^o, \geq_d \rangle$ *is a cpo.*

Proof. Let $S \subseteq List^o$ be a chain. We have to show that S has a least upper bound. If S has finite cardinality, the claim is trivial. Suppose S has an infinite number of elements. By the definition of $\geq_d$ and since S is a chain, if S contains an infinite list l, then l is the least upper bound of S. Suppose otherwise, i.e. all elements in S are finite lists of natural numbers or finite lists of natural numbers ending with o. We denote the length of a list $l \in S$ by $length(l) \in \mathbb{N}$. We can order the lists in S following the total order, and we denote by l^i the i-th list in S with such ordering, i.e. $i_1 \leq i_2$ if and only if $l^{i_1} \leq l^{i_2}$. Notice that if $length(l^i) = length(l^{i+1})$, then l^i must end with o and l^{i+1} with a natural number. In particular, $length(l^{i+2})$ is strictly greater than $length(l^i)$. Let $l \in S$ be a list and $n \leq length(l)$, we denote by l_n the n-th element of l. We define an infinite list $L := [L_j]_{j \in \mathbb{N}}$ where

$$\forall j \in \mathbb{N}: L_j := l^{2j+2}_{length(l^{2j+2})-1}.$$

Notice that, for all $i \in \mathbb{N}$, the first $\min(length(l^i), length(l^{i+1})) - 1$ elements of l^i and l^{i+1} coincide. Hence, it is not hard to see that for any $l \in S$ we have $l \geq_d L$ by construction. Let U be an upper bound for S, i.e. U is an infinite sequence of natural numbers such that $l \geq_d U$ for all $l \in S$. By the definition of the order, for all $l \in S$, the first $length(l) - 1$ elements of l and U coincide. Hence, it is easy to see that $L = S$. In particular, L is the least upper bound of S.

The order $\geq_d$ can be extended in the standard, pointwise way to $(List^o)^{\mathbb{N}^2}$. We define the construction space $\mathcal{C}_{co\_Fib} = \langle (List^o)^{\mathbb{N}^2}, \geq_d \rangle$. The inversion of the definedness order, often used for recursion, mimics the order inversion between inductive and coinductive definitions (hence the term *corecursion*).

Finally, we define the operator $O_{co\_Fib}: \mathcal{C}_{co\_Fib} \to \mathcal{C}_{co\_Fib}$ by sending a function $f \in \mathcal{C}_{co\_Fib}$ to

$$O_{co\_Fib}(f): \mathbb{N}^2 \to List^o : (x, y) \mapsto [x \mid f(y, x + y)].$$

By the definition of $\geq_d$, it is easy to see that O_{co_Fib} is a monotone operator. Moreover, the desired function co_Fib is the least fixpoint of the operator O_{co_Fib}. This coincides with the limit of the increasing sequence $f_0, f_1, \ldots$ constructed by iterating O_{co_Fib} on the bottom element $\bot_{\mathcal{C}_{co\_Fib}}$ of $\mathcal{C}_{co\_Fib}$, i.e., the function $\bot_{\mathcal{C}_{co\_Fib}}: \mathbb{N}^2 \to List^o$ sending every tuple to o. We report here the images of the functions in the first iterations of the process, depending on $(x, y) \in \mathbb{N}^2$:

$f_0(x,y) = \bot_{\mathcal{C}_{co\_Fib}}(x,y) = o$ $\quad f_2(x,y) = [x, y \mid o]$

$f_1(x,y) = [x \mid o]$ $\quad\quad\quad\quad\quad\quad\; f_3(x,y) = [x, y, x + y \mid o]$

3.3 Definitions with Custom-Designed Cpo's

In this third and last subsection, we present a final example of a constructive definition of a function. Even though this definition deviates from the standard (co)recursive account, it can indeed be formalized using our proposed framework. Just like Example 5, the definition illustrated here falls under the company controls domain: in this case, we want to define the number of shares of a company that another company controls.

Example 9 (Controlled shares). If x and y are two companies, we say that x controls n shares of y if n is the sum of the shares of y owned by x or any company z of which x controls more than half of the shares.

We can formally define the desired function $Csh\colon C^2 \to [0,1]$ as follows:

$$\left\{ \forall x \forall y : Csh(x,y) := \sum_{z \in \{x\} \cup \{u \mid Csh(x,u) > 0.5\}} Sh(z,y). \right\}$$

where $Sh\colon C^2 \to [0,1]$ is still the function mapping a pair of companies (x,y) to the fraction of shares of y owned by x. The exact space is the set of functions from S^2 to the interval $[0,1]$. The construction process is more complex than before: we now need to be able to decide whether $Csh(x,y) > 0,5$ is true before $Csh(x,y)$ is determined.

We can think about this construction process as a gradual refinement of each tuple's image. At the beginning of the process, we have no information about the image of Csh except that $Csh(x,y) \in [0,1]$ for all $x,y \in C$. At every rule application we get new information on the lower bounds of the images of elements of C^2. Since the upper bounds remain constant equal to 1, we may as well identify the interval in which an image is contained with its lower bound. Now, the choice for a construction space $\mathcal{C}_{Csh}$ becomes clear, namely we consider the cpo of functions from C^2 to $[0,1]$, with the pointwise extension of the standard order $\leq$ on real numbers.

Proposition 9. $\mathcal{C}_{Csh}$ *is a cpo.*

Proof. Since $([0,1], \leq)$ is a cpo, the proof is analogous to the proof of Proposition 6.

Finally, we can consider the monotone operator $O_{Csh}\colon \mathcal{C}_{Csh} \to \mathcal{C}_{Csh}$, which maps a function $f\colon C^2 \to [0,1]$ to $O_{Csh}(f)$, defined by

$$O_{Csh}(f)(x,y) := \sum_{z \in \{x\} \cup \{u \mid f(x,u) > 0,5\}} Sh(x,y).$$

Proposition 10. *The operator O_{Csh} is monotone.*

Proof. Let $f,g\colon C^2 \to [0,1]$ such that $f \leq_L g$. In particular, for all $x \in C$, we have $\{u \mid f(x,u) > 0,5\} \subseteq \{u \mid g(x,u) > 0,5\}$. Since $Sh(z,y) \geq 0$ for all $z,y \in C$, we have $O_{Csh}(f) \leq_L O_{Csh}(g)$, as desired.

As anticipated, we can start the recursion from the bottom element of $\mathcal{C}_{Csh}$, namely the function f_0 sending every pair of companies to 0. By iteratively applying the operator O_{Csh} we get an increasing sequence of functions $f_0 \leq_{\mathcal{C}_{Csh}} f_1 \leq_{\mathcal{C}_{Csh}} f_2 \leq_{\mathcal{C}_{Csh}} \cdots$, whose limit is the desired defined function Csh and coincides with the least fixpoint of O_{Csh}. Notice that at any step t of the construction process, for each pair $(x,y) \in C^2$, the image $f_t(x,y)$ may not be the correct value of $Csh(x,y)$. Only in the last step, when the fixpoint is reached, certainty is reached of the correct value $Csh(x,y)$, for all pairs (x,y) at once. This is much unlike previous examples. This type of construction, using increasingly more precise bounds, lies at the basis of bound-founded ASP [4,8].

4 Conclusion and Future Work

We investigated a heterogeneous set of monotone constructive definitions, coming from different domains and never brought together before, in a uniform framework. Our analysis confirms the power of fixpoint theory for abstract formalization, but also points to a key distinguishing factor: the construction space, the set of objects that serve as approximations of the object being defined. We propose the general term *monotone constructive definitions* for a class of definitions that includes recursive and inductive definitions and developed a framework that clearly emphasises how different types of definitions can be classified according to different types of construction spaces. This is a crucial step towards the development of knowledge representation languages that include a variety of constructive definitions. Our framework suggests such language requires a formal syntax for definitions such that one can automatically and uniformly derive a suitable exact space, semantic operator and construction space. As shown by the examples, while the first two are straightforward, the latter may be non-trivial. We have illustrated different types of definitions by example, allowing us to handpick the most convenient, natural construction space. The challenge is that a uniformly derived construction space needs to be strong enough to handle all considered definitions, and the defined object should coincide with the one obtained with the handpicked construction space. This paper offers an important first step towards solving this issue by classifying different types of definitions based on the kind of construction space they require. This means identifying the correct type of definition will be an essential part of the syntax of the considered knowledge representation language.

By no means do we claim our list of types of constructive definitions to be exhaustive. Other types of constructive definitions not considered here are nested inductive and coinductive definitions where multiple objects are defined in a hierarchy of inductive and coinductive definitions [25,30] or non-monotone "iterated" inductive definitions which have been researched in mathematical logic [7,18,23]. In iterated inductive definitions, e.g., over a well-founded order, multiple objects are defined in terms of other defined objects on a lower or equal level. Once all objects on some level are well-defined, their values can be used to derive the value of any object on a higher level. This is the natural principle of *stratification*. It has been argued that this principle is implemented by the well-founded semantics of logic programming [11,14,15]. Thus, the declarative logic underlying logic programming can be seen as a logic of this type of constructive definition.

In this paper, we focused on *monotone* constructive definitions. Non-monotone *inductive* definitions have been studied intensively, including in a fixpoint-theoretic setting (known as *Approximation Fixpoint Theory (AFT)* [12]). In the terminology of the current paper, dealing with this non-monotonicity requires switching to a different construction space (a space of approximations). A natural next question we wish to tackle is whether this framework can also be of use for studying non-monotone *recursive* definitions.

As a final remark, we argued that constructive definitions are an important form of human knowledge. Of course, many other types of knowledge are important as well. Contrary to the languages of logic and functional programming, which support mainly definitions used as programs, expressive KR languages should offer language constructs for expressing a broad range of knowledge. In this respect, an example is the logic FO(·), which extends FO with among others an expressive rule-based language construct for definitional knowledge, inspired by logic programming [9,10].

References

1. Abramsky, S., Jung, A.: Domain theory. In: Handbook of Logic in Computer Science, vol. 3, pp. 1–168. Oxford University Press, Inc., New York (1994). http://dl.acm.org/citation.cfm?id=218742.218744
2. Aczel, P.: An introduction to inductive definitions. In: Studies in Logic and the Foundations of Mathematics, vol. 90, pp. 739–782. Elsevier (1977)
3. Aczel, P.: The type theoretic interpretation of constructive set theory: inductive definitions. In: Studies in Logic and the Foundations of Mathematics, vol. 114, pp. 17–49. Elsevier (1986)
4. Aziz, R.A.: Bound founded answer set programming. CoRR abs/1405.3367 (2014). http://arxiv.org/abs/1405.3367
5. Bogaerts, B., Vennekens, J., Denecker, M.: Safe inductions and their applications in knowledge representation. Artif. Intell. **259**, 167–185 (2018). https://doi.org/10.1016/j.artint.2018.03.008. http://www.sciencedirect.com/science/article/pii/S000437021830122X
6. Brouwer, L.E.J.: Over de grondslagen der wiskunde. Maas & van Suchtelen (1907)
7. Buchholz, W., Feferman, S., Pohlers, W., Sieg, W.: Iterated Inductive Definitions and Subsystems of Analysis: Recent Proof-Theoretical Studies. Lecture Notes in Mathematics, vol. 897. Springer, Heidelberg (1981). https://doi.org/10.1007/BFb0091894
8. Cabalar, P., Fandinno, J., Schaub, T., Schellhorn, S.: Lower bound founded logic of here-and-there. In: Calimeri, F., Leone, N., Manna, M. (eds.) JELIA 2019. LNCS (LNAI), vol. 11468, pp. 509–525. Springer, Cham (2019). https://doi.org/10.1007/978-3-030-19570-0_34
9. De Cat, B., Bogaerts, B., Bruynooghe, M., Janssens, G., Denecker, M.: Predicate logic as a modeling language: the IDP system. In: Declarative Logic Programming: Theory, Systems, and Applications, pp. 279–323 (2018). https://doi.org/10.1145/3191315.3191321
10. Denecker, M.: Extending classical logic with inductive definitions. In: Lloyd, J., et al. (eds.) CL 2000. LNCS (LNAI), vol. 1861, pp. 703–717. Springer, Heidelberg (2000). https://doi.org/10.1007/3-540-44957-4_47
11. Denecker, M., Bruynooghe, M., Marek, V.: Logic programming revisited: logic programs as inductive definitions. ACM Trans. Comput. Log. **2**(4), 623–654 (2001)
12. Denecker, M., Marek, V., Truszczyński, M.: Approximations, stable operators, well-founded fixpoints and applications in nonmonotonic reasoning. In: Minker, J. (ed.) Logic-Based Artificial Intelligence. The Springer International Series in Engineering and Computer Science, vol. 597, pp. 127–144. Springer, Boston (2000). https://doi.org/10.1007/978-1-4615-1567-8_6
13. Denecker, M., Ternovska, E.: Inductive situation calculus. Artif. Intell. **171**(5–6), 332–360 (2007)

14. Denecker, M., Ternovska, E.: A logic of nonmonotone inductive definitions. ACM Trans. Comput. Log. **9**(2), 14:1–14:52 (2008). https://doi.org/10.1145/1342991.1342998
15. Denecker, M., Vennekens, J.: The well-founded semantics is the principle of inductive definition, revisited. In: Baral, C., De Giacomo, G., Eiter, T. (eds.) KR, pp. 1–10. AAAI Press (2014). http://www.aaai.org/ocs/index.php/KR/KR14/paper/view/7957
16. Doets, K., Eijck, J.: The Haskell Road to Logic, Maths and Programming. College Publications (2004)
17. Downen, P., Ariola, Z.M.: Classical (co)recursion: programming. CoRR abs/2103.06913 (2021). https://arxiv.org/abs/2103.06913
18. Feferman, S.: Formal theories for transfinite iterations of generalised inductive definitions and some subsystems of analysis. In: Kino, A., Myhill, J., Vesley, R. (eds.) Intuitionism and Proof Theory, pp. 303–326. North Holland (1970)
19. Hudak, P.: Conception, evolution, and application of functional programming languages. ACM Comput. Surv. **21**(3), 359–411 (1989). https://doi.org/10.1145/72551.72554
20. Kemp, D.B., Stuckey, P.J.: Semantics of logic programs with aggregates. In: Saraswat, V.A., Ueda, K. (eds.) ISLP, pp. 387–401. MIT Press (1991)
21. Kleene, S.C.: On the forms of the predicates in the theory of constructive ordinals. Am. J. Math. **66**, 41–58 (1944)
22. Kozen, D., Silva, A.: Practical coinduction. Math. Struct. Comput. Sci. **27**(7), 1132–1152 (2017)
23. Martin-Löf, P.: Hauptsatz for the intuitionistic theory of iterated inductive definitions. In: Fenstad, J. (ed.) Second Scandinavian Logic Symposium, pp. 179–216 (1971)
24. Moschovakis, Y.N.: Elementary Induction on Abstract Structures. North-Holland Publishing Company, Amsterdam; New York (1974)
25. Paulson, L.C.: A fixedpoint approach to (co) inductive and (co) datatype definitions. In: Proof, Language, and Interaction, pp. 187–212 (2000)
26. Roberts, E.S.: Thinking Recursively. Wiley, Hoboken (1986)
27. Rubio-Sanchez, M.: Introduction to Recursive Programming. CRC Press, Boca Raton (2017)
28. Rusu, V., Nowak, D.: Defining corecursive functions in Coq using approximations. In: Ali, K., Vitek, J. (eds.) 36th European Conference on Object-Oriented Programming, ECOOP 2022, 6–10 June 2022, Berlin, Germany. LIPIcs, vol. 222, pp. 12:1–12:24. Schloss Dagstuhl - Leibniz-Zentrum für Informatik (2022). https://doi.org/10.4230/LIPIcs.ECOOP.2022.12
29. Scott, D.: Data types as lattices. In: Müller, G.H., Oberschelp, A., Potthoff, K. (eds.) International Summer Institute and Logic Colloquium. Lecture Notes in Mathematics, vol. 499, pp. 579–651. Springer, Heidelberg (1975). https://doi.org/10.1007/BFb0079432
30. Simon, L., Mallya, A., Bansal, A., Gupta, G.: Coinductive logic programming. In: Etalle, S., Truszczyński, M. (eds.) ICLP 2006. LNCS, vol. 4079, pp. 330–345. Springer, Heidelberg (2006). https://doi.org/10.1007/11799573_25
31. Smyth, M.B., Plotkin, G.D.: The category-theoretic solution of recursive domain equations. SIAM J. Comput. **11**(4), 761–783 (1982)
32. Soare, R.I.: Recursively Enumerable Sets and Degrees - A Study of Computable Functions and Computability Generated Sets. Perspectives in Mathematical Logic. Springer, Heidelberg (1987)
33. Spector, C.: Inductively defined sets of natural numbers. In: Infinitistic Methods (Proc. 1959 Symposium on Foundation of Mathematis in Warsaw), pp. 97–102. Pergamon Press, Oxford (1961)
34. Tarski, A.: A lattice-theoretical fixpoint theorem and its applications. Pac. J. Math. (1955)

Partial Boolean Functions for QBF Semantics

Allen van Gelder(✉)

University of California, Santa Cruz, CA 95064, USA
avg@cs.ucsc.edu

Abstract. Long-distance resolution for quantified boolean formula (QBF) solving was introduced in 2002 by Zhang and Malik, but has been controversial for the following decade, because it derives and uses tautologous clauses. Balabanov and Jiang (2012) gave a set of proof rules (called LDQ-resolution) that include long-distance resolution with conditions, but did not attach any "meaning" (i.e., semantic interpretation) to the derived tautologous clauses. Egly, Lonsing and Widl (2013) showed that the QBF certificate extraction algorithm of Goutltiaeva, Van Gelder and Bacchus (2011) could be applied with correct results to LDQ refutations. These results and others have brought LDQ-resolution back into the mainstream. This paper introduces partial boolean functions (pbfs) and develops a semantic interpretation for universal literals in QBF clauses. It develops LDP-resolution, which is refutationally complete for QBF formulas in prenex conjunction normal form (PCNF). LDP-resolution is shown to derive only logical consequences.

Keywords: Quantified boolean formulas · Partial boolean functions · Long-distance resolution

1 Introduction

Quantified boolean formulas (QBFs) provide a natural expression of many AI problems on finite domains, such as planning, hypothetical reasoning, counterfactual reasoning, game analysis, quantified constraint satisfaction, and various forms of verification. Practical tools for QBF reasoning are of interest and value in a variety of constraint-satisfaction problems.

Long-distance resolution for quantified boolean formula (QBF) solving (see Sect. 2 for definitions) was introduced more than a decade ago [23], but has been controversial for the following decade, because it derives and uses tautologous clauses. Balabanov and Jiang gave a set of proof rules (called *LDQ-resolution*) that include long-distance resolution with conditions [2]. They claimed that any LDQ-resolution refutation can be transformed into a Q-resolution refutation. Further work showed that LDQ-resolution can be implemented efficiently in a search-based QBF solver and may generate exponentially shorter refutations than Q-resolution on a certain family of formulas [8]. Peitl *et al.* [15] showed that it may be combined with the *standard dependency scheme* [13,17]. Beyersdorff and Blinkhorn used the idea of fully exhibiting model trees to show that it may be combined with the *RRS dependency scheme* [3]. Blinkhorn and Beyersdorff showed that combining formula restriction with the *RRS dependency scheme* can exponentially shorten refutations in a variant of a well-studied QBF family, even without long-distance resolution [5].

None of this work addressed the question of whether clauses or similar formulas derived by these proof systems were ***logical consequences*** of the original QBF; they only considered whether derived formulas preserved the *truth value* of a closed QBF. A derived formula is a logical consequence of an original formula if conjoining the derived formula with the original formula does not reduce the set of model trees.

Definition 1. We say a system is *strategy sound* if it only derives logical consequences of the original formula. The qualifier "strategy" is included because many papers consider a system to be "sound" if it does not return an incorrect truth value. That is, only one bit of the output is considered important. □

It is known in the literature that the Q-resolution proof system, dating from 1995, derives only logical consequences. Recent years have seen the introduction of several QBF proof systems that might have exponentially shorter refutations than are possible with Q-resolution on certain formula families. This paper introduces **partial boolean functions** (**pbfs**) and develops a semantic interpretation for universal literals in QBF clauses based on pbfs. A special case of pbfs is described, called **partial Herbrand functions** (**phfs**). It introduces a long-distance resolution system using phfs called LDP-resolution that is "strategy sound," i.e., it derives only logical consequences of the original QBF formula. LDP applies to QBF formulas in prenex conjunction normal form (PCNF).

It is then shown that some well known systems may derive formulas that are **not** logical consequences, although they do not change the truth value of the entire original formula.

The importance of this distinction becomes apparent if the original formula is a subformula in the overall logic for a large project: A logical consequence may be added to the original subformula without changing its set of model trees, and may be useful in its own right in other parts of the project. It further demonstrates that universal reductions based on several more aggressive dependency schemes found in the literature [20,22], even without any form of long-distance resolution, lead to derived clauses that are *not* necessarily logical consequences.

QBF can be formulated as a decidable fragment of full first-order logic. Future work may include study of other decidable fragments to see if partial Herbrand functions or partial Skolem functions have a useful role as replacements for total functions. For example, Samer has considered quantified constraint-satisfaction problems in which variables take values in finite domains [16].

2 Basic Definitions and Notation

This paper uses standard notation as much as possible, but contains many specific definitions and notations that are used throughout the paper. Page limits force us to assume that the reader is familiar with the common definitions related to QBFs in the literature.

This paper uses capitalized Greek letters Ψ, Φ, and Θ to denote QBFs. We use 0 and 1 for truth values of literals and use *true* and *false* for semantic values of formulas.

We follow certain notational conventions for boolean variables and literals (signed variables) to make reading easier: Lowercase letters near the **beginning** of the alphabet (e.g., b, c, d, e) denote existential literals, while lowercase letters near the **end** of the alphabet (e.g., u, v, w, x) denote universal literals, while **middle** letters (e.g., p, q, r) are of ambiguous quantifier type. Quantifier types are implied frequently throughout the paper without restating this convention.

In contexts where a literal is expected, p might denote a positive or negative literal, while $\overline{p}$ denotes the negation of p. To emphasize that p stands for a **variable**, rather than a *literal*, the notation $|p|$ is used. Clauses may be written as $[p, q, \overline{r}]$; $[]$ denotes the empty clause; $[1]$ denotes an existentially tautologous clause.

For this paper we require the **original QBF** (to be solved) to be **closed** (i.e., with no free variables), in **prenex conjunctive normal form** (**PCNF**), and have a **matrix** of **non-tautologous original clauses** that are **completely reduced**, which means that every *universal* literal in the clause that is inner to every *existential* literal in the same clause is removed. Derived formulas may be in a slightly more general prenex form.

Perhaps the most constructive way to think of QBF semantics is as a two-player game between the E-player, who sets existential variables, and the A-player, who sets universal variables. Repeatedly, the outermost unset variable in the prenex is set to 0 or 1 by the appropriate player. The E-player tries to make the matrix eventually evaluate to *true*, which requires *every* clause to be satisfied, whereas the A-player tries to make the matrix eventually evaluate to *false*, which is accomplished by *some* clause being falsified.

Practical proof systems focus on proving that the A-player has a winning strategy for the original QBF Φ, which is synonymous with Φ being *false*. When the A-player needs to choose a value for an outermost unset universal variable u, the values of all *existential* variables outer to u have been set. Therefore the choice made by the A-player can be thought of as a boolean function of these outer existential variables, say $u(e_1, e_2, \ldots)$. Such functions are called **Herbrand functions** for u.

A winning A-strategy is a complete set of Herbrand functions, one for each universal variable, such that setting the universal variables according to these functions during a play of the game ensures that Φ evaluates to *false* no matter how the E-player chooses. Now suppose each occurrence of a universal variable u in the matrix of Φ is replaced by the corresponding Herbrand function

$u(e_1, e_2, \ldots)$. The resulting Φ has entirely existential variables and is *false* if and only if it is unsatisfiable.

Historically, Herbrand functions are total boolean functions but in practice partial functions are often sufficient to specify a winning A-strategy. This paper introduces a theory based on partial boolean functions and partial Herbrand functions that avoids certain logical problems that arise when total Herbrand functions are used. Several technical definitions are needed.

Definition 2. For this paper the prenex is *totally ordered*. The ***qdepth***, or ***scope***, of a variable is its alternation depth, with 1 outermost. Notations: $p \prec q$ means $qdepth(p) < qdepth(q)$; $p \preceq q$ means $qdepth(p) \leq qdepth(q)$; $p \prec\!\prec q$ means p precedes q in the totally ordered prenex.

A few special operations involving a set of literals S are defined, assuming the prenex $\vec{Q}$ is known by the context.

$$\mathsf{exist}(S) = \{\text{the existential literals in } S\} \tag{1}$$
$$\mathsf{univ}(S) = \{\text{the universal literals in } S\} \tag{2}$$
$$(S \prec q) = \{\text{the literals in } S \text{ outer to } q\} \tag{3}$$
$$(q \prec S) = \{\text{the literals in } S \text{ inner to } q\} \tag{4}$$

Depending on context, S might be a clause, a prenex, a partial assignment, or some other logical expression. □

Definition 3. A ***prenex-ordered assignment*** is a *total* assignment that is represented by a sequence of literals that are assigned 1 and are in prenex order. □

A strategy for the E-player can be represented as an E-assignment tree, which we now describe. (See also [18,19]). We say that a ***branch*** is a path from the root node to some leaf node, represented as a sequence of tree edges, which are labeled with a prenex-ordered assignment.

Although it seems artificial at first, it is very useful to represent a tree as the nonempty set of its branches. A ***branch prefix*** is a path (sequence of edges) from the root node that might terminate before reaching a leaf. The same branch prefix can occur in many branches, and represents a tree node; the empty branch prefix is the tree root.

Definition 4. Let a QBF $\Phi = \vec{Q}.\mathcal{F}$ be given, with $k = |\mathsf{univ}(\Phi)|$. An ***E-assignment tree*** T for Φ is a set of exactly 2^k prenex-ordered assignments for Φ that defines a tree and satisfies these constraints on branching:

1. Let (σ, e) denote a branch prefix in T. Then *no* branch of T has the prefix $(\sigma, \overline{e})$.
2. Let (σ, u) denote a branch prefix in T. Then *some* branch of T has the prefix $(\sigma, \overline{u})$.

A ***model tree*** M for Φ is an E-assignment tree in which each branch τ makes $\mathcal{F}$ *true*, i.e., $\tau \models \mathcal{F}$ in the usual propositional sense.

Although the wording is different, if M is a model tree by this definition, it is also a model tree by definitions found in other papers [19]. Every model tree M represents a winning E-strategy in the obvious way.

Definition 5. Departing from the CNF formalism, we denote the ***if-then-else*** boolean operator on three parameters (or gate with three inputs) by $ite(C, T, F)$. Here C, T, and F are themselves boolean formulas. For any partial assignment τ:

if $C\lceil_\tau = 1$, then $ite(C, T, F)\lceil_\tau = T\lceil_\tau$;
if $C\lceil_\tau = 0$, then $ite(C, T, F)\lceil_\tau = F\lceil_\tau$.

See Definition 12 for abbreviated forms of ite denoted by $if(C, T)$ and $cp(T)$. □

The ite operator is basic to many programming languages and is popular for circuit design, due to its nice properties. For example, it can simulate unary and binary boolean formulas, e.g., $(A \wedge B) \equiv ite(A, B, 0)$, $xor(A, B) \equiv ite(A, \neg(B), B)$. It is famous as the foundation for binary decision diagrams (BDDs) developed in the pioneering work of Bryant [6]. Recently it has played a prominent role in works on QBF [7,12]. For this paper, ite and its formulation in terms of restriction are important for the development of partial Herbrand functions in derived clauses.

We assume the reader is familiar with the *Q-resolution* proof system, with two operations, *resolution* and *universal reduction*. Q-resolution is *refutationally* complete [11], but not *inferentially* complete. That is, $\Phi \models C$ might hold for a clause C, yet neither C nor a clause that subsumes C is derivable by Q-resolution. Also, Q-resolution prohibits derivation of a tautologous clause.

Definition 6. *Resolution* is defined as usual for propositional clauses, but the notation used is more specific than usual, to facilitate analysis of proofs. Let clauses $C_1 = [\overline{p}, \alpha]$ and $C_2 = [p, \beta]$, where α and β are (possibly empty) literal sequences containing neither p nor $\overline{p}$. Then the ***resolvent*** is: $\mathsf{res}_p(C_1, C_2) = \alpha \cup \beta$.

Variable $|p|$ and its literals are called the ***clashing variable*** and ***clashing literals***. Note that the second operand is always the clause that contains p. This paper considers only existential clashing literals, but considers proof systems that permit tautologous resolvents under certain conditions. □

Definition 7. A universal literal u is said to be ***tailing*** in a clause C if its *qdepth* is greater than that of any existential literal in C, i.e., $(u \prec \mathsf{exist}(C)) = \emptyset$.

Universal reduction removes tailing universal literals from clauses. If u is *tailing* in $C_3 = [\alpha, u]$, then the ***reduced clause*** is: $\mathsf{unrd}_u(C_3) = \alpha$.

In addition $\mathsf{unrd}_*(C_3)$ denotes the *completely reduced* clause that results from applying unrd successively on all tailing universal literals in C_3. □

Lemma 8. Resolution and universal reduction are strategy-sound operations.

Proof: Straightforward application of the definitions. ∎

Definition 9. We now define an extension of resolution that makes it a total function so that operations in a resolution proof remain well defined after a restriction is applied to the leaf clauses [9,21].. As with *Q-resolution*, the clashing literal is required to be existential. To distinguish from standard resolution we call the new operator **total resolution** and denote it as "trex".

As a technicality, the clause [1] represents a clause with the literal 1 in it, due to a restriction; it is treated as the widest possible clause and always evaluates to *true*. If normal resolution is defined, trex is the same. The other cases are summarized next.

If the clashing literal is present in one operand and not the other, and neither operand is [1], the resolvent is the operand *without* a clashing literal. If one operand is [1] and the other contains its clashing literal, or both operands are [1], the resolvent is [1]. If one operand is [1] and the other *lacks* its clashing literal, the other clause is the resolvent. Finally, if both operands are regular clauses that lack their clashing literal, the clause with fewer literals becomes the resolvent; ties are broken by some deterministic procedure. These cases explain the importance of specifying the clashing literal in the notation. □

Lemma 10. If $D = \text{trex}_p(C_1, C_2)$, then

$$\text{res}_p(C_1\lceil_\tau, C_2\lceil_\tau) \subseteq D\lceil_\tau. \tag{5}$$

Proof: Straightforward application of the definitions. ∎

2.1 Boolean If-Then-Else

Definition 11. Departing from the CNF formalism, we denote the *if-then-else* boolean operator on three parameters (or gate with three inputs) by $ite(C, T, F)$. Here $C, T,$ and F are themselves boolean formulas. For any partial assignment τ:

if $C\lceil_\tau = 1$, then $ite(C, T, F)\lceil_\tau = T\lceil_\tau$;
if $C\lceil_\tau = 0$, then $ite(C, T, F)\lceil_\tau = F\lceil_\tau$. □

This operator is basic to many programming languages and is popular for circuit design, due to nice properties. For example, it can simulate unary and binary boolean formulas, e.g., $(A \wedge B) \equiv ite(A, B, 0)$, $xor(A, B) \equiv ite(A, \neg(B), B)$, et al. It is famous as the foundation for binary decision diagrams (BDDs) developed in the pioneering work of Bryant [6]. Recently it has played a prominent role in works on QBF [7,12].

The *ite* operator has several implementations using two-parameter boolean operators, but it is often most natural as a three-parameter operator. Well known representations in DNF and CNF are:

$$\textbf{DNF: } ite(C, T, F) \equiv (C \wedge T) \vee (\neg(C) \wedge F) \tag{6}$$
$$\textbf{CNF: } ite(C, T, F) \equiv (\neg(C) \vee T) \wedge (C \vee F). \tag{7}$$

3 Partial Boolean Functions

This section introduces *partial boolean functions*, which play a central role in the interpretation of QBF derivations. For this discussion, let **V** be a fixed nonempty set of k propositional variables: $\mathbf{V} = \{q_i \mid 1 \leq i \leq k\}$.

Recall that the mathematical definition of a boolean function F on **V** is a $(k+1)$-ary relation (i.e., set of rows) in which each *total* assignment τ to the variables of **V** appears in the first k columns of exactly one row of F, and $F(\tau)$ appears in column $(k+1)$, called the **output column**. This relation has 2^k rows and is often called a *truth table*. Although this seems like a cumbersome way to represent a function it is the natural lead-in to partial functions. For example, set operations like union and intersection are not useful on total functions because the result is not a function except in trivial cases. However, these operations prove to be very useful on partial functions.

q_1	q_2	q_3	F_1		q_1	q_2	q_3	G_1		q_1	q_2	q_3	G_2
1	0	0	1		0	0	0	0		0	0	0	0
1	0	1	1		1	0	0	0		1	0	1	1

Fig. 1. Pbfs discussed in Example 14.

Definition 12. A *partial boolean function* (**pbf**) on **V** is a (not necessarily proper) subset of the rows of some total boolean function on **V**. The *empty pbf* has no rows and is denoted by $\Diamond$. With some abuse of notation we also write $\Diamond$ for the "value" of a pbf at total assignments that do not correspond to any row in the pbf.

We use the *ite* abbreviations $if(C, T) = ite(C, T, \Diamond)$ and $cp(T) = if(1, T)$.

Two pbfs F and G are said to be **consistent** if the union of their rows is a pbf; that is, there is no total assignment τ such that $F(\tau) = 0$ and $G(\tau) = 1$, or vice versa. Consistency extends to sets of pbfs in the natural way. □

Definition 13. *Restriction* on a pbf (or total boolean function) f defined over a set of variables **V** is denoted as $f\lceil_\sigma$ and can be defined as an operation on the relation that defines f, as follows: In general σ is a partial assignment to **V**, represented as an ordered set of *literals*, $p_1, \ldots, p_m$. It suffices to describe the operation for $m = 1$, as the complete restriction is the same as applying it one variable at a time. So, let $\sigma = (p)$, a single literal with $|p| \in \mathbf{V}$ and let column j in the relational table for f represent $|p|$. **(1)** If p is the positive literal of $|p|$, let f_0 be the selection of all rows of f in which column j is 1; if p is the negative literal select the rows of f that have 0 in column j. **(2)** Let f_1 be a copy of f_0 except that all the entries in column j are inverted. **(3)** Then $f\lceil_p = (f_0 \cup f_1)$. Actually, the union is disjoint and this relation is called the *cylindrification* of f_0.

Thus, in our terminology, $f\lceil_\sigma$ is defined on the same set of variables as f, although its value is independent of the values of variables assigned by σ. □

Example 14. Let $\mathbf{V} = \{q_1, q_2, q_3\}$; Let $F = (q_1 \wedge \neg(q_2))$ and $G = (q_2 \vee q_3)$ be total boolean functions. The relational tables for three pbfs are shown in Fig. 1, where F_1 is a pbf of F and G_1 and G_2 are pbfs of G. Note that F_1 and G_1 **are not consistent**, whereas F_1 and G_2 are consistent. Of course, G_1 and G_2 must be consistent, because they are subsets of the same total function. □

Example 15. Let $\mathbf{V} = \{q_1, q_2, q_3\}$. Let $F = (q_1 \wedge \neg(q_2))$ and let $G = (q_2 \vee q_3)$, and let pbfs F_1, G_1, and G_2 be as discussed in Example 14 and shown in Fig. 1. Then $(F_1)\lceil_{\overline{q_1}} = \Diamond$. Restrictions on q_1 are shown in Fig. 2. The standard definition of restriction (Definition 13) gives $F\lceil_{q_1} = \neg(q_2)$ and $G\lceil_{q_1} = G$. □

q_1	q_2	q_3	$(F_1)\lceil_{q_1}$
0	0	0	1
0	0	1	1
1	0	0	1
1	0	1	1

q_1	q_2	q_3	$(G_1)\lceil_{q_1}$
0	0	0	0
1	0	0	0

q_1	q_2	q_3	$(G_2)\lceil_{q_1}$
0	0	1	1
1	0	1	1

Fig. 2. Restrictions on pbfs discussed in Example 15.

Lemma 16. Let F be a pbf defined on $\mathbf{V}$ and let G be an extension of F to a total boolean function. Let σ be a partial assignment on $\mathbf{V}$. Then $G\lceil_\sigma$ is a total boolean function, and it is an extension of $F\lceil_\sigma$. ∎

Definition 17. Boolean operators can be extended to accept partial boolean functions (pbfs) as parameters. The input pbfs should all be defined on the same set of propositional variables $\mathbf{V}$. The idea is that for every extension of the input pbfs to total boolean functions the output should be an extension of the output pbf on the same $\mathbf{V}$.

The main idea is that for a boolean operator whose parameters are pbfs instead of total functions, the output is defined for an assignment σ if and only if the input pbfs have sufficient known information at σ so that only one output value is possible no matter how the input pbfs are extended to total functions.

Most identities for unary and binary boolean operators are obvious, such as $\neg(\Diamond) \equiv \Diamond$, $(1 \wedge \Diamond) \equiv \Diamond$, $(0 \wedge \Diamond) \equiv 0$, $(1 \vee \Diamond) \equiv 1$, $(0 \vee \Diamond) \equiv \Diamond$, etc.

Some cases where the output of the ternary operator *ite* is defined may not be immediately obvious:

$$ite(\Diamond, 0, 0) \equiv 0 \text{ and}$$
$$ite(\Diamond, 1, 1) \equiv 1.$$

The *restriction* operator can be applied to pbfs with the same rules as for total boolean functions; that is, if G is an extension of pbf F to a total boolean function, then $G\lceil_\sigma$ is an extension of $F\lceil_\sigma$. This can be shown by straightforward induction.

Other cases are easily evaluated from these examples. □

4 Syntax of LDP-Resolution

The term *long-distance resolution* was coined by Zhang and Malik [23] for resolution that accepted pairs of complementary universal literals in its resolvents. Balabanov and Jiang gave a precise condition for allowing long-distance resolution [2]. We call operations that satisfy their conditions *LD-resolution*. What they call *LDQ-resolution* consists of LD-resolution and universal reduction. This section introduces **LDP-resolution**, an operation that integrates long-distance resolution with partial Herbrand functions.

Definition 18. A *mixed variable* in a clause is a universal variable with both positive and negative literals in the same clause. (We simply list both literals on the mixed variable.) □

Definition 19. This paper uses a form of *nonclausal resolution*, which is a generalization of propositional clause resolution due to Manna and Waldinger [14]. We use the following specific notation:

$$\mathsf{res}_e(F_0, F_1) = F_0 \lceil_e \vee F_1 \lceil_{\overline{e}} \tag{8}$$

The clausal case has $\overline{e} \in F_0$ and $e \in F_1$. For this paper e and $\overline{e}$ are always existential literals, called the *clashing literals*. Note that the second operand is always the clause that contains e.

The general case is only useful if the resolvent is not tautologously true, $F_0 \lceil_e$ properly subsumes F_0 and $F_1 \lceil_{\overline{e}}$ properly subsumes F_1; otherwise the resolvent is weaker than an operand.

A logically equivalent definition is:

$$\mathsf{res}_e(F_0, F_1) = ite(e, F_0, F_1) \tag{9}$$

This is clear from Definition 11 and shows that the resolvent expressions do not actually depend on e. □

Definition 20. The *LDP-resolution criteria* apply when the operands are QBF clauses and the resolvent contains u and $\overline{u}$. Say the clashing variable was $|e|$. We say $|u|$ is a *mixed variable*. Unlike most papers, we do not "merge" such complementary pairs, for reasons explained in Sect. 4. LDP permits mixed $|u|$ in a resolvent **if (A)** all u-literals came from a single resolution operand, **or if (B)** ($|e| \prec |u|$) holds, i.e., $|e|$ is outer to $|u|$ in the prenex. (see Definition 2). Mixed existential variables are never permitted. □

It is not useful to regard LDP resolvents (or LD resolvents) with mixed variables as logical formulas because they are tautologous; their value lies in using them as a *notation* in solver implementations [23].

5 Partial Herbrand Functions

The intuitive meaning of partial Herbrand functions (phfs) are at the core of LDP methodology. We explain this before proceeding to the technical definitions. We have an original closed PCNF $\Phi = \vec{Q}.\mathcal{F}$ and possibly some derived clauses. For clause $C \in \mathcal{F}$ and a universal literal $u \in C$, the A-player (trying to falsify *some* clause) thinks:

> The value of literal u matters for C only if the assignment to existential literals outer to u (call it τ) falsifies ($\text{exist}(C) \prec u$), and then the useful value is $u = 0$. (10)

This explains the base definitions

$$phf(u, C) = \begin{cases} 0 & \text{if all literals in } (\text{exist}(C) \prec u) \\ & \text{are assigned } 0 \\ \diamond & \text{otherwise.} \end{cases} \quad (11)$$

If u is a negative literal, i.e., $\overline{|u|}$; then $phf(u, C)$ tells when variable $|u|$ should be 1.

The "intended meaning" of $phf(u, C)$ in (10) is generalized in a natural way by the phfs of derived clauses, but now with the goal to detect whether the value of u matters to the A-player anywhere in the proof of a derived clause D. For this purpose we may restrict attention to the support of D, which we denote as $\mathcal{G}$. Let M be any model tree for $\Psi = \vec{Q}.\mathcal{G}$. The analogous goal of the A-player is to show that some branch of M, call it τ, falsifies D. If such τ exists in M, it shows D is *not* a logical consequence of Ψ.

Two kinds of partial Herbrand functions (phfs) are defined for LDP-resolution. The definitions depend implicitly on the original formula Φ. To avoid excessive verbiage we use "phf" to refer to either $phf()$ or $phfm()$. The name *phf* indicates that the first parameter is a universal *literal* and the output is 0 or $\diamond$.

The name *phfm* indicates that the first parameter is a universal *variable* (v in (12) below), and the output is 0, 1, or $\diamond$. Although phfms are not part of any LDP derivation, they facilitate *verifying* that a derivation of the empty clause is correct, which is considered important in some circles.

In all cases *phfm* is defined naturally in terms of *phf* as follows:

$$phfm(v, D) = (\neg phf(\overline{v}, D) \cup phf(v, D)). \quad (12)$$

Recall that "$\cup$" is not the same as "or" for pbfs. It is easily shown that "$\cup$" is consistent in Eq. (12) in all the cases of Definition 21 where it is used, below. Note that $phfm(v, D)$ is *not* defined recursively.

Phfs for derived clauses are defined recursively in terms of the operand(s) of the proof operation that derived the clause. For any form of universal reduction on v, the phf is copied from the operand if $|v| \neq |u|$, or is $\diamond$ if $|u|$ itself is being reduced.

The general formula for phfs of resolvents, where $D = \mathsf{res}_e(D_0, D_1)$, is:

$$phf(u, D) = ite(e, phf(u, D_0), phf(u, D_1)) \qquad (13)$$

where u may be a positive or negative literal. Then compute $phfm(|u|, D)$ with Eq. (12). Some of the phfs in an *ite* may be $\Diamond$.

An important identity for *all* resolution cases is:

$$phf(u, D) \equiv phf(u, D_0)\lceil_e \vee phf(u, D_1)\lceil_{\overline{e}}. \qquad (14)$$

This shows that $phf(u, D)$ does not actually depend on the clashing literal e, although the *ite* operands in Eq. 13 do. Therefore it does not matter if $u \prec e$; the phf only depends on existential variables outer to u, as required for Herbrand functions.

It is important to remember that in the examples a pair of literals u, $\overline{u}$ is shorthand for the pbf $(phf(u, D) \cup phf(\overline{u}, D))$, which must be consistent because the only boolean value taken by phfs is 0. In implementations, a pair of pointers to the phfs probably suffices.

Resolution simplifies in several special cases, detailed below. The complicated cases are when u is mixed in D and/or when D was derived by LDS-resolution with a clashing variable e such that $u \prec e$ but $indep(u, e)$ holds.

Definition 21. Let $D = \mathsf{res}_e(D_0, D_1)$. For each universal variable $|u| \in vars(D)$:

1. If $|e|$ depends on $|u|$ (i.e., $|u| \prec |e|$ and $indep(|u|, |e|)$ does *not* hold), and literals on $|u|$ occur in only one operand, then the phfs in the operands cannot depend on e. Copy the phfs of the operand that contains literals on $|u|$. This includes the case that $|u|$ is mixed. One of the phfs is $\Diamond$ unless $|u|$ is mixed. The *phfm* may use Eq. (12), but it will be the same as in the operand that had the literals on $|u|$.
2. If $|u| \prec |e|$ and u occurs in both operands, then $|u|$ cannot be mixed, and

$$phf(u, D) = (phf(u, D_0) \cap phf(u, D_1)) \qquad (15)$$

   because $0 \vee \Diamond = \Diamond$. Then compute $phfm(|u|, D)$ with Eq. (12).
3. If $|e| \prec u$ and u occurs in only one operand, then copy the phfs and phfm of that operand. This includes the case that $|u|$ is mixed.
4. If $|e| \prec u$ and u occurs in both operands, and u is *not* mixed in D, then $phf(u, D)$ is given by Eq. 13. One of the phfs in the *ite* may be $\Diamond$. Then compute $phfm(|u|, D)$ with Eq. (12).
5. If $|e| \prec |u|$ and $|u|$ *is* mixed in D, then $phf(u, D)$ is given by Eq. 13. Then compute $phfm(|u|, D)$ with Eq. (12). Some of the phfs in an *ite* may be $\Diamond$.

□

6 Logical Implications in QBF Derivations

This section considers the logical properties of derivations from closed PCNFs. The main result of this section is Theorem 25, which states that LDP-resolution derives only logical consequences; it follows from the more technical Lemma 24. We begin with a helpful definition.

Definition 22. Let Π_D be an LDP-resolution derivation of clause D from Φ. Let τ be a prenex-ordered assignment for Φ. Then τ is said to **effectively falsify** D if: **(1)** For each literal $p \in D$ that is *not* mixed, $\tau(p) = 0$ (p may be universal or existential). **(2)** For each *mixed* universal variable $v \in vars(D)$, define $\sigma(v)$ to be the subset of τ that assigns values to *existential* variables outer to v. Then $phf(v, D)\lceil_{\sigma(v)} = \Diamond$ or $\tau(v) = phfm(v, D)\lceil_{\sigma(v)}$.

That is, τ assigns the variable v the value obtained by evaluating $phfm(v, D)$ at the point $(\mathsf{exist}(\tau) \prec v)$ if that value is defined. If $phfm(v, D)\lceil_{\sigma(v)} = \Diamond$, then both values for $\tau(v)$ allow τ to effectively falsify D. □

Example 23. If D is a resolvent in LDQ-resolution and universal variable u is in $vars(D)$, then for certain choices of operands and some partial assignment τ that only assigns values to $(\mathsf{exist}(D) \prec u)$ (i.e., existential variables outer to u), it is possible that $(\mathsf{exist}(D) \prec u)\lceil_\tau = 0$ and $phf(u, D)$ is not defined at τ.

An example with $\{a, b, c\} \prec u$ uses the following clauses. The cases $u \prec e$ and $e \prec u$ are both considered.

$$D_1 = [a, \overline{e}],$$
$$D_3 = [b, c, u, e, f],$$
$$D_4 = [\overline{c}, f],$$
$$D_2 = \mathsf{res}_c(D_4, D_3) = [b, u, e, f],$$
$$D_5 = \mathsf{res}_e(D_1, D_3) = [a, b, c, u, f].$$

If $u \prec e$, let $D = \mathsf{res}_e(D_1, D_2) = [a, b, u, f]$. Then

$$phf(u, D_3) = if\left((\overline{b} \wedge \overline{c}), 0\right),$$
$$phf(u, D_2) = if\left((\overline{c}), phf(u, D_3)\right),$$
$$phf(u, D) = cp(phf(u, D_2)) = if\left(\overline{c}, phf(u, D_3)\right).$$

Let $\tau = \{\overline{a}, \overline{b}, c\}$. Then $phf(u, D)\lceil_\tau = \Diamond$.

Next consider $e \prec u$ and let $D = \mathsf{res}_c(D_4, D_5)$. Here D is the same as above, but

$$phf(u, D_3) = if\left((\overline{b} \wedge \overline{c}) \wedge \overline{e}, 0\right),$$
$$phf(u, D_5) = if(\overline{e}, phf(u, D_3)),$$
$$phf(u, D) = if(\overline{c}, phf(u, D_5)).$$

The same τ as above produces $phf(u, D)\lceil_\tau = \Diamond$.

For this example, it is not really problematical that $phf(u, D)$ is not defined at τ because $D\lceil_\tau$ is subsumed by $D_4\lceil_\tau$ and D_4 does not contain u. □

Lemma 24. Let clause D be derived by LDP-resolution (see Definition 4) from an original QBF $\Phi = \overrightarrow{Q}.\mathcal{F}$. Let S_D be the *support* of D in Π_D, that is, the clauses used in its derivation. Let T be an E-assignment tree for Φ such that some branch $\tau \in T$ effectively falsifies D. Then there is some branch $\tau_f \in T$ that falsifies some original clause $C_j \in S_D$.

Proof: (Sketch) Adopt the method of Goultiaeva *et al.* [9] for extracting certificates (see also [21]). ∎

Theorem 25. Any non-tautologous clause D derived by LDP-resolution from PCNF $\Phi = \overrightarrow{Q}.\mathcal{F}$ is logically implied by Φ.

Proof: Contrapositive of Lemma 24. ∎

Corollary 26. If PCNF Φ has an LDP-resolution refutation, then $\Phi = \text{false}$. ∎

7 Conclusion

The proof system LDP-resolution was introduced and shown to derive only logical consequences, when the derived clause has no mixed phfs. It uses a novel version of long-distance resolution with partial Herbrand functions to avoid tautologous universal literals. Certain aggressive dependency schemes are shown in Appendix B to derive clauses that are *not* logical consequences in some cases.

Future work should develop more integration of QBF reasoning with applications. More than a one-bit result is needed in practice. Incremental formulas have been productive in the propositional domain.

Partial Herbrand functions and partial Skolem functions might be useful in other fragments of first-order logic. Quantified constraint satisfaction problems in which variables have finite domains are one possibility. Modeling procedurally defined functions with preconditions is another possibility, such as `first` and `rest` in Common Lisp, which require a non-empty list for their parameter.

Appendices

A Logical Operators on Pbfs

Boolean operators can be extended to accept partial boolean functions (pbfs) as parameters. The input pbfs should all be defined on the same set of propositional variables **V**. The idea is that for every extension of the input pbfs to total boolean functions the output should be an extension of the output pbf on the same **V**.

Definition 27. Let **V** be a fixed nonempty set of k propositional variables. Let op be an m-ary boolean operator (m may be 0) with parameters P_i, $1 \leq i \leq m$, where the P_i are pbfs on **V**. Then $op(P_1, \ldots, P_m)$ is the pbf P on **V** defined as follows:

For each total assignment σ on **V** there is a $b \in \{0, 1\}$ such that the following two statements are equivalent:

1. For every extension of P_i to a total boolean function on $\mathbf{V}$, call it F_i,
$$op(F_1(\sigma), \ldots, F_m(\sigma)) = b.$$
2. P contains exactly one row of the form (σ, b).

In other words, $P(\sigma)$ is defined if any only if the input pbfs have sufficient known information at σ to determine the output of op at σ. □

It suffices to consider the cases that one or more operands are the empty pbf $\Diamond$. Also we adopt the convention that an output value of $\Diamond$ means there is no matching row in the output pbf. The following relationships are straightforward applications of Definition 27:

$$cp(\Diamond) \equiv \Diamond \tag{16}$$

$$\neg(\Diamond) \equiv \Diamond \tag{17}$$

$$\bigwedge_{1 \leq i \leq m} (P_i(\sigma)) = \begin{cases} 0 & \text{if some } P_i(\sigma) = 0 \text{;} \\ 1 & \text{if every } P_i(\sigma) = 1 \text{;} \\ \Diamond & \text{otherwise.} \end{cases} \tag{18}$$

$$\bigvee_{1 \leq i \leq m} (P_i(\sigma)) = \begin{cases} 1 & \text{if at least one } P_i(\sigma) = 1 \text{;} \\ 0 & \text{if every } P_i(\sigma) = 0 \text{;} \\ \Diamond & \text{otherwise.} \end{cases} \tag{19}$$

Also, $xor(\Diamond, P) \equiv \Diamond$ for all pbfs P.

The extension of *ite* and other operators on three or more parameters requires care. By application of Definition 27:

$$ite(\Diamond, 0, 0) = 0 \tag{20}$$

$$ite(\Diamond, 1, 1) = 1 \tag{21}$$

Referring back to Sect. 2, use of the DNF in (6) fails to give (21), and use of the CNF in (7) fails to give (20). The solution is to close a CNF that represents the operator on total boolean functions under resolution (see Definition 28 below). The next theorem shows that the resulting CNF correctly represents the corresponding operator on pbfs. Closure of a DNF under consensus (also called *term resolution*) accomplishes the same thing by a similar argument.

Definition 28. A propositional CNF $\mathcal{F} = \{C_j\}$ is said to be ***closed under resolution*** if for every pair of resolvable clauses $C_i \in \mathcal{F}$ and $C_j \in \mathcal{F}$, The resolvent is either tautologous or is subsumed by a clause already in $\mathcal{F}$. A propositional DNF $\mathcal{G} = \bigwedge_j (T_j)$ is said to be ***closed under consensus*** if for every pair of resolvable terms $T_i \in \mathcal{G}$ and $T_j \in \mathcal{G}$, The consensus is either contradictory or is subsumed by a term already in $\mathcal{G}$. □

Theorem 29. Let $\mathbf{V}$ be a fixed nonempty set of k propositional variables. Let the boolean operator $op(P_1, \ldots, P_m)$ be defined or represented for total boolean functions on $\mathbf{V}$ as parameters by a propositional CNF $\mathcal{F}$ that treats the P_i as boolean variables and has no other variables. We abbreviate $\{P_i \mid 1 \leq i \leq m\}$ to $\{P_i\}$. Then $\mathcal{F}$ represents the correponding operator for pbfs on $\mathbf{V}$ as parameters (as defined in Definition 27) if and only if $\mathcal{F}$ is closed under resolution.

Proof: ($\Rightarrow$) Assume for purposes of contradiction that $\mathcal{F}$ is not closed under resolution. There is some non-tautologous clause $D = \text{res}_p(C_i, C_j)$ such that $C_i \in \mathcal{F}$, $C_j \in \mathcal{F}$, and D is not subsumed by any clause in $\mathcal{F}$. Consider the partial assignment on $\{P_i\}$ defined by $\sigma = \neg(D)$. Treat σ as a pbf on $\{P_i\}$ (i.e., every variable not assigned by σ is $\Diamond$). Every extension of σ to a total assignment τ on $\{P_i\}$ causes $\mathcal{F}\lceil_\tau = 0$ (because either $C_i\lceil_\tau = 0$ or $C_j\lceil_\tau = 0$), so $\mathcal{F}(\sigma)$ should be 0. However, every clause $C \in \mathcal{F}$ contains a literal not mentioned in σ (or D would be subsumed), so $C(\sigma) = \Diamond$ or $C(\sigma) = 1$ by (19). In particular, $C_i(\sigma) = \Diamond$ and $C_j(\sigma) = \Diamond$, so $\mathcal{F}(\sigma) = \Diamond$ by (18). This contradicts the hypothesis that $\mathcal{F}$ represents op for pbfs.

($\Leftarrow$) By hypothesis, $\mathcal{F}$ is closed under resolution. Let σ be a partial assignment on $\{P_i\}$ and let τ range over extensions of σ to total assignments on $\{P_i\}$. If there is some $C_j \in \mathcal{F}$ such that $C_j\lceil_\sigma \neq 1$, then there is some τ such that $C_j\lceil_\tau = 0$. Therefore, if $op(\tau) = 1$ for all τ, then $op(\sigma) = 1$.

Now assume $op(\tau) = 0$ for all τ. It remains to show that $op(\sigma) = 0$. This is immediate if $C_j\lceil_\sigma = 0$ for any $C_j \in \mathcal{F}$, so assume this is not the case. Define:

$$\mathcal{G} = \{C_j \mid C_j \in \mathcal{F} \text{ and } C_j\lceil_\sigma = \Diamond\}. \tag{22}$$

If $\mathcal{G}$ has no clauses, $op(\sigma)$ must be 1, contradicting the hypothesis that all $op(\tau) = 0$, so assume $\mathcal{G}$ has some clauses. In this case $\mathcal{G}\lceil_\sigma$ must be unsatisfiable or else there would be some total extension τ that satisfies $\mathcal{G}$ and also satisfies $\mathcal{F}$. Let Π be a resolution refutation of $\mathcal{G}\lceil_\sigma$. Removing the restriction converts Π to a resolution derivation on $\mathcal{G}$ of some clause $D \subseteq \neg(\sigma)$. But D is also derivable from $\mathcal{F}$, which is closed under resolution. Therefore some $C_j \in F$ subsumes D, and $C_j\lceil_\sigma = 0$, so $op(\sigma) = 0$. ∎

Corollary 30. Let **V** be a fixed nonempty set of k propositional variables. Let the boolean operator $op(P_1, \ldots, P_m)$ be defined or represented for total boolean functions on **V** as parameters by a propositional DNF $\mathcal{G}$ that treats the P_i as boolean variables and has no other variables. We abbreviate $\{P_i \mid 1 \leq i \leq m\}$ to $\{P_i\}$. Then $\mathcal{G}$ represents the correponding operator for pbfs on **V** as parameters (as defined in Definition 27) if and only if $\mathcal{G}$ is closed under consensus (also called term resolution).

Proof: Define $\mathcal{F} = \neg(\mathcal{G})$ and apply Theorem 29 to $\mathcal{F}$. ∎

Corollary 31. Let **V** be a fixed nonempty set of k propositional variables. Let C, T, F be pbfs on **V**. Then

$$ite(C, T, F) \equiv [\neg(C), T] \wedge [C, F] \wedge [T, F] \tag{23}$$
$$ite(C, T, F) \equiv (C \wedge T) \vee (\neg(C) \wedge F) \vee (T \wedge F) \tag{24}$$
$$ite(\Diamond, T, F) \equiv (T \cap F) \tag{25}$$

Note that the rightmost clause in (23) and the rightmost term in (24) have no effect when C is a total boolean function. ∎

B Derivations That Are Not Logical Consequences

Slivovsky and Szeider [20] define the ***rrs*** dependency scheme (denoted D^{rrs}). A universal literal u is called *rrs-tailing* if no existential literal in the same clause depends on it in D^{rrs}. Such universal literals are deleted from that clause. Their paper shows that Q-resolution with universal reductions based on rrs-tailing is refutationally sound. The rrs scheme is strictly more aggressive[1] than the *standard dependency scheme* (D^{std}) that was introduced and studied theoretically in the pioneering work of Samer and Szeider [17] and is also refutationally sound.

More aggressive dependency schemes such as Samer's generalized triangle dependencies [16], *strict* standard dependencies [22], and reflexive resolution-path dependencies [20] may derive clauses that are not logically implied by Φ. An example with a mere four clauses shows that universal reduction can delete some model trees.

Example 32. Let $\Phi = \vec{Q}.\mathcal{F}$ be given in Clause-literal matrix form below on the left, followed by two of several model trees, M_1 and M_2: :

Φ	$\exists b$	$\exists c$	$\forall u$	$\exists f$	$\exists g$
C_1		c	u	f	
C_2	b		$\overline{u}$		g
C_3	$\overline{b}$			f	
C_4		$\overline{c}$			g

M_1:

	$\overline{u}$	f	g
$\overline{b}$ $\overline{c}$			
	u	$\overline{f}$	$\overline{g}$

M_2:

	$\overline{u}$	f	$\overline{g}$
$\overline{b}$ $\overline{c}$			
	u	$\overline{f}$	g

Clauses $D_1 = [\overline{b}, c]$ and $D_2 = [b, \overline{c}]$ are derivable by Q-resolution, so they are logically implied by Φ. Under the standard dependency scheme f and g depend on u, so the only independence is that implied by the prenex order.

In the more aggressive dependency schemes mentioned above, f and g are considered to be independent of u. Under these schemes $D_3 = [b, g]$ and $D_4 = [c, f]$ are derivable by universal reduction from C_2 and C_1, respectively. Adding D_3 and D_4 to Φ is *safe* (its truth-value does not change); M_1 is still a model tree. However, M_2 is *not* a model tree for either $\Phi \cup \{D_3\}$ or $\Phi \cup \{D_4\}$, so neither clause is a logical consequence. □

References

1. Balabanov, V., Jiang, J.H., Janota, M., Widl, M.: Efficient extraction of QBF (counter)models from long-distance resolution proofs. In: Proceedings of AAAI, pp. 3694–3701 (2015)
2. Balabanov, V., Jiang, J.R.: Unified QBF certification and its applications. Formal Methods Syst. Des. **41**, 45–65 (2012)

---

[1] The expression "D_1 is *more aggressive* than D_2" means that the *independence* found by D_1 is a superset of that found by D_2.

3. Beyersdorff, O., Blinkhorn, J.: Dependency schemes in QBF calculi: semantics and soundness. In: Rueher, M. (ed.) CP 2016. LNCS, vol. 9892, pp. 96–112. Springer, Cham (2016). https://doi.org/10.1007/978-3-319-44953-1_7
4. Bjorner, N., Janota, M., Klieber, W.: On conflicts and strategies in QBF. In: LPAR-20 (2015)
5. Blinkhorn, J., Beyersdorff, O.: Shortening QBF proofs with dependency schemes. In: Gaspers, S., Walsh, T. (eds.) SAT 2017. LNCS, vol. 10491, pp. 263–280. Springer, Cham (2017). https://doi.org/10.1007/978-3-319-66263-3_17
6. Bryant, R.: Graph-based algorithms for Boolean function manipulation. IEEE Trans. Comput. C **35**(8), 677–691 (1986)
7. Bubeck, U., Kleine Büning, H.: Nested Boolean functions as models for quantified Boolean formulas. In: Järvisalo, M., Van Gelder, A. (eds.) SAT 2013. LNCS, vol. 7962, pp. 267–275. Springer, Heidelberg (2013). https://doi.org/10.1007/978-3-642-39071-5_20
8. Egly, U., Lonsing, F., Widl, M.: Long-distance resolution: proof generation and strategy extraction in search-based QBF solving. In: McMillan, K., Middeldorp, A., Voronkov, A. (eds.) LPAR 2013. LNCS, vol. 8312, pp. 291–308. Springer, Heidelberg (2013). https://doi.org/10.1007/978-3-642-45221-5_21
9. Goultiaeva, A., Van Gelder, A., Bacchus, F.: A uniform approach for generating proofs and strategies for both true and false QBF formulas. In: Proceedings of IJCAI (2011)
10. Beyersdorff, O., Chew, L., Janota, M.: On unification of QBF resolution-based calculi. In: Csuhaj-Varjú, E., Dietzfelbinger, M., Ésik, Z. (eds.) MFCS 2014. LNCS, vol. 8635, pp. 81–93. Springer, Heidelberg (2014). https://doi.org/10.1007/978-3-662-44465-8_8
11. Kleine Büning, H., Karpinski, M., Flögel, A.: Resolution for quantified Boolean formulas. Inf. Comput. **117**, 12–18 (1995)
12. Klieber, W., Janota, M., Marques-Silva, J., Clarke, E.: Solving QBF with free variables. In: Schulte, C. (ed.) CP 2013. LNCS, vol. 8124, pp. 415–431. Springer, Heidelberg (2013). https://doi.org/10.1007/978-3-642-40627-0_33
13. Lonsing, F., Biere, A.: Integrating dependency schemes in search-based QBF solvers. In: Strichman, O., Szeider, S. (eds.) SAT 2010. LNCS, vol. 6175, pp. 158–171. Springer, Heidelberg (2010). https://doi.org/10.1007/978-3-642-14186-7_14
14. Manna, Z., Waldinger, R.: A deductive approach to program synthesis. TOPLAS **2**, 90–121 (1980)
15. Peitl, T., Slivovsky, F., Szeider, S.: Long distance q-resolution with dependency schemes. In: Proceedings of SAT (2016)
16. Samer, M.: Variable dependencies of quantified CSPs. In: Cervesato, I., Veith, H., Voronkov, A. (eds.) LPAR 2008. LNCS (LNAI), vol. 5330, pp. 512–527. Springer, Heidelberg (2008). https://doi.org/10.1007/978-3-540-89439-1_36
17. Samer, M., Szeider, S.: Backdoor sets of quantified Boolean formulas. JAR **42**, 77–97 (2009)
18. Samulowitz, H., Davies, J., Bacchus, F.: Preprocessing QBF. In: Benhamou, F. (ed.) CP 2006. LNCS, vol. 4204, pp. 514–529. Springer, Heidelberg (2006). https://doi.org/10.1007/11889205_37
19. Samulowitz, H., Bacchus, F.: Dynamically partitioning for solving QBF. In: Theory and Applications of Satisfiability Testing (SAT), pp. 215–229 (2007)
20. Slivovsky, F., Szeider, S.: Soundness of Q-resolution with dependency schemes. Theor. Comp. Sci. **612**, 83–101 (2016)

21. Van Gelder, A.: Input distance and lower bounds for propositional resolution proof length. In: Theory and Applications of Satisfiability Testing (SAT) (2005)
22. Gelder, A.: Variable independence and resolution paths for quantified Boolean formulas. In: Lee, J. (ed.) CP 2011. LNCS, vol. 6876, pp. 789–803. Springer, Heidelberg (2011). https://doi.org/10.1007/978-3-642-23786-7_59
23. Zhang, L., Malik, S.: Conflict driven learning in a quantified Boolean satisfiability solver. In: Proceedings of ICCAD, pp. 442–449 (2002)

Author Index

A
Austhof, Bethany 1
Awasthi, Pranjal 10

B
Biswas, Arindam 206
Bogaerts, Bart 218
Bounan, Samuel 50

C
Cortes, Corinna 10

D
Denecker, Marc 218
Dutt, Mousumi 206

E
Edelkamp, Stefan 50

H
Häntschel, Tim 150

J
Jégou, Philippe 94
Jena, Sangram K. 68

K
Kuroki, Manabu 165

M
Maity, Anukul 206
Mansour, Yishay 10

Mao, Anqi 107
Mohri, Mehryar 10, 107

O
Olasz, Csaba 182

P
Pollaci, Samuele 218

R
Rahm, Erhard 150
Reyzin, Lev 1, 136

S
Schäfer, Martin 150
Shingaki, Ryusei 165
Subramani, K. 68, 81

T
Tu, Duan 136

U
Uhrich, Benjamin 150

V
van Gelder, Allen 236
Vanbesien, Linde 218
Velasquez, Alvaro 68

W
Wojciechowski, Piotr 81

Z
Zhong, Yutao 107

SPRINGER NATURE

GPSR Compliance

The European Union's (EU) General Product Safety Regulation (GPSR) is a set of rules that requires consumer products to be safe and our obligations to ensure this.

If you have any concerns about our products, you can contact us on ProductSafety@springernature.com

In case Publisher is established outside the EU, the EU authorized representative is:

Springer Nature Customer Service Center GmbH
Europaplatz 3
69115 Heidelberg, Germany

The manufacturer's authorised representative in the EU is Springer Nature Customer Service Centre GmbH, Europaplatz 3, 69115 Heidelberg, Germany. If you have any concerns regarding our products, please contact ProductSafety@springernature.com

Printed and bound by CPI Group (UK) Ltd, Croydon, CR0 4YY

26/03/2026

02078984-0002